Textbook of Biotechnology

Textbook of Biotechnology

Edited by **Lydell Norris**

SYRAWOOD
PUBLISHING HOUSE

New York

Published by Syrawood Publishing House,
750 Third Avenue, 9th Floor,
New York, NY 10017, USA
www.syrawoodpublishinghouse.com

Textbook of Biotechnology
Edited by Lydell Norris

International Standard Book Number: 978-1-68286-194-3 (Hardback)

Printed in the United States of America.

Contents

Preface

The main aim of this book is to educate learners and enhance their research focus by presenting diverse topics covering this vast field. This is an advanced book which compiles significant studies by distinguished experts. This book addresses successive solutions to the challenges arising in the area of application, along with it; the book provides scope for future developments.

Biotechnology is a rapidly growing discipline that emphasises on modifying and using biological organisms for producing various things. Its applications extend to medicine, engineering, agriculture and other disciplines. This book includes some of the vital pieces of work being conducted across the world, on different topics related to biotechnology and also includes some crucial developments such as advancements in cancer and stem cell research, genomics and proteomics, use of biotechnology in gene therapy, pharmaceuticals, and biosensors, etc. The students and academicians will find this book full of innovative insights.

It was a great honour to edit this book, though there were challenges, as it involved a lot of communication and networking between me and the editorial team. However, the end result was this all-inclusive book covering diverse themes in the field.

Finally, it is important to acknowledge the efforts of the contributors for their excellent chapters, through which a wide variety of issues have been addressed. I would also like to thank my colleagues for their valuable feedback during the making of this book.

Editor

Extracellular synthesis of silver nanoparticles by the *Bacillus* strain CS 11 isolated from industrialized area

Vidhya Lakshmi Das · Roshmi Thomas ·
Rintu T. Varghese · E. V. Soniya · Jyothis Mathew ·
E. K. Radhakrishnan

Abstract Biological synthesis of silver nanoparticles using microorganisms has received profound interest because of their potential to synthesize nanoparticles of various size, shape and morphology. In the current study, synthesis of silver nanoparticles by a bacterial strain (CS 11) isolated from heavy metal contaminated soil is reported. Molecular identification of the isolate showed it as a strain of *Bacillus* sp. On treating the bacteria with 1 mM $AgNO_3$, it was found to have the ability to form silver nanoparticles extracellularly at room temperature within 24 h. This was confirmed by the visual observation and UV–Vis absorption at 450 nm. Further characterization of nanoparticles by transmission electron microscopy confirmed the size of silver nanoparticles in 42–92 nm range. Therefore, the current study is a demonstration of an efficient synthesis of stable silver nanoparticle by a *Bacillus* strain.

Keywords Silver nanoparticle · Biosynthesis · Extracellular · *Bacillus* sp.

V. L. Das and R. Thomas contributed equally to the work.

V. L. Das · R. Thomas · J. Mathew · E. K. Radhakrishnan (✉)
School of Biosciences, Mahatma Gandhi University,
PD Hills (PO), Kottayam 686 560, Kerala, India
e-mail: radhakrishnanek@mgu.ac.in

R. T. Varghese · E. V. Soniya
Plant Molecular Biology, Rajiv Gandhi Centre
for Biotechnology, Thycaud (PO), Poojappura,
Thiruvananthapuram 695 014, Kerala, India

Introduction

Nanotechnology involving synthesis and applications of nanomaterials is a rapidly growing field with significant applications in various areas (Duran et al. 2005). The attraction of silver nanoparticles (AgNPs) is mainly because of its application in therapeutics, biomolecular detection, catalysis and also as antimicrobial agents (Sadhasivam et al. 2010; Shrivastava et al. 2009; Wei et al. 2008; Christopher et al. 2011). Microbial synthesis of nanoparticles is eco-friendly and has significant advantages over other processes since it takes place at relatively ambient temperature and pressure (Gade et al. 2008; Mukherjee et al. 2008; Wei et al. 2012). As the size and shape of nanoparticles can also be controlled in microbial synthesis (Narayanan and Sakthivel 2010), screening of unexplored microorganisms for AgNPs synthesizing property is very important.

Microbial synthesis of metal nanoparticles can take place either intracellularly or extracellularly (Ahmad et al. 2003, 2007; Jain et al. 2011; Kalishwaralal et al. 2010; Pugazhenthiran et al. 2009; Saifuddin et al. 2009). Intracellular synthesis of nanoparticles requires additional steps such as ultrasound treatment or reactions with suitable detergents to release the synthesized nanoparticles (Babu et al. 2009; Kalimuthu et al. 2008). At the same time extracellular biosynthesis is cheap and it requires simpler downstream processing. This favours large-scale production of silver nanoparticles to explore its potential applications. Because of this, many studies were focussed on extracellular methods for the synthesis of metal nanoparticles (Duran et al. 2005). When the culture supernatant of *Bacillus megaterium* was treated with aqueous solutions of Ag^+ ions, within few minutes it formed silver nanoparticles (AgNPs) extracellularly (Saravanan et al. 2011).

Studies using culture supernatants of bacteria like *Pseudomonas proteolytica, Pseudomonas meridiana, Arthrobacter kerguelensis, Bacillus indicus,* etc., were also proven its property to form extracellular nanoparticles very effectively (Shivaji et al. 2011). Studies on reduction of Ag+ ions to AgNPs by *Staphylococcus aureus* also highlight the potential of extracellular method of nanoparticle formation (Nanda and Saravanan 2009).

Long and continuous microbial interactions with various metals can be greatly expected among bacterial isolates from industrialized area. This provides high possibility of identifying interesting microbial strains with nanoparticle forming properties. In the current study, soil microorganisms from industrialized area were screened for its potential to form silver nanoparticles. The selected strain was subjected to identification by molecular methods which resulted in its identification as a *Bacillus* sp. and very interestingly the isolate was found to have the ability to synthesize silver nanoparticles both intracellularly and extracellularly. The reduction of silver ions was checked by visual inspection as well as by measuring its UV–visible absorption. Further characterization by transmission electron microscopy confirmed the synthesis of stable silver nanoparticles with size range of 42–94 nm.

Materials and methods

Isolation of metal-resistant bacteria

Soil samples were collected from heavy metal contaminated areas of Cochin, Kerala, India, and were used as the source material for bacterial isolation. The samples were serially diluted in sterile 0.8 % NaCl and then plated onto nutrient agar (Babu and Gunasekaran 2009). The colonies obtained were further subcultured on nutrient agar supplemented with 1 mM concentration of filter-sterilized AgNO$_3$. This was further incubated at room temperature for 48 h and the plates were observed for bacterial growth. The isolated colonies were subcultured and obtained in the form of pure culture and from this one of the strains CS 11 was selected randomly to explore its potential to form silver nanoparticles.

Molecular identification

Molecular identification of the isolated strain was carried out by 16S rDNA sequence-based method. Total genomic DNA was isolated from a selected strain for PCR. Quality of the isolated DNA was checked by agarose gel electrophoresis and was further quantified using UV–Vis spectrophotometer (Hitachi U5100). The genomic DNA was then used as template for PCR using the primers 16SF (5′-AgA gTTTgA TCM Tgg CTC-3′) and 16SR (5′-AAg gAggTg WTC CAR CC-3′). The PCR was carried out in a total volume of 50 µL containing 50 ng of genomic DNA, 20 pmol of each primer, 1.25 Units of *Taq* DNA polymerase, 200 µM of each dNTPs and 1× PCR buffer as components. The PCR was performed for 35 cycles in a Mycycler™ (Bio-Rad, USA) with the initial denaturation for 3 min at 94 °C, cyclic denaturation for 30 s at 94 °C, annealing for 30 s at 58 °C and extension for 2 min at 72 °C with a final extension of 7 min at 72 °C. After the PCR, the reaction products were analysed by agarose gel electrophoresis. The product was then gel purified and was further subjected to sequencing PCR using the Big Dye Terminator Sequence Reaction Ready Mix (Applied Biosystem). After the reaction, product was purified, precipitated and was used for sequence run in the DNA sequencer ABI 310 Genetic Analyser. The sequence data of 16S rDNA thus obtained was further aligned using BioEdit program. This sequence was then used for BLAST analysis. The 16S rDNA sequence of CS 11 was also used for phylogenetic analysis using neighbor-joining method in MEGA5 (Tamura et al. 2011).

Synthesis of silver nanoparticles

For silver nanoparticle biosynthesis studies, the selected bacterial isolate was inoculated in to 250-ml Erlenmeyer flask containing 100 ml sterile nutrient broth. The cultured flasks were incubated in a rotating shaker set at 200 rpm for 48 h at room temperature. After this the culture was centrifuged at 12,000 rpm for 10 min. The biomass and supernatant were separated and used separately for the synthesis of silver nanoparticles. The supernatant was used for studying extracellular production of silver nanoparticles by mixing it with filter-sterilized AgNO$_3$ solution at 1 mM final concentration. At the same time, bacterial biomass was taken for intracellular synthesis. For this approximately 2 g of wet biomass was resuspended in 100 ml of 1 mM aqueous solution of AgNO$_3$ in a 250-ml Erlenmeyer flask. All the reaction mixtures were incubated on rotating shaker (200 rpm) at room temperature for a period of 72 h in light. Heat-killed samples with AgNO$_3$ were also incubated along with experimental samples as control. Visual observation was conducted periodically to check for the nanoparticle formation. Further characterization was conducted for nanoparticle generated through extracellular methods.

Characterization of silver nanoparticle

The optical characteristics of the synthesized silver nanoparticles were analysed using UV–Vis spectrophotometer. For this, nanoparticle containing samples were subjected to absorption analysis at 200–700 nm range using UV–Vis spectrophotometer (Hitachi U5100). The silver nanoparticles synthesized by the supernatant were further subjected to TEM analysis. This was used to get an insight into size and morphology of the formed silver nanoparticles. Samples for TEM analysis were prepared on carbon-coated copper TEM grids. The films on the TEM grids were allowed to stand for 2 min, then extra solution was removed and the grid was allowed to dry prior to measurement. TEM measurements were recorded using a JEOL-JEM-1011 instrument at 80 kV.

Results

Among the 11 bacterial isolates (CS 1–CS 11) purified from soil samples, CS 11 was randomly selected for the study and was found to have the ability to form silver nanoparticles as observed by change in colour of the reaction. The selected isolate was further subjected to molecular identification by 16S rDNA sequencing-based method. The sequence data were subjected to BLAST analysis and the result showed its maximum identity of 99 % to various *Bacillus* sp. mainly *Bacillus cereus* (Fig. 1). The 16S rDNA sequence of the isolate was submitted to NCBI under the accession number JN835219.

The biosynthesis of silver nanoparticles using both biomass and supernatant were separately investigated primarily through the observation of colour change of the experimental samples in the presence of 1 mM AgNO$_3$. A colour change from pale yellow to brown occurred for both bacterial biomass and supernatant within 24 h of incubation in the presence of light and is shown in Fig. 2. The positive result as observed by the formation of brown colour was maintained throughout the 72-h period of observation. At the same time, experimental control containing heat-killed biomass or supernatant with silver nitrate showed no colour change. This suggests the colour change observed in the bacterial biomass and the supernatant samples was due to the formation of silver nanoparticles. Even though the colour change was observed for samples containing both biomass and supernatant, further experiments were continued only for the extracellular samples. This is because of the comparative advantages of extracellular synthesis over intracellular method.

The colour change observed for the extracellular samples were further confirmed by UV–Vis spectral analysis as part of primary confirmation. Silver nanoparticles are known to have an intense absorption peak in UV absorption spectra due to its surface plasmon excitation. For the culture supernatant of CS 11, an absorption peak was observed at 450 nm and is an indication of formation of

Fig. 1 The phylogenetic analysis of the 16S rDNA sequence of the soil bacterial isolate obtained in the study along with other selected sequences from database. The analysis was conducted using neighbor-joining method in MEGA5. Among the various species of *Bacillus*, the isolate formed cluster with *Bacillus cereus*

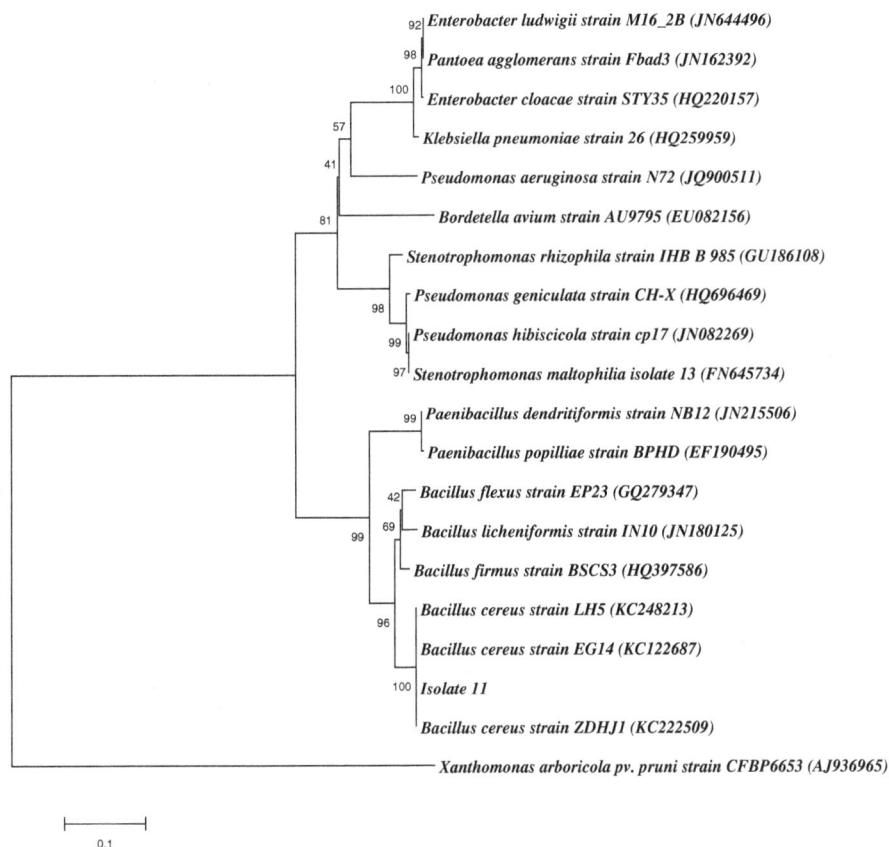

92 *Enterobacter ludwigii* strain M16_2B (*JN644496*)
98 *Pantoea agglomerans* strain Fbad3 (*JN162392*)
100 *Enterobacter cloacae* strain STY35 (*HQ220157*)
Klebsiella pneumoniae strain 26 (*HQ259959*)
57
41 *Pseudomonas aeruginosa* strain N72 (*JQ900511*)
81 *Bordetella avium* strain AU9795 (*EU082156*)
Stenotrophomonas rhizophila strain IHB B 985 (*GU186108*)
98 *Pseudomonas geniculata* strain CH-X (*HQ696469*)
99 *Pseudomonas hibiscicola* strain cp17 (*JN082269*)
97 *Stenotrophomonas maltophilia* isolate 13 (*FN645734*)
99 *Paenibacillus dendritiformis* strain NB12 (*JN215506*)
Paenibacillus popilliae strain BPHD (*EF190495*)
42 *Bacillus flexus* strain EP23 (*GQ279347*)
69 *Bacillus licheniformis* strain IN10 (*JN180125*)
99 *Bacillus firmus* strain BSCS3 (*HQ397586*)
96 *Bacillus cereus* strain LH5 (*KC248213*)
Bacillus cereus strain EG14 (*KC122687*)
100 *Isolate 11*
Bacillus cereus strain ZDHJ1 (*KC222509*)
Xanthomonas arboricola pv. *pruni* strain CFBP6653 (*AJ936965*)

0.1

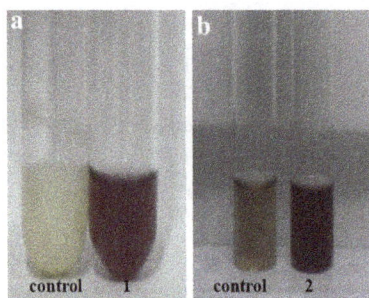

Fig. 2 Visual observation of the biosynthesis of silver nanoparticles by biomass and supernatant of *Bacillus* sp. CS 11 selected for the study. **a1** bacterial biomass with AgNO₃ solution (colour change from *pale yellow* to *brown*) and control heat-killed biomass with AgNO₃ solution (no colour change). **b2** culture supernatant with AgNO₃ solution (colour change from *pale yellow* to *brown*) and control heat-killed supernatant with AgNO₃ solution (no colour change)

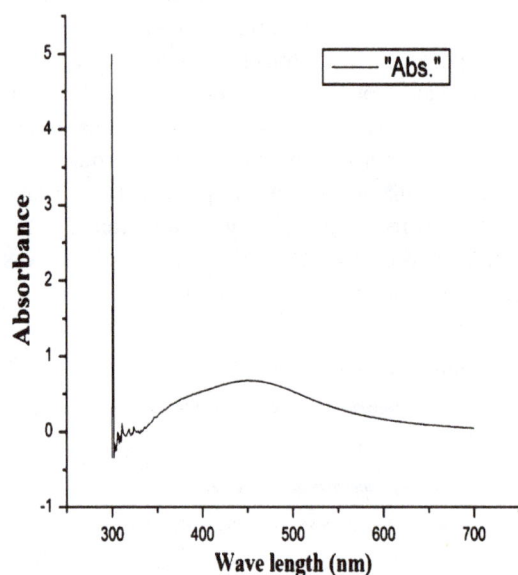

Fig. 3 The UV–Vis absorption spectrum of silver nanoparticles synthesized by supernatant of *Bacillus* sp CS 11. The absorption spectrum of silver nanoparticles exhibited a strong broad peak at 450 nm and observation of such a band is assigned to surface plasmon resonance of the particles

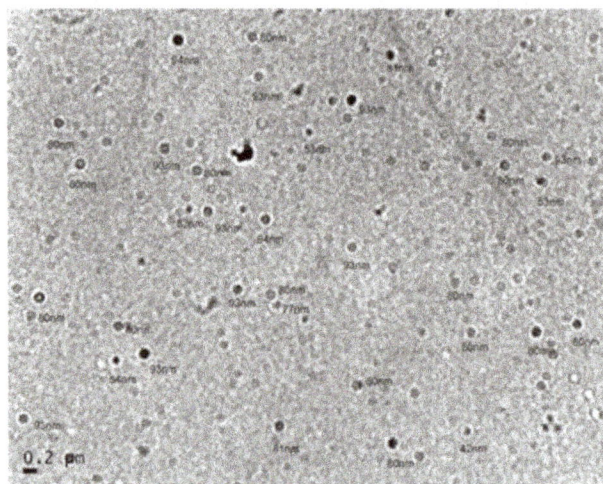

Fig. 4 TEM analysis of the silver nanoparticles synthesized by *Bacillus* sp. CS 11 used in the study. TEM image indicates the size controlled synthesis of silver nanoparticles with the particle size ranging from 42 to 94 nm

Discussion

The biological agents in the form of microbes are efficient candidates for the synthesis of nanoparticles. These biogenic nanoparticles are cost-effective, simpler to synthesize and the method is greener in approach. Silver nanoparticle had drawn much attention because of their extensive application to new technologies in chemistry, electronics, medicine and biotechnology. In the current study, a bacterial strain isolated from heavy metal contaminated soil samples was found to have resistance to silver nitrate. Very interestingly, the selected *Bacillus* strain showed the ability to synthesize silver nanoparticles by both extracellular and intracellular synthesis mechanisms. This was observed by the change in colour from pale yellow to brown. Observation on colour change is a method generally used for screening microbial isolates for silver nanoparticle synthesis (Kalimuthu et al. 2008). The excitation of surface plasmon vibration in the silver nanoparticles was considered as the basis for formation of brown colour. Similar observation was previously reported for the supernatant of *Bacillus megaterium*, where a pale yellow to brown colour was formed due to the reduction of aqueous silver ions to silver nanoparticles (Saravanan et al. 2011). This supports the fact that change in colour as observed in the experiment can be considered as an indication of silver nanoparticles formation. This was further confirmed by UV–Vis spectroscopy which measures the absorption spectra of silver nanoparticles formed due to the collective excitation of conduction electrons in the metal. Thus, methods based on UV–Vis spectroscopy have been shown to be an effective technique for the analysis of nanoparticles (Sastry et al. 1998). UV–Vis spectra of silver

silver nanoparticles (Fig. 3). In addition, these absorption spectra for the silver nanoparticles were obtained within 24 h. The colour change and UV absorption data analysis thus confirms the reduction of AgNO₃ to silver nanoparticles by the culture supernatant of CS 11.

For further confirmation of nanoparticles formed by bacterial supernatant, the samples were subjected to TEM analysis. The result revealed that the silver nanoparticles were generally spherical in shape and in the size range of 42–94 nm (Fig. 4). The size of the nanoparticles can have direct effect on its physico-chemical properties and this can vary among the particles formed by different groups of microorganisms.

nanoparticles synthesized by the selected isolate showed an absorption band at 450 nm by treating the supernatant with AgNO₃. Presence of such peak, assigned to a surface plasmon, was also well documented for silver nanoparticles as reported in the case of *Neurospora crassa* (Longoria et al. 2011).

The mechanism behind the extracellular synthesis of nanoparticles using microbes is not fully known. But it is considered that the enzymes like nitrate reductase secreted by microbes help in the bioreduction of metal ions to metal nanoparticles (Duran et al. 2005). This was reported in *Bacillus licheniformis* where nitrate reductase secreted by the bacteria was found to be responsible for the reduction of Ag⁺ to nanoparticles (Kalimuthu et al. 2008). Detailed analysis of TEM data proved that the silver nanoparticles formed were not in direct contact with each other. This can be taken as a proof of good stabilization of the nanoparticles which can be due to the proteins secreted by microorganisms. The size ranges of silver nanoparticles produced by the CS 11 (42–94 nm) fall closer to the size of silver nanoparticles produced by other bacteria (Gurunathan et al. 2009).

In conclusion, this work demonstrates the silver nanoparticles synthesizing property of bacteria isolated from industrialized area. The isolate CS 11 selected in the study was found to have the potential to form silver nanoparticles extracellularly at room temperature within 24 h. The synthesized silver nanoparticles were characterized by UV–Vis spectroscopy and confirmed by TEM. The nanoparticles formed by the isolate were found to be stable with size range of 42–92 nm which indicate its potential applications.

Acknowledgments The authors gratefully acknowledge School of Chemical Sciences, Mahatma Gandhi University, Kottayam, Kerala, India, for the help and support for the analysis of samples and also DBT-RGYI (Department of Biotechnology-Rapid Grant Young Investigator) Programme for the instrumentation facility.

Conflict of interest The authors declare that they have no conflict of interest in the publication.

References

Ahmad A, Mukherjee P, Senapati S, Mandal D, Khan MI, Kumar R, Sastry M (2003) Extracellular biosynthesis of silver nanoparticles using the fungus *Fusarium oxysporum*. Coll Surf B 28:313–318

Ahmad R, Minaeian S, Shahverdi HR, Jamalifar H, Nohi A (2007) Rapid synthesis of silver nanoparticles using culture supernatants of Enterobacteria: A novel biological approach. Process Biochem 42:919–923

Christopher P, Xin H, Linic S (2011) Visible-light-enhanced catalytic oxidation reactions on plasmonic silver nanostructures. Nat Chem 3:467–472

Duran N, Priscyla D, Marcato PD, Alves O, De Souza G, Esposito E (2005) Mechanistic aspects of biosynthesis of silver nanoparticles by several *Fusarium oxysporum* strains. J Nanobiotechnol 3:1–7

Gade AK, Bonde P, Ingle AP, Marcato PD, Duran N, Rai MK (2008) Exploitation of *Aspergillus niger* for synthesis of silver nanoparticles. J Biobase Mater Bioenergy 2:243–247

Ganesh Babu MM, Gunasekaran P (2009) Production and structural characterization of crystalline silver nanoparticles from *Bacillus cereus* isolate. Coll Surf B 74:191–195

Gurunathan S, Kalishwaralal K, Vaidyanathan R, Deepak V, Pandian SRK, Muniyandi JH, Hariharan N, Eom SH (2009) Biosynthesis, purification and characterization of silver nanoparticles using *Escherichia coli*. Coll Surf B 74:328–335

Jain N, Bhargava A, Majumdar S, Tarafdar JC, Panwar J (2011) Extracellular biosynthesis and characterization of silver nanoparticles using *Aspergillus flavus* NJP08: a mechanism perspective. Nanoscale 3:635–641

Kalimuthu K, Babu RS, Venkataraman D, Bilal M, Gurunathan S (2008) Biosynthesis of silver nanocrystals by *Bacillus licheniformis*. Coll Surf B 65:150–153

Kalishwaralal K, Deepak V, Pandian SRK, Kottaisamy M, BarathManiKanth S, Kartikeyan B, Gurunathan S (2010) Biosynthesis of silver and gold nanoparticles using *Brevibacterium casei*. Coll Surf B 77:257–262

Longoria EC, Nestor ARV, Borja MA (2011) Biosynthesis of silver, gold and bimetallic nanoparticles using the filamentous fungus *Neurospora crassa*. Coll Surf B 83:42–48

Mukherjee P, Roy M, Mandal BP, Dey GK, Mukherjee PK, Ghatak J, Tyagi AK, Kale SP (2008) Green synthesis of highly stabilized nanocrystalline silver particles by a non-pathogenic and agriculturally important fungus *T. asperellum*. Nanotechnology 19:103–110

Nanda A, Saravanan M (2009) Biosynthesis of silver nanoparticles from *Staphylococcus aureus* and its antimicrobial activity against MRSA and MRSE. Nanomedicine 5:452–456

Narayanan KB, Sakthivel N (2010) Biological synthesis of metal nanoparticles by microbe. Adv. Coll Interface Sci 156:1–13

Pugazhenthiran N, Anandan S, Kathiravan G, Prakash NKU, Crawford S, Ashokkumar M (2009) Microbial synthesis of silver nanoparticles by *Bacillus* sp. J Nanopart Res 11:1811–1815

Sadhasivam S, Shanmugam P, KyuSik Yun K (2010) Biosynthesis of silver nanoparticles by *Streptomyces hygroscopicus* and antimicrobial activity against medically important pathogenic microorganism. Coll Surf B 81:358–362

Saifuddin N, Wong CW, NurYasumira AA (2009) Rapid biosynthesis of silver nanoparticles using culture supernatant of bacteria with microwave irradiation. J Chem 6:61–70

Saravanan M, Vemu AK, Barik SK (2011) Rapid biosynthesis of silver nanoparticles from *Bacillus megaterium* (NCIM 2326) and their antibacterial activity on multi drug resistant clinical pathogens. Coll Surf B 88:325–331

Sastry M, Patil V, Sainkar SR (1998) Electrostatically controlled diffusion of carboxylic acid derivatized silver colloidal particles in thermally evaporated fatty amine films. J. Phys Chem B 102:1404–1410

Shivaji S, Madhu S, Singh S (2011) Extracellular synthesise of antibacterial silver nanoparticles using psychrophilic bacteria. Process Biochem 49:830–837

Shrivastava S, Bera T, Singh SK, Singh G, Ramachandrarao Dash P (2009) Characterization of antiplatelet properties of silver nanoparticles. ACS Nano 3:1357–1364

Tamura K, Peterson D, Peterson N, Stecher G, Nei M, Kumar S
    (2011) MEGA5: Molecular evolutionary genetics analysis using
    maximum likelihood, evolutionary distance, and maximum
    parsimony methods. Biol Evol 28:2731–2739
Wei H, Chen C, Han B, Wang E (2008) Enzyme colorimetric assay
    using unmodified silver nanoparticles. Anal Chem 80:7051–7055

Wei X, Luo M, Li W, Yang L, Liang X, Xu L, Kong P, Liu H (2012)
    Synthesis of silver nanoparticles by solar irradiation of cell-free
    Bacillus amyloliquefaciens extracts and AgNO$_3$. Bioresour
    Technol 103:273–278

2

Bacteria viability assessment after photocatalytic treatment

Yanling Cai · Maria Strømme · Ken Welch

Abstract The aim of the present work was to evaluate several methods for analyzing the viability of bacteria after antibacterial photocatalytic treatment. Colony-forming unit (CFU) counting, metabolic activity assays based on resazurin and phenol red and the Live/Dead® *Bac*Light™ bacterial viability assay (Live/Dead staining) were employed to assess photocatalytically treated *Staphylococcus epidermidis* and *Streptococcus mutans*. The results showed conformity between CFU counting and the metabolic activity assays, while Live/Dead staining showed a significantly higher viability post-treatment. This indicates that the Live/Dead staining test may not be suitable for assessing bacterial viability after photocatalytic treatment and that, in general, care should be taken when selecting a method for determining the viability of bacteria subjected to photocatalysis. The present findings are expected to become valuable for the development and evaluation of photocatalytically based disinfection applications.

Keywords Photocatalysis · Bacterial viability · Live/Dead staining · Metabolic activity assays · CFU counting

Introduction

Photocatalysis of titanium dioxide (TiO_2) has been widely investigated and successfully applied in a wide variety of applications such as solar cells (Nah et al. 2010), disinfection, anti-fouling and self-cleaning surfaces (Chen and Poon 2009; Robertson et al. 2012; Sanchez et al. 2012). When the anatase crystalline form of TiO_2 is irradiated with light having a wavelength less than 385 nm, an electron–hole pair is generated as electrons are excited above the material's band gap of 3.2 eV. TiO_2 can also be doped to change the band gap energy and thereby enable the photocatalytic process under visible light (Chatterjee and Dasgupta 2005; Jie et al. 2012; Sheng et al. 2009). The excited electrons can react with oxygen to produce a superoxide ion ($O_2{}^{\cdot-}$), while the positive holes can react with H_2O or OH^- to produce hydroxyl radicals ($\cdot OH$). Further reactions can generate other reactive oxygen species (ROS) like hydroxyl peroxide (H_2O_2) and singlet oxygen ($^1O_2$) (Chen and Poon 2009; Fujishima et al. 2008). The ROS generated by TiO_2 photocatalysis have been proved to provide an antibacterial effect by many researchers (Welch et al. 2010; Li et al. 2008; Sanchez et al. 2012; Robertson et al. 2012). This disinfection ability of photocatalytic materials is due to the high redox reaction ability of the photocatalytic products, and the primary mechanism is thought to be the destruction of the cell membrane or cell wall causing leakage or structural damage of the cell (Maness et al. 1999). Research has demonstrated killing of viruses, Gram-positive and Gram-negative bacteria and even cancer cells by the photocatalysis of TiO_2 (Li et al. 2008; Blake et al. 1999; Welch et al. 2010).

There is a growing interest in applying TiO_2 photocatalysis to disinfection and antibacterial applications (Allahverdiyev et al. 2011; Robertson et al. 2012; Sanchez et al. 2012; Welch et al. 2010; Lilja et al. 2012). To produce reliable results from research, it is critical to have accurate and high-throughput methods for screening

Y. Cai · M. Strømme (✉) · K. Welch (✉)
Division for Nanotechnology and Functional Materials, Department of Engineering Sciences, The Ångström Laboratory, Uppsala University, Box 534, 75121 Uppsala, Sweden
e-mail: maria.stromme@angstrom.uu.se

K. Welch
e-mail: ken.welch@angstrom.uu.se

bacterial viability after photocatalytic treatment. When assessing bacterial viability and, in particular, bacteria in biofilm form, it is often necessary to use several methods in concert to get reliable results. Currently, methods widely used in bacterial viability analysis include indirect methods based on further culture of bacterial samples or direct methods based on molecular probes.

Colony-forming unit (CFU) counting is a conventional indirect method for assessing viability based on cell counting. Given the assumption that each viable bacterium grows and forms a colony, CFU counting method provides advantages like sensitivity (very low concentrations of living bacteria can be determined) and only counts viable cells. However, CFU counting is not a reliable method for bacteria forming clumps or chains, and especially biofilms, which are the prevalent growth form of most bacteria found in nature (Bettencourt et al. 2010). Furthermore, CFU counting after serial dilution and plating is a labor-intensive and time-consuming process, which hinders its application in high-throughput experiments.

A group of indirect methods for quantifying live bacteria, and even biofilms, is based on the detection of metabolic activity. A number of different indicators are used for this purpose, including resazurin (Sandberg et al. 2009), fluorescein diacetate (FDA) (Diaper et al. 1992), tetrazolium salt (XTT) (Belanger et al. 2011) and pH indicators like phenol red (Pantanella et al. 2008). These metabolic activity tests rely on the production of detectable signals resulting from a reaction between the indicator and the metabolite intermediate (e.g., NADPH) or product (e.g., lactic acid) (Peeters et al. 2008). The intensities of the detectable signals are assumed to be proportional to the viability of the bacterial samples, which depend on both the number of bacteria and metabolic rate of the bacteria. An advantage of metabolic activity detection is the ability to avoid or minimize sample manipulation, which makes these methods more suitable for high-throughput screening (Belanger et al. 2011). A limitation of metabolic activity detection is the uncertainty arising from the variation in innate metabolic rates of different bacteria. For example, different strains of the same bacterial species or the same bacteria strain in planktonic or biofilm form may have different growth rates (Welch et al. 2012; Mah and O'Toole 2001; Donlan 2001).

Molecular probe assays are direct methods for bacterial viability detection that do not require further culturing. Cell membrane integrity is typically considered a criterion of cell viability and is, thus, used in molecular probe assays. There are many commercially available kits based on fluorescent dyes, such as the Live/Dead® BacLight™ bacterial viability assay kit (Live/Dead staining) containing SYTO 9 and propidium iodide dyes (Berney et al. 2007; Bar et al. 2009), the redox activity assay based on CTC or

RedoxSensor™ Green reagent (Asadishad et al. 2011) and the BacLight™ Bacterial Membrane Potential Kit based on $DiOC_2(3)$ (Lisle et al. 1999). The Live/Dead staining is a widely used method and utilizes both SYTO 9, which has a green fluorescence emission and stains both live and dead bacterial DNA, and propidium iodide, which has a red fluorescence emission and penetrates only damaged cell membranes. When the fluorescence is measured directly (e.g., with a microplate reader) or combined with flow cytometry, bacterial viability can be detected rapidly and accurately (Berney et al. 2007), while when assessed with fluorescent microscopy or laser scanning confocal microscopy (LSCM), regions of varying viability can be differentiated with imagery (Wierzchos et al. 2004).

In this work, we evaluated and compared several methods for analyzing bacteria treated with TiO_2 photocatalysis, including CFU counting, metabolic activity assays based on resazurin and phenol red and Live/Dead staining. The methods were applied on two different bacterial strains: *Staphylococcus epidermidis* and *Streptococcus mutans*. *S. epidermidis* was chosen because it is a common cause of infections on skin-penetrating implants (Collinge et al. 1994; Mahan et al. 1991), and such infections are of interest to prevent using photocatalysis (Lilja et al. 2012). *S. mutans* was chosen because it is one of the initial colonizers in the formation of dental plaque and plays an important role in acid production leading to the development of dental cavity (Banas 2004). Applications of photocatalysis in dental materials could be used, for example, to reduce secondary dental caries following dental restoration (Welch et al. 2010).

Materials and methods

Bacterial strains and culture medium

Two bacterial strains, *S. epidermidis* (CCUG 18000A) and *S. mutans* (NCTC 10449), were employed to evaluate the conformity of the different viability quantification methods after TiO_2 photocatalytic treatment. *S. epidermidis* was employed in planktonic form, while *S. mutans* was employed in both planktonic and biofilm form, depending on the quantification method. *S. epidermidis* was inoculated in 20 mL cation-adjusted Mueller–Hinton (MH) Broth (Fluka, Sigma-Aldrich Chemie GmbH, Steinheim, Germany) and cultured at 37 °C under agitation to late log phase. *S. mutans* was inoculated into Brain–Heart Infusion (BHI) broth (Fluka, Sigma-Aldrich Chemie GmbH, Steinheim, Germany) culture medium and cultured overnight at 37 °C.

Before the photocatalytic treatment, bacteria were collected by centrifugation (4,000 rpm, 10 min, EBA 30 centrifuge, Hettich, Tuttlingen, Germany) and re-suspended in sterile deionized water to achieve the desired concentration of bacteria for the tests involving planktonic bacteria. Bacterial concentration was determined by optical density measurements. The *S. mutans* biofilm preparation procedure is described below in the section for LSCM.

Photocatalytic test surfaces

In this study, resin-based nanocomposite disks comprising a dental adhesive containing TiO_2 nanoparticles, hereafter referred to as NP adhesives, were used as a standard photocatalytic surface. The photocatalytic nanoparticles used in this work were P25 TiO_2 nanoparticles (lot number 4166031598, Evonik Industries (previously Degussa) AG, Germany), which consist of anatase and rutile crystalline phases of TiO_2 in a ratio of about 3:1. The average sizes of anatase and rutile elementary particles are 25 nm and 85 nm, respectively (Ohno et al. 2001; Kirchnerova et al. 2005). NP adhesives have been proved to possess sufficient photocatalytic activity for achieving bacteria (Welch et al. 2010) and even biofilm elimination (Cai et al. 2013).

The light-cured dental adhesive resin was made by mixing 2, 2-bis [4-(2-hydroxy-3- methacryloxypropoxy) phenyl]-propane (BisGMA, Polysciences Europe GmbH, Eppelheim, Germany) and 2-hydroxyethyl methacrylate (HEMA, Sigma-Aldrich, Schnelldorf, Germany) in a 55/45 wt/wt ratio. Photoinitiator and coinitiators were added as follows: 0.5 mol % camphorquinone (CQ); 0.5 mol % 2-(dimethylamino) ethyl methacrylate (DMAEMA); 0.5 mol % ethyl-4-(dimethylamino) benzoate (EDMAB); and 1 wt % diphenyliodonium hexafluorophosphate (DPIHP) (all from Sigma-Aldrich, Steinheim, Germany).

The NP adhesive disks were made by mixing 20 wt % P25 TiO_2 nanoparticles with the adhesive resin. The disks were cast in circular Teflon molds (diameter 8 mm, thickness 1 mm) and light-cured with 460 nm light for 30–40 s (BlueLEX GT1200, Monitex, Taiwan) under N_2 flow. Sample disks were randomly grouped for the different viability test methods.

Antibacterial treatments

Prior to antibacterial treatment, NP adhesive disks were first sterilized and cleaned in an ultrasonic bath of 70 % ethanol for 30 min. The disks were then washed twice with sterile deionized water and air dried at room temperature.

Antibacterial treatment for comparison of viability assessment of planktonic bacteria with CFU counting, metabolic activity assays and Live/Dead staining

The bactericidal effect of photocatalytic treatment as a function of UV-A dose was evaluated with CFU counting,

metabolic activity assays and Live/Dead staining combined with fluorescent intensity measurements. For antibacterial tests with both planktonic *S. epidermidis* and *S. mutans*, 10 µL of bacterial suspension (bacterial population $\sim 10^7$) was spread on each NP adhesive disk using a pipette tip. The disks with bacteria were irradiated with a high-power UV-A diode ($\lambda = 365$ nm, NSCU033B(T), Nichia, Japan). A collimating lens ensured an even UV-A light intensity of 15 mW/cm^2 over the irradiated area (UV light meter, UV-340, Lutron), and the treatment times were varied to provide UV-A doses ranging from 0 to 13.6 J/cm^2. The 0 J/cm^2 dose refers to control disks that were not exposed to UV-A light and were included to provide a reference level for determining the log reduction in viability of the samples subjected to UV-A irradiation. Four disks at each UV-A dose and for each bacteria strain were irradiated. The disks were inspected for moisture loss on the surface so that any bactericidal effect due to desiccation would be minimized. After photocatalytic treatment, each disk was immediately put into a well in a 48-well plate containing 100 µL of sterile water. The 48-well plate was then fixed to an incubating orbital shaker (Talboys, Troemner, USA) and shaken at 500 rpm for 2 min to re-suspend the bacteria from the disk surfaces. The sample disks were removed from the wells and bacterial viability was subsequently analyzed.

From the 100 µL of bacteria suspension of *S. epidermidis* after each test, 10 µL was taken for CFU counting, 10 µL for the metabolic assay incorporating resazurin and 50 µL for fluorescence intensity measurements following Live/Dead staining. From the 100 µL of bacteria suspension of *S. mutans* after each test, 10 µL was taken for the metabolic assay incorporating resazurin, 10 µL for the metabolic assay incorporating phenol red and 50 µL for fluorescence intensity measurements following Live/Dead staining.

Antibacterial treatment for comparison of viability assessment of planktonic bacteria with CFU counting and flow cytometry

To further assess Live/Dead staining, a comparison between CFU counting and flow cytometry was performed. Ten microliters of planktonic *S. epidermidis* bacterial suspension with a bacterial population of 10^8 was spread on an NP adhesive disk using a pipette tip and illuminated with a UV-A dose of 42 J/cm^2 to ensure a strong bactericidal effect. After the photocatalytic treatment, bacteria were re-suspended from the disk surface into 2 mL of sterile water. Ten microliters of bacterial suspension was taken for CFU counting, while the remainder was taken for Live/Dead staining and subsequent analysis with flow cytometry. To provide a control sample, a suspension of untreated planktonic *S. epidermidis* (10^8 CFU in 2 mL) was analyzed with flow cytometry after Live/Dead staining.

*Antibacterial treatment for assessing viability of an
S. mutans biofilm with laser scanning confocal microscopy
(LSCM)*

The viability of 16-h-old *S. mutans* biofilm was assessed after photocatalytic treatment using LSCM. Three NP adhesive disks were first incubated with *S. mutans* (10^6 CFU/mL) in BHIS broth for 4 h at 37 °C. The disks were then cultured in fresh BHIS broth in an orbital shaking incubator (100 rpm, Talboy) at 37 °C for 16 h to induce biofilm formation. For photocatalytic treatment, a UV-A irradiation dose of 40 J/cm^2 was applied to one of the NP adhesive disks coated with the 16-h-old biofilm. The other biofilm-coated NP adhesive disks were used as a live control and a dead control in which the bacteria were killed by immersing in 70 % ethanol.

Methods for analyzing bacterial viability after photocatalytic treatment

Six methods were used for assessing bacterial viability after photocatalytic treatment: CFU counting, metabolic activity assays based on resazurin and phenol red and Live/ Dead staining viability assays combined with fluorescent intensity measurements, flow cytometry and LSCM.

CFU counting

Ten microliters of *S. epidermidis* bacterial suspension was taken from the bacterial suspension after the photocatalytic treatment for CFU counting. A dilution series was performed to achieve a suitable amount of bacteria on the LB agar plates (Sigma-Aldrich, Steinheim, Germany) for counting. The LB agar plates were cultured at 37 °C overnight and the resulting CFUs on the agar plates were imaged with a digital microscope (Dino Lite, Netherlands) and counted with the aid of the software Dotcount (developed by Martin Reuter, MIT, MA, USA). CFU counting was not used with *S. mutans* as testing showed a significant tendency for *S. mutans* cells to aggregate and form clumps of several cells, resulting in a gross underestimation of viability when using CFU counting.

Metabolic activity assay based on resazurin

The metabolic activity assay based on resazurin is an indirect method used to evaluate the viability of bacteria by measuring the accumulation of resorufin (pink in color and highly fluorescent), which is the reaction product of resazurin (blue in color and non-fluorescent) and reductive metabolic intermediates. Ten microliters of bacterial suspension was taken from the 100 μL bacterial suspension after the

photocatalytic treatment for resazurin bacterial viability testing. For *S. epidermidis*, the 10 μL of bacterial suspension was added to 200 μL of MH broth containing resazurin (1.25 μg/mL) in a 96-well plate. A tenfold dilution series of living *S. epidermidis* bacterial suspension, from 10^7 to 10 CFUs/well, was also prepared and placed in the 96-well plate to calibrate the number of surviving bacteria after photocatalytic treatment on the sample disks. The 96-well plate containing the resazurin assay was incubated at 37 °C, and color change (from blue to pink) due to reduction of non-fluorescent resazurin to pink resorufin by the bacterial metabolic activity was automatically recorded with a digital camera every 10 min. The initial number of surviving bacteria in the test wells was determined by comparing the time for color change to the *S. epidermidis* calibration series. The viability of *S. mutans* was evaluated with the same procedure used with *S. epidermidis* except that the assay culture media was BHI broth with 2.5 μg/mL resazurin instead of MH broth with 1.25 μg/mL resazurin.

Metabolic activity assay based on phenol red

The metabolic activity assay based on phenol red is an indirect method used to evaluate the amount of viable bacteria, which is related to a pH change in a culture medium containing the bacteria resulting from the accumulation of metabolic acid products. In this study, the phenol red assay was only used to determine the viability of *S. mutans* since they readily produce acidic metabolites. The assay changes color from red to yellow, due to accumulation of lactic acid, which is a sucrose metabolic byproduct produced by *S. mutans*. The assay culture media consisted of BHI broth plus 2 % sucrose and 25 mg/L of the pH indicator phenol red (BHIS–PR broth). The pH of the BHIS–PR broth was adjusted to 7.10 before autoclaving. The same batch of BHIS–PR broth was used for both the calibration curve and viability testing to avoid variances caused by difference in broth media (All chemicals were obtained from Sigma-Aldrich, Steinheim, Germany).

Ten microliters of bacterial suspension was taken from 100 μL of bacterial suspension after photocatalytic treatment and added to 1.5 mL of BHIS–PR broth in a 48-well plate (Nunclon® Δ Multidishes, Thermo Fisher Scientific, Germany). A calibration concentration series of *S. mutans* ranging from 10^7 to 10 CFUs/well was also cultured parallel with the photocatalytic samples. The 48-well plate was incubated at 37 °C and the color of the wells containing the culture medium was automatically recorded every 10 min with a digital camera. The initial number of surviving bacteria in the test wells was determined by comparing the time of color change to the *S. mutans* calibration series.

Fifty microliters of bacterial suspension was taken from the 100 μL bacterial suspension after photocatalytic treatment for assessment of viability with Live/Dead staining (Live/Dead® *Bac*Light™ bacterial viability assay kit, L13152, Invitrogen, Eugene, USA). For both *S. epidermidis* and *S. mutans*, calibration curves were performed according to the product instructions. For each sample, fluorescence intensity was measured at an emission wavelength at 530 nm (green) and 620 nm (red) using an excitation wavelength at 485 nm (Infinite 200 microplate reader, Tecan, Switzerland). The ratio of green/red fluorescence intensities was calculated and compared to the calibration curves from samples of known viability to determine the viability of the photocatalytically treated bacteria.

Live/Dead staining combined with flow cytometery

After the photocatalytic treatment, approximately 2 mL of both untreated and treated *S. epidermidis* suspension was stained with Live/Dead stain according to the product instructions and the viability of the sample was determined with multi-laser analytical flow cytometry (LSR II, BD Biosciences).

Live/Dead staining combined with LSCM

After photocatalytic treatment, the *S. mutans* biofilm on the surface of the NP adhesive was stained with the Live/Dead staining kit according to the product instructions. A control NP adhesive disk with live 16-h-old *S. mutans* biofilm and a control NP adhesive disk with 16-h-old *S. mutans* biofilm treated with 70 % ethanol to kill the biofilm were also stained with the Live/Dead staining kit. The viability of the biofilm samples was assessed by imaging the samples with an LSCM (LSM 510 META, Carl Zeiss MicroImaging GmbH, Jena, Germany) using an excitation wavelength of 488 nm.

Statistical tests

The Student's *t* test was employed to determine if statistically significant differences existed between measured bactericidal effects using the different viability assessment methods.

Results

Viability measurements of planktonic *S. epidermidis* as a function of UV-A dose

Figure 1 shows the bacterial viability of planktonic *S. epidermidis* after photocatalytic antibacterial treatment, as

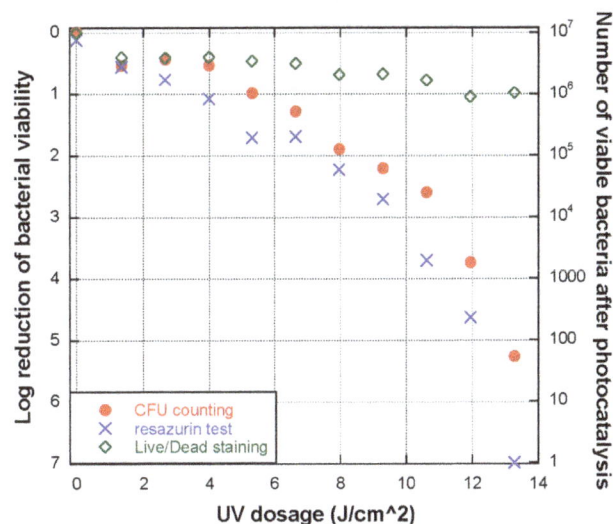

Fig. 1 Bacterial viability of planktonic *S. epidermidis* after photocatalytic antibacterial treatment, measured with CFU counting, metabolic activity assay incorporating resazurin and Live/Dead staining. Each data point is the average of four tests; the standard deviations are within 0.63 log

measured with CFU counting, metabolic activity assay incorporating resazurin and Live/Dead staining with fluorescent intensity measurements.

The resazurin metabolic assay and CFU counting show a similar trend in that an increasing dose of UV-A irradiation leads to a greater antibacterial effect. It can be observed from Fig. 1 that an *S. epidermidis* population of $\sim 10^7$ CFUs on the Φ 8 mm NP adhesive disks can be disinfected by photocatalysis with UV-A dose of 13.6 J/cm^2 since a reduction of greater than 5 log was achieved. However, Live/Dead staining provided a much higher measure of bacterial viability at higher UV doses (less than 1 log reduction at UV doses of 4–13.6 J/cm^2). For UV doses greater than 5 J/cm^2, Live/Dead staining showed a statistically higher viability than both CFU counting and the resazurin assay (Student's *t* test, $p < 0.005$).

Viability measurements of planktonic *S. mutans* as a function of UV-A dose

Figure 2 shows the quantification of viable planktonic *S. mutans* after photocatalytic treatment with UV-A doses ranging from 0 to 13.6 J/cm^2. For assessing *S. mutans* viability, three methods were employed: metabolic activity assay based on phenol red, metabolic activity assay based on resazurin and Live/Dead staining.

The metabolic assays incorporating phenol red and resazurin provided similar measures of bacterial viability, which indicate that an increasing dose of UV-A irradiation leads to an increasing antibacterial effect, as expected. From Fig. 2, it can be observed that an *S. mutans*

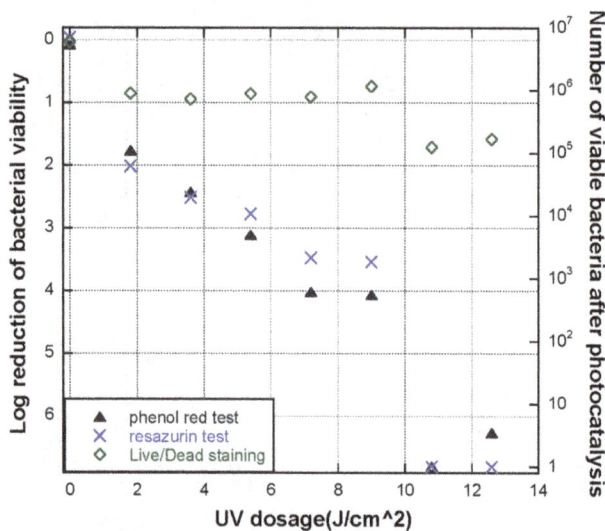

Fig. 2 Bacterial viability of planktonic *S. mutans* after the photocatalytic antibacterial treatment, measured with metabolic activity assays incorporating phenol red and resazurin, respectively, and Live/Dead staining. Each data point is the average of four tests; the standard deviations are within 0.88 log

population of $\sim 10^7$ CFUs on Φ 8 mm NP adhesive disks can be disinfected (>5 log reduction) by photocatalysis with a UV-A dose of 10 J/cm^2 or greater. As with the viability measurements of *S. epidermidis*, Live/Dead staining indicates a much higher bacterial viability than the metabolic assays, see Fig. 2. For all tested non-zero UV doses, Live/Dead staining showed a statistically higher viability than both the resazurin and phenol red assays (Student's *t* test, $p < 0.005$). With a UV-A dose of 13.6 J/

cm^2, Live/Dead staining shows less than a 2 log reduction of bacteria, while both metabolic activity assays show more than 6 log bacterial reduction.

Bacterial viability evaluation of photocatalytically treated planktonic *S. epidermidis* based on Live/Dead staining and flow cytometry

Figure 3 shows flow cytometry analysis of a control *S. epidermidis* sample (a) and an *S. epidermidis* sample after being subject to a UV-A dose of 42 J/cm^2 (b). Both samples were stained with the Live/Dead stain kit prior to analysis with flow cytometry. In Fig. 3a, in which the *S. epidermidis* sample was not treated with UV-A light, it could be observed that 16 % of the cells were non-viable while 75 % were active. Figure 3b shows that after a UV-A irradiation dose of 42 J/cm^2, 51 % of *S. epidermidis* population was non-viable while 39 % was alive. However, a CFU counting analysis of the same sample of photocatalytically treated *S. epidermidis* displayed in Fig. 3b showed that only 7 of 10^6 bacteria survived. The results from CFU counting are in contrast to the Live/Dead staining results when combined with flow cytometry as shown in Fig. 3b.

Bacterial viability evaluation of photocatalytically treated *S. mutans* biofilm based on Live/Dead staining and LSCM

Figure 4a shows an LSCM image of a photocatalytically treated *S. mutans* biofilm after a UV-A irradiation dose of 40 J/cm^2. Figure 4b shows an LSCM image from an

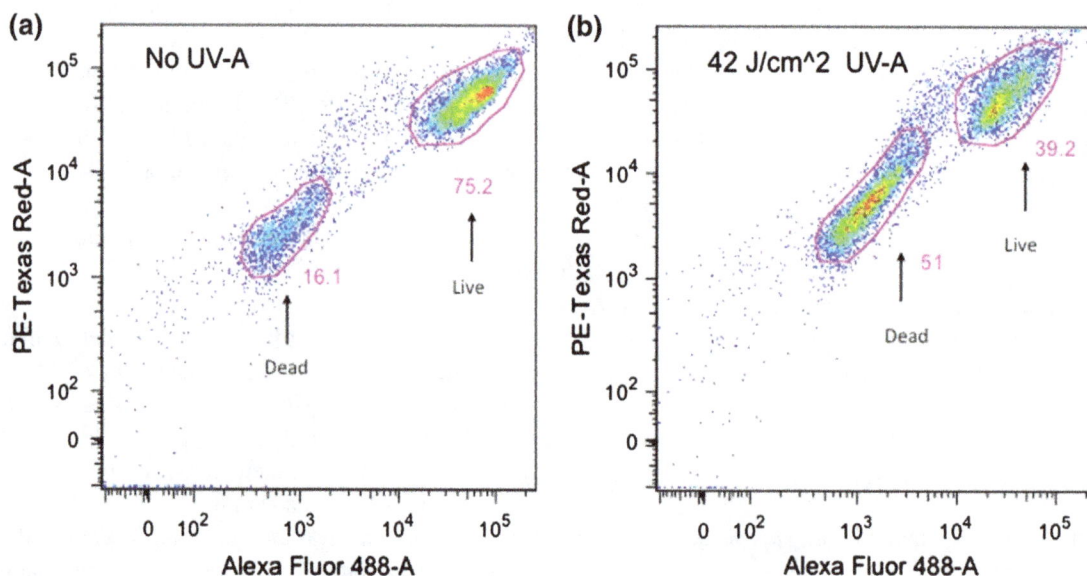

Fig. 3 Planktonic *S. epidermidis* viability assessed with flow cytometry utilizing Live/Dead staining. **a** A control sample of *S. epidermidis* without photocatalytic treatment. **b** The viability of an *S. epidermidis* sample subjected to a UV-A irradiation dose of 42 J/cm^2 on an NP adhesive disk

Fig. 4 *S. mutans* biofilm with Live/Dead staining and imaged with LSCM. The *green* signal is due to the dye SYTO9, indicating alive cells while the *red* signal is due to propidium iodide which marks the dead cells. **a** Photocatalytically treated biofilm with a UV-A irradiation dose of 40 J/cm^2; **b** control of dead biofilm; **c** control of living biofilm

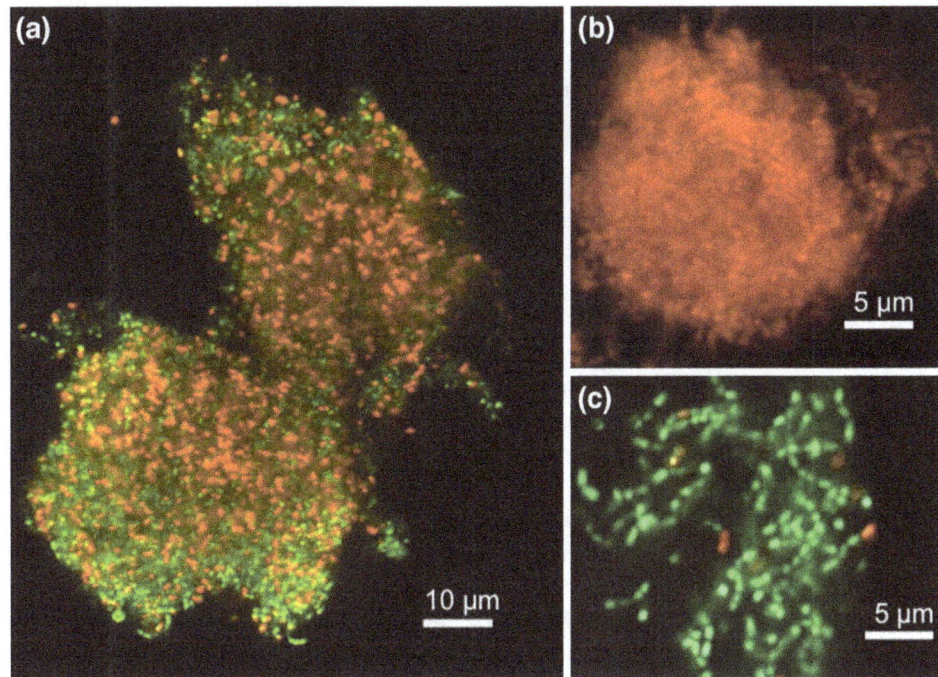

S. mutans biofilm treated with 70 % ethanol and Fig. 4c displays an LSCM image of an untreated *S. mutans* biofilm. From the figures, green and/or red signals can be observed, which represent living and dead bacteria, respectively. Thus, Fig. 4b indicates that the 70 % ethanol treatment effectively killed the biofilm since no green signal was observed. Conversely, Fig. 4a shows qualitatively that a large part of the photocatalytically treated biofilm is alive.

Discussion

In this work, different methods were used in the analysis of *S. epidermidis* and *S. mutans* bacterial viability after photocatalysis treatment. An important issue raised from the above results is the disagreement of Live/Dead staining data compared to both CFU counting and the two types of metabolic activity assays. CFU counting, the resazurin assay and the phenol red assay all showed the same tendency of bacterial viability to decrease with a corresponding increase in UV-A light irradiation, whereas Live/Dead staining indicated a much higher level of viability in the bacteria samples subjected to photocatalytic treatment. This tendency for Live/Dead staining to indicate a higher viability in photocatalytically treated bacteria compared to other methods can be observed in Figs. 1 and 2 where Live/Dead staining was quantified through fluorescent intensity measurements in a multiplate reader and in Fig. 3 where Live/Dead staining was combined with flow cytometry.

From Fig. 4, it can even be seen that LSCM with Live/Dead staining showed a high degree of viability in an *S. mutans* biofilm that had been subjected to photocatalytic treatment with a high UV-A dose. It has been previously shown that a similar UV-A dose on *S. mutans* biofilm cultured on NP adhesives has a potent bactericidal effect (Cai et al. 2013). In these tests, a metabolic activity assay incorporating phenol red was used to assess viability and showed a 5 log reduction in viability.

The reason for the discrepancy between the Live/Dead staining results and other methods could be related to the criteria for bacterial viability utilized by the different methods. For example, CFU counting examines the number of viable bacteria that can form colonies on a broth agar plate, while metabolic activity assays assess the accumulation of metabolic product or intermediate, which depends on both the number and metabolic rate of bacteria. As mentioned previously. CFU counting can sometimes provide an underestimation of viability if the bacterial cells aggregate, and this was the reason *S. mutans* was not assessed with CFU counting. On the other hand, when comparing the CFU results with the resazurin assay for planktonic *S. epidermidis*, similar estimations of viability were found, where the CFU counting results indicated slightly higher viability. Live/Dead staining is based on assessment of the bacterial membrane integrity with the help of two nucleic acid dyes, SYTO 9 and propidium iodide. SYTO9 can permeate the cell membrane of both dead and living cells, while propidium iodide can only permeate damaged cell membranes, resulting in dead bacteria producing a red fluorescence signal and live bacteria producing a green signal. It is generally accepted that ROS generated during TiO_2 photocatalysis attack the bacterial cell wall and/or membrane and are responsible for killing the bacteria (Maness et al. 1999). However, it appears that even though the ROS attack causes a reduction or total loss of normal cellular function, the membrane integrity (as probed by propidium iodide) may not be significantly affected. Regardless of the mechanism that gives rise to the higher measure of viability when using Live/Dead staining, the results in this study suggest that this method may not be suitable for the analysis of bacterial viability following photocatalytic treatments.

Returning to Fig. 1, it is interesting to note that viability assessed with the resazurin assay consistently showed a slightly lower viability than that determined from CFU counting, for tests involving UV-A irradiation. A possible explanation for this could be related to the recovery of some bacteria subjected to the photocatalytic treatment. It is known that bacteria that are sub-lethally injured due to ROS exposure can recover under optimum environmental conditions (Rizzo 2009). This would result in a delay of growth and division and consequently exhibit itself as a lower signal in a metabolic assay due to the lower/delayed metabolic activity of the affected cells. However, CFU counting would not necessarily distinguish between a healthy cell and a damaged cell that recovered from its injuries if both result in a countable colony at a later time point. Support for this hypothesis was found by observing the agar plates containing the *S. epidermidis* samples during the incubation time prior to CFU counting. While colonies formed by a control sample of healthy *S. epidermidis* not subjected to UV-A light appeared on the agar plate at approximately the same time and were of the same size at the time of counting, colonies formed from samples subjected to UV-A irradiation appeared visibly at different times during the incubation period and were of different sizes at the time of counting.

When choosing an appropriate method for assessing viability in antibacterial testing, it is important to consider the mechanism by which the method probes viability. Often a combination of methods is required to give a more certain indication of viability. Each method has unique criteria for determining bacterial viability. CFU counting shows the number of living bacteria; metabolic activity assays show the multiplication and metabolic rate of an amount of living bacteria; and molecular probe methods examine the membrane integrity. The sensitivity of the various methods is also an important issue to consider in practice. For example, CFU counting is suitable for examining very low concentrations of living bacteria, but is only reliable for assessing bacterial populations where one can be certain that individual cells can be well separated from each other on the plate. Metabolic activity assays are also applicable for sample showing more than 6 log reduction of viability, as demonstrated in this study, and because the technique can largely avoid sample manipulation (Pantanella et al. 2008), it is suitable for assessing the viability of both planktonic and biofilm forms of bacteria. For the antibacterial tests not involving photocatalysis, Live/Dead staining has been used to analyze viability, visualize both viability and distribution of live and dead cells and analyze samples containing multiple bacterial species.

Conclusions

Multiple methods were compared for the assessment of bacterial viability after photocatalytic treatment. The results of CFU counting and metabolic activity assays incorporating resazurin and phenol red showed good agreement with each other, while tests based on the Live/Dead staining differed significantly, showing a much higher viability. Our results suggest that the use of Live/Dead staining may not be applicable to the assessment of bacterial viability following antibacterial photocatalytic treatments. The present findings are expected to become valuable for the development and evaluation of photocatalytically based sterilization applications in, e.g., medicine and dentistry.

Acknowledgments We greatly acknowledge the Carl Trygger Foundation, Göran Gustafsson Foundation, Swedish Research Council, Vinnova and Swedish Foundation For Strategic Research for financially supporting this work.

Conflict of interest The authors declare that they have no conflict of interest.

References

Allahverdiyev AM, Abamor ES, Bagirova M, Rafailovich M (2011) Antimicrobial effects of TiO_2 and Ag_2O nanoparticles against drug-resistant bacteria and leishmania parasites. Future Microbiol 6(8):933–940

Asadishad B, Ghoshal S, Tufenkji N (2011) Method for the direct observation and quantification of survival of bacteria attached to negatively or positively charged surfaces in an aqueous medium. Environ Sci Technol 45(19):8345–8351

Banas JA (2004) Virulence properties of *Streptococcus mutans*. Front Biosci 9:1267–1277

Bar W, Bade-Schumann U, Krebs A, Cromme L (2009) Rapid method for detection of minimal bactericidal concentration of antibiotics. J Microbiol Methods 77(1):85–89

Belanger PA, Beaudin J, Roy S (2011) High-throughput screening of microbial adaptation to environmental stress. J Microbiol Methods 85(2):92–97

Berney M, Hammes F, Bosshard F, Weilenmann HU, Egli T (2007) Assessment and interpretation of bacterial viability by using the LIVE/DEAD *Bac*Light kit in combination with flow cytometry. Appl Environ Microbiol 73(10):3283–3290

Bettencourt P, Pires D, Carmo N, Anes E (2010) Application of confocal microscopy for quantification of intracellular mycobacteria in macrophages. FORMATEX Research Center, Badajoz

Blake DM, Maness PC, Huang Z, Wolfrum EJ, Huang J, Jacoby WA (1999) Application of the photocatalytic chemistry of titanium dioxide to disinfection and the killing of cancer cells. Sep Purif Method 28(1):1–50

Cai YL, Strømme M, Melhus Å, Engqvist H, Welch K (2013) Photocatalytic inactivation of biofilms on bioactive dental adhesives. J Biomed Mater Res B (in press)

Chatterjee D, Dasgupta S (2005) Visible light induced photocatalytic degradation of organic pollutants. J Phototch Photobio C 6(2–3): 186–205

Chen J, Poon CS (2009) Photocatalytic construction and building materials: from fundamentals to applications. Build Environ 44(9):1899–1906

Collinge CA, Goll G, Seligson D, Easley KJ (1994) Pin tract infections: silver vs uncoated pins. Orthopedics 17(5):445–448

Diaper JP, Tither K, Edwards C (1992) Rapid assessment of bacterial viability by flow-cytometry. Appl Microbiol Biotechnol 38(2): 268–272

Donlan RM (2001) Biofilm formation: a clinically relevant microbiological process. Clin Infect Dis 33(8):1387–1392

Fujishima A, Zhang XT, Tryk DA (2008) TiO$_2$ photocatalysis and related surface phenomena. Surf Sci Rep 63(12):515–582

Jie H, Lee HB, Chae KH, Huh MY, Matsuoka M, Cho SH, Park JK (2012) Nitrogen-doped TiO$_2$ nanopowders prepared by chemical vapor synthesis: band structure and photocatalytic activity under visible light. Res Chem Intermediat 38(6):1171–1180

Kirchnerova J, Cohen MLH, Guy C, Klvana D (2005) Photocatalytic oxidation of *n*-butanol under fluorescent visible light lamp over commercial TiO$_2$ (Hombicat UV100 and Degussa P25). Appl Catal A-Gen 282(1–2):321–332

Li QL, Mahendra S, Lyon DY, Brunet L, Liga MV, Li D, Alvarez PJJ (2008) Antimicrobial nanomaterials for water disinfection and microbial control: potential applications and implications. Water Res 42(18):4591–4602

Lilja M, Welch K, Astrand M, Engqvist H, Strømme M (2012) Effect of deposition parameters on the photocatalytic activity and bioactivity of TiO$_2$ thin films deposited by vacuum arc on Ti-6Al-4V substrates. J Biomed Mater Res B 100B(4):1078–1085

Lisle JT, Pyle BH, McFeters GA (1999) The use of multiple indices of physiological activity to access viability in chlorine disinfected *Escherichia coli* O157: H7. Lett Appl Microbiol 29(1):42–47

Mah TFC, O'Toole GA (2001) Mechanisms of biofilm resistance to antimicrobial agents. Trends Microbiol 9(1):34–39

Mahan J, Seligson D, Henry SL, Hynes P, Dobbins J (1991) Factors in pin tract infections. Orthopedics 14(3):305–308

Maness PC, Smolinski S, Blake DM, Huang Z, Wolfrum EJ, Jacoby WA (1999) Bactericidal activity of photocatalytic TiO$_2$ reaction: toward an understanding of its killing mechanism. Appl Environ Microbiol 65(9):4094–4098

Nah YC, Paramasivam I, Schmuki P (2010) Doped TiO$_2$ and TiO$_2$ nanotubes: synthesis and applications. Chem Phys Chem 11(13):2698–2713

Ohno T, Sarukawa K, Tokieda K, Matsumura M (2001) Morphology of a TiO$_2$ photocatalyst (Degussa, P-25) consisting of anatase and rutile crystalline phases. J Catal 203(1):82–86

Pantanella F, Valenti P, Frioni A, Natalizi T, Coltella L, Berlutti F (2008) BibTimer Assay, a new method for counting *Staphylococcus* spp. in biofilm without sample manipulation applied to evaluate antibiotic susceptibility of biofilm. J Microbiol Methods 75(3):478–484

Peeters E, Nelis HJ, Coenye T (2008) Comparison of multiple methods for quantification of microbial biofilms grown in microtiter plates. J Microbiol Methods 72(2):157–165

Rizzo L (2009) Inactivation and injury of total coliform bacteria after primary disinfection of drinking water by TiO$_2$ photocatalysis. J Hazard Mater 165(1–3):48–51

Robertson PKJ, Robertson JMC, Bahnemann DW (2012) Removal of microorganisms and their chemical metabolites from water using semiconductor photocatalysis. J Hazard Mater 211:161–171

Sanchez B, Sanchez-Munoz M, Munoz-Vicente M, Cobas G, Portela R, Suarez S, Gonzalez AE, Rodriguez N, Amils R (2012) Photocatalytic elimination of indoor air biological and chemical pollution in realistic conditions. Chemosphere 87(6):625–630

Sandberg ME, Schellmann D, Brunhofer G, Erker T, Busygin I, Leino R, Vuorela PM, Fallarero A (2009) Pros and cons of using resazurin staining for quantification of viable *Staphylococcus aureus* biofilms in a screening assay. J Microbiol Methods 78(1):104–106

Sheng GD, Li JX, Wang SW, Wang XK (2009) Modification to promote visible-light catalytic activity of TiO$_2$. Prog Chem 21(12):2492–2504

Welch K, Cai YL, Engqvist H, Strømme M (2010) Dental adhesives with bioactive and on-demand bactericidal properties. Dent Mater 26(5):491–499

Welch K, Cai YL, Strømme M (2012) A method for quantitative determination of biofilm viability. J Funct Biomater 3(2):418–431

Wierzchos J, De los Rios A, Sancho LG, Ascaso C (2004) Viability of endolithic micro-organisms in rocks from the McMurdo Dry Valleys of Antarctica established by confocal and fluorescence microscopy. J Microsc 216:57–61

3

Mycoremediation of Benzo[a]pyrene by *Pleurotus ostreatus* in the presence of heavy metals and mediators

Sourav Bhattacharya · Arijit Das ·
Kuruvalli Prashanthi · Muthusamy Palaniswamy ·
Jayaraman Angayarkanni

Abstract Benzo[a]pyrene is considered as a priority pollutant because of its carcinogenic, teratogenic and mutagenic effects. The highly recalcitrant nature of Benzo[a]pyrene poses a major problem for its degradation. White-rot fungi such as *Pleurotus ostreatus* can degrade Benzo[a]pyrene by enzymes like laccase and manganese peroxidase. The present investigation was carried out to determine the extent of Benzo[a]pyrene degradation by the PO-3, a native isolate of *P. ostreatus,* in the presence of heavy metals and ligninolytic enzyme mediators. Modified mineral salt medium was supplemented with 5 mM concentration of different heavy metal salts and ethylene-diaminetetraacetic acid. Vanillin and 2,2'-azinobis-(3-ethylbenzothiazoline-6-sulfonate) (1 and 5 mM) were used to study the effect of mediators. Results indicated that *P. ostreatus* PO-3 degraded 71.2 % of Benzo[a]pyrene in the presence of copper ions. Moderate degradation was observed in the presence of zinc and manganese. Both biomass formation and degradation were severely affected in the presence of all other heavy metal salts used in the study. Copper at 15 mM concentration supported the best degradation (74.2 %), beyond which the degradation progressively reduced. Among the mediators, 1 mM 2,2'-azinobis-(3-ethylbenzothiazoline-6-sulfonate) supported 78.7 % degradation and 83.6 % degradation was observed under the influence of 5 mM vanillin. Thus, metal ion like copper is essential for better biodegradation of Benzo[a] pyrene. Compared to synthetic laccase mediator like 2,2'-azinobis-(3-ethylbenzothiazoline-6-sulfonate), natural mediator such as vanillin may play a significant role in the degradation of aromatic compounds by white-rot fungi.

Keywords Benzo[a]pyrene · *Pleurotus ostreatus* · Degradation · Heavy metals · Mediators

S. Bhattacharya (✉) · M. Palaniswamy
Department of Microbiology, Karpagam University,
Coimbatore 641021, Tamil Nadu, India
e-mail: sourav3011@rediffmail.com

A. Das
Department of Microbiology, Genohelix Biolabs,
Centre for Advanced Studies in Biosciences, Jain University,
Bangalore 560019, Karnataka, India

K. Prashanthi
Department of Biotechnology, Genohelix Biolabs,
Centre for Advanced Studies in Biosciences, Jain University,
Bangalore 560019, Karnataka, India

J. Angayarkanni
Department of Microbial Biotechnology, Bharathiar University,
Coimbatore 641046, Tamil Nadu, India

Introduction

Benzo[a]pyrene (BaP), a representative of high molecular weight polycyclic aromatic hydrocarbon (HMW PAH), consists of five fused benzene rings and is of environmental concern since it behaves as a potent teratogen, mutagen and carcinogen. Its high molecular weight and extremely low water solubility reduce its bioavailability, thus making it resistant to microbial degradation (Rentz et al. 2008). Together with other PAHs, BaP is commonly formed under natural conditions like volcanic eruptions and forest fires and by anthropogenic activities like the pyrolysis and incomplete combustion of fossil fuels (Li et al. 2009).

Different strategies for removal of BaP and other highly recalcitrant compounds from contaminated sites include chemical oxidation, photolysis and biodegradation. Among all these clean up strategies, involvement of microorganisms is a subject of intense research and gains a cutting edge advantage for being less expensive and environment friendly (Chatterjee et al. 2008).

In nature there exists a great diversity of microorganisms capable of PAH degradation. Bacteria are capable of degrading the low molecular weight PAH, such as naphthalene, acenaphthene and phenanthrene and relatively few genera have been observed to degrade the HMW PAHs, such as BaP (Juhasz and Naidu 2000). However, several white-rot fungi (*Phanerochaete chrysosporium*, *Coriolus versicolor*, *Stropharia coronilla*, *Pleurotus ostreatus*, *Irpex lacteus* and *Bjerkandera adusta*) that synthesizes extracellular lignin modifying enzymes like lignin peroxidases (LiP), manganese peroxidases (MnP), laccases and other oxidases can oxidize BaP and similar HMW PAHs with up to six aromatic rings (Pointing 2001).

Though many studies on microbial degradation of BaP have been performed, literature on the effect of simultaneous existence of different pollutants groups on the extent of BaP bioremediation is scanty. PAH contaminated sites near industrial land are often accompanied by the presence of high levels of heavy metals (Sun et al. 2010). Heavy metals may be toxic for white-rot fungi and may have a negative effect on the activity of their ligninolytic enzymes (Bamforth and Singleton 2005). Occurrence of both organic and inorganic contaminants on the same site can therefore challenge the effectiveness of bioremediation technologies (Roy et al. 2005).

Among the oxidases produced by white-rot fungi, laccases have relatively low redox potential (0.4–0.8 V), belong to a group of polyphenol oxidases containing copper atoms in the catalytic site and usually called multicopper oxidases (Palmer et al. 1999). Laccases are responsible for lignin degradation in nature. This enzyme lacks substrate specificity and is thus capable of degrading a wide range of non ligninolytic compounds including certain xenobiotics like industrial colored waste water, oil and oil products, chlorinated compounds and PAHs (Abo-State et al. 2011).

Laccase limitations, owing to their relative low redox potential, have been overcome using redox mediators like 2,2′-azinobis-(3-ethylbenzothiazoline-6-sulfonate) (ABTS) or 1-hydroxybenzotriazole (HBT) which improves the oxidation of PAHs by laccase (Johannes and Majcherczyk 2000). However, the use of these synthetic mediators makes the process expensive and may result in the development of toxic end products. In contrast, some products generated from lignin degradation could act as natural redox mediators of laccase. The use of natural compounds provides a low cost, ecofriendly and toxicity less process (Moldes et al. 2008).

Therefore the objective of the study is to evaluate the effect of heavy metals on the extent of BaP degradation and the potential of synthetic and natural mediators to promote BaP transformation by the native isolate, *P. ostreatus* PO-3.

Materials and methods

Chemicals and reagents

HPLC grade BaP standard (98 % pure) was procured from Spectrochem Pvt. Ltd., Mumbai, India. Other fine chemicals used were procured from SRL Chemicals, India and were of the highest purity and analytical grade.

Effect of heavy metals and mediators

To study the effect of heavy metals and mediators on the degradation of BaP (10 μg/ml of the mineral salt medium), the PO-3 isolate of *P. ostreatus* from our previous study was used (Bhattacharya et al. 2012a). The optimized nutrients and surfactant from our earlier study were added to the modified mineral salt medium (devoid of $CuSO_4$ and $MnSO_4$) with the following composition (g/L): $(NH_4)_2HPO_4$, 05; KH_2PO_4, 0.8; K_2HPO_4, 0.3; $MgSO_4 \cdot 7H_2O$, 0.3; $CaCl_2 \cdot 2H_2O$, 0.055; Thiamine, 1 ml (2 mg/ml). The optimized physical parameters were also maintained (Bhattacharya et al. 2012b).

The effect of heavy metals and ethylenediaminetetraacetic acid (EDTA) on BaP degradation and biomass formation was studied by incorporating salts of different heavy metals ($CuSO_4$, $ZnSO_4$, $MnSO_4$, $AgNO_3$, $CdCl_2$, and $HgCl_2$) at 5 mM concentration to the modified mineral salt medium. The effect of different concentrations (5, 10, 15, 20, 25, 30 and 50 mM) of selected heavy metal salt that resulted in highest BaP degradation and fungal growth was analyzed. ABTS and vanillin (1 and 5 mM) were used as mediators in the study. Broth without the heavy metals or the mediators served as the control.

Analytical methods

Extraction of residual BaP

Following incubation, the extraction of residual BaP was performed using a modified method of Capotorti et al. (2004). The concentrated extract was subjected to high performance liquid chromatography analysis.

High performance liquid chromatography analysis

The condensed sample was subjected to filtration using 0.25 μ nitrocellulose membrane filter. The working standard solution of BaP (concentration of 5 μg/ml) was prepared using 80:20 (v/v) of acetonitrile: water. 20 μl of the eluate containing 0.1 μg of the standard BaP was injected into the HPLC system (Waters, USA, model number-2487, with dual λ absorbance UV detector and binary pump system, model number-1525). A reverse phase, C-18 column (150 × 4.6 mm) was used. The mobile phase used was acetonitrile: water (80:20 v/v). The flow rate was maintained at 1 ml/min. The concentration of the BaP standard solution was determined at 254 nm. Area under the absorbance peak was used to estimate the percentage of degradation using a formula:

$[(C_i - C_f)/C_i]*100$, where C_i is the initial concentration of BaP and C_f is the final concentration of BaP.

Statistical analysis

The effect of each parameter was studied in triplicate and the data are graphically presented as the mean ± S.D. of triplicates ($n = 3$). All the graphs have been prepared using Microsoft Excel 2007.

Results and discussion

Mycoremediation refers to fungal degradation or transformation of hazardous organic contaminants to less toxic compounds (Sasek et al. 2003). *P. ostreatus* is a white-rot basidiomycete fungus whose ligninolytic enzyme machinery consists of laccase and manganese peroxidase. It is often considered a model organism for the degradation of xenobiotics because of its ability to colonize soil and remains unaffected by the presence of indigenous microflora, as well as its excellent performance in biodegradation studies. Several isoenzymes of laccase from *P. ostreatus* have been reported and many have been purified for their potential applications. A range of compounds, including aromatic substrates of laccase that regulate the enzyme activity is well documented (Palmieri et al. 1997).

Effect of heavy metals on BaP degradation

Heavy metals often act as important modulators for enzyme activity. Many of these heavy metals are present in the environment either naturally (Cu^{2+}) or may gain their entry as a result of anthropogenic activities (Cd^{2+}, Hg^{2+}, Pb^{2+}). Polluted sites often contain high concentrations of heavy metals. Co-contamination with heavy metals may be a road block for in situ biodegradation of xenobiotics like PAHs or polychlorinated biphenyls (Koeleman et al. 1999).

Maximum degradation and biomass formation by *P. ostreatus* PO-3 resulted in the presence of Cu^{2+} ions (71.2 % and 63 mg of biomass/50 ml of broth). Moderate degradation resulted from the presence of Zn^{2+} and Mn^{2+}. Both biomass formation and degradation were severely

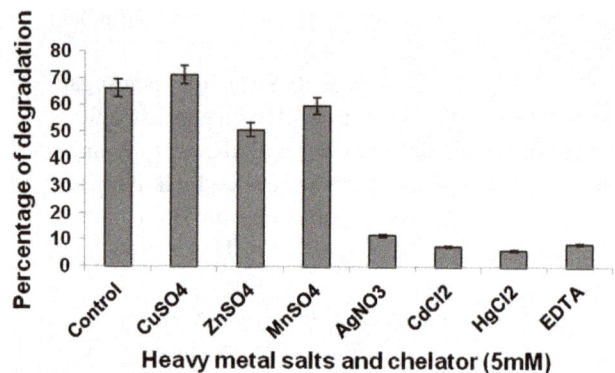

Fig. 1 Effect of heavy metal salts and chelator on degradation of BaP by *P. ostreatus*

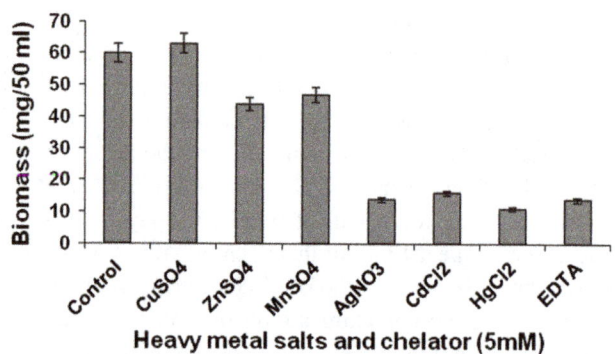

Fig. 2 Effect of heavy metal salts and chelator on *P. ostreatus* biomass formation

affected in the presence of Ag^{2+}, Cd^{2+}, Hg^{2+} and EDTA (Figs. 1, 2).

The present result demonstrating the increase in the extent of BaP degradation by *P. ostreatus* PO-3 following the addition of Cu^{2+} may be attributed to the increase in the laccase activity, which may in turn be due to both increased laccase production and the stabilization of the enzyme in an extracellular environment. Copper acting as a strong laccase inducer in *P. ostreatus* was reported earlier (Palmieri et al. 2000).

While characterizing the laccase from *Myrothecium verrucaria* NF-05, the resultant large extent activation of laccase activity in the presence of Cu^{2+} might be caused by the filling of type-2 copper binding sites with Cu^{2+} (Sulistyaningdyah et al. 2004).

Cadmium is a non-essential heavy metal. The presence of cadmium severely affected *P. ostreatus* PO-3 biomass formation probably due to the induction of oxidative stress. BaP degradation was also hindered by the presence of cadmium, since the sensitivity of fungi toward Cu^{2+} and Cd^{2+} changes with time. Laccase induction only occurs when cadmium is added during the later stages of growth. Parallel to our result, when cadmium was added earlier than 12 days, the activation of laccase was decreased.

Ag^{2+}, Hg^{2+} and Pb^{2+} also decreased the activity of laccase (Baldrian and Gabriel 2002). Mercury has high affinity for thiol groups in proteins, which might have lead to the inactivation of enzymes involved in the degradation (Baldrian et al. 2000).

Effect of different concentration of Cu^{2+} on BaP degradation

Since copper is a biogenic metal present in the environment, its concentration may have an important role in enzyme performance and stability. Copper requirement by microorganisms is usually satisfied by very low concentrations of the metal (1–10 mM). Copper present in higher concentration proves to be extremely toxic to microbial cells (Tychanowicz et al. 2006), although reports of copper tolerance by some fungi exist (De Groot and Woodward 1999).

Result from the present study shows that greater biomass was formed in lower concentrations as compared to higher concentrations of Cu^{2+}. Appreciable biomass and BaP degradation were observed up to the presence of 15 mM of Cu^{2+} supporting 74.2 % BaP degradation. However, higher concentrations of copper progressively decreased biomass and BaP degradation by *P. ostreatus* PO-3 (Fig. 3).

An early research reports similar finding, where higher Cu^{2+} concentration resulted in the reduced biomass formation. The probable reason for reduction in biomass formation at elevated level of Cu^{2+} may be attributed to the induced oxidative stress similar to that of cadmium (Baldrian et al. 2000). Similarly, Tychanowicz et al. (2006) reported that in the culture where copper was present at 10.0–25.0 mM concentrations, laccase activity increased up to eightfold when compared to the control cultures.

Effect of mediators on BaP degradation

It is well quoted in literature that many species of white-rot fungi, including *Pleurotus* sp, produce laccases that can efficiently degrade HMW PAHs such as BaP (Juhasz and Naidu 2000). The laccase-catalyzed reaction depends on monoelectronic oxidation, which transforms the substrates to corresponding reactive radicals. With the incorporation of special compounds called mediators that act as a single electron donor and activator of the enzymes, oxidation rates of the enzyme can be enhanced (Morozova et al. 2007).

Synthetic mediators like ABTS and HBT are commonly used in degradation studies involving laccases and other lignin-modifying enzymes. Several natural phenolic compounds produced in the process of lignin depolymerization like vanillin, acetovanillone, acetosyringone, syringaldehyde, 2,4,6-trimethylphenol, *p*-coumaric acid, ferulic acid, and sinapic acid may also act as alternative mediators for the laccase mediated bioremediation of HMW PAHs by white-rot fungi (Cañas et al. 2007).

In our study, both ABTS (artificial mediator) and vanillin (natural mediator) demonstrated positive effect on the degradation as compared to that of the control (Fig. 4). ABTS proved to be superior at 1 mM concentration (78.7 % and 42 mg of biomass/50 ml of broth). Vanillin supported better degradation (83.6 %) at 5 mM concentration. Both the mediators affected the biomass formation (Fig. 5).

Similar result was observed previously, wherein lignin-related phenolic compounds acted as laccase mediators and resulted in a significant degradation of PAH. Vanillin, acetovanillone and ferulic acid significantly promoted anthracene and BaP transformation by *Pycnoporus cinnabarinus* laccase (Cañas et al. 2007). The probable reason could be the structural similarities of these recalcitrant aromatic pollutants to that of lignin network.

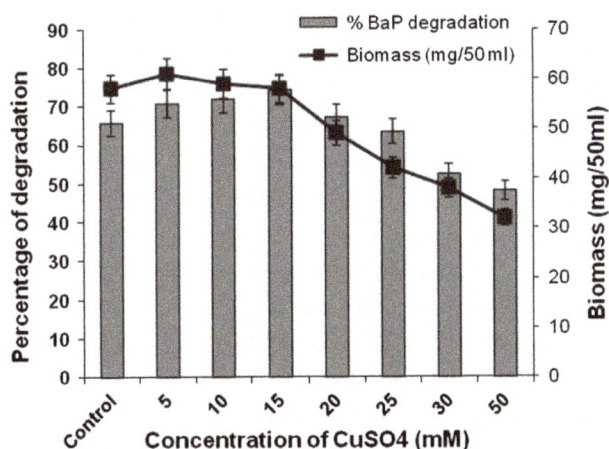

Fig. 3 Effect of different concentrations of $CuSO_4$ on BaP degradation and *P. ostreatus* biomass formation

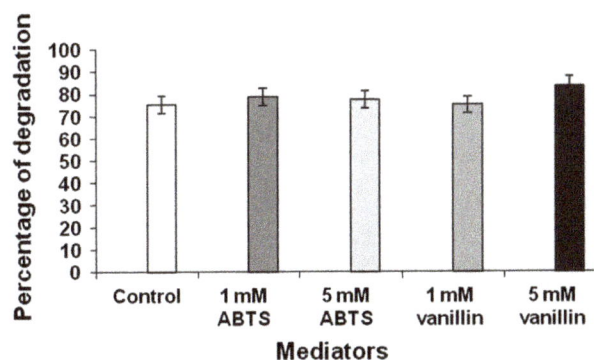

Fig. 4 Effect of ABTS and vanillin on degradation of BaP by *P. ostreatus*

Earlier, *Pycnoporus cinnabarinus* laccase without mediator oxidized only 12 % of BaP in 24 h. The presence of the natural compounds like vanillin, acetovanillone, 2,4,6-trimethylphenol and *p*-coumaric acid notably

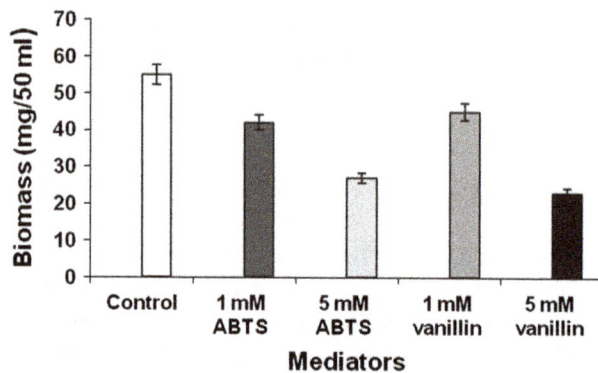

Fig. 5 Effect of ABTS and vanillin on *P. ostreatus* biomass formation

promoted transformation of BaP by laccase (Camarero et al. 2005). Positive effect has also been found by joining the artificial mediator ABTS and vanillin during penta-chlorophenol transformation by laccase (Jeon et al. 2008).

Of the phenols tested, 4-hydroxybenzaldehyde and vanillin were most inhibitory to *Pleurotus sajor-caju* growth, and cultures supplemented with these two compounds also exhibited large increase in laccase-specific activity when compared to the unsupplemented controls (Lo et al. 2001).

Naturally occurring phenols such as syringaldehyde, vanillic acid, vanillin, and *p*-coumaric acid proved to be important laccase mediators effectively transforming cyprodinil up to 70 % (Kang et al. 2002). The positive effect of the phenolic compounds derived from the lignin polymer and constituents of humic acids on the oxidative transformation of widespread pollutants such as chlorinated phenols had already been demonstrated (Park et al. 1999).

Fig. 6 HPLC chromatogram of BaP **a** control, **b** after degradation

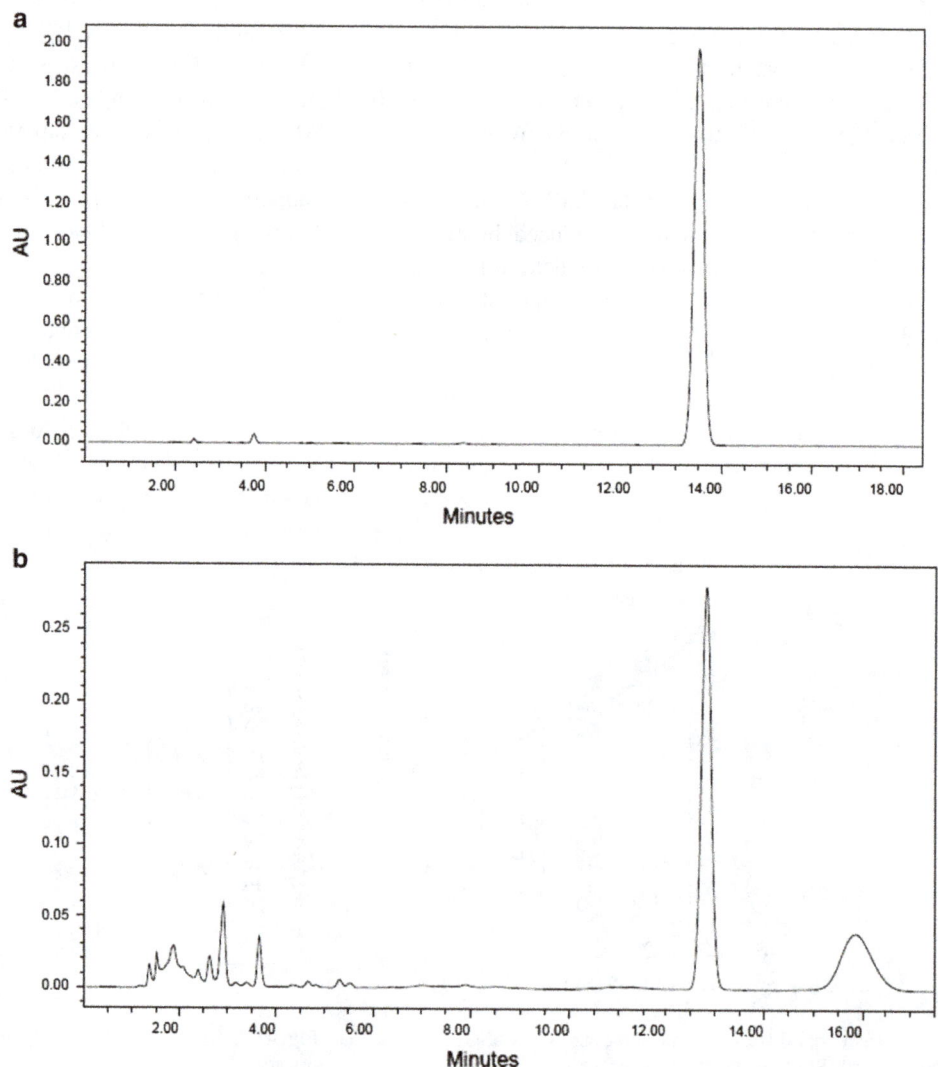

HPLC analysis reveals biodegradation of BaP by *P. ostreatus* PO-3 under optimized media and growth conditions (Fig. 6a, b).

Conclusions

Various metals may be present in a contaminated site alongside the PAHs. The results showed varied levels of BaP degradation in the presence of heavy metals. Improved biodegradation in case of copper availability can be correlated to the activation and stabilization of the produced laccase by *P. ostreatus* PO-3.

Natural mediators may play an important role in the degradation of PAH by white-rot fungi. In future, lignin-derived mediator like vanillin may act as an alternative to synthetic mediator like ABTS. Besides improving the working potential of the ligninolytic enzymes, these natural mediators are released in large amount as a result of the microbial degradation of lignocellulose and are present in humus as common secondary plant metabolites. Vanillin and similar mediators would thus effectively reduce the cost of the bioremediation process.

Acknowledgments We wish to extend our sincere gratitude to the managements of Karpagam University, Jain University and Bharathiar University for their encouraging support. Our special thanks to Dr. R. Chenraj Jain, Chairman of Jain Group of Institutions, Dr. N. Sundararajan, Vice-Chancellor of Jain University and Dr. S. Sundara Rajan, Director of Genohelix Biolabs, A Division of Centre for Advanced Studies in Biosciences, Jain University, Bangalore for providing us with the laboratory facilities required for this research work.

Conflict of interest The authors declare that they have no conflict of interest.

References

Abo-State MAM, Khatab O, Abo-El Nasar A, Mahmoud B (2011) Factors affecting laccase production by *Pleurotus ostreatus* and *Pleurotus sajor-caju*. World Appl Sci J 14:1607–1619

Baldrian P, Gabriel J (2002) Copper and cadmium increase activity in *Pleurotus ostreatus*. FEMS Microbiol Lett 206:69–74

Baldrian P, Der Weische CW, Gabriel J, Nerud F, Zadrazil F (2000) Influence of cadmium and mercury on activities of ligninolytic enzymes and degradation of polycyclic aromatic hydrocarbons by *Pleurotus ostreatus* in soil. Appl Environ Microbiol 66:2471–2478

Bamforth SM, Singleton I (2005) Bioremediation of polycyclic aromatic hydrocarbons: current knowledge and future directions. J Chem Technol Biotechnol 80:723–736

Bhattacharya S, Angayarkanni J, Das A, Palaniswamy M (2012a) Mycoremediation of Benzo[a]pyrene by *Pleurotus ostreatus* isolated from Wayanad district in Kerala, India. Int J Pharm Bio Sci 2:84–93

Bhattacharya S, Angayarkanni J, Das A, Palaniswamy M (2012b) Evaluation of physical parameters involved in mycoremediation of Benzo[a]pyrene by *Pleurotus ostreatus*. J Pure Appl Microbio 6:1721–1726

Camarero S, Ibarra D, Martínez MJ, Martínez AT (2005) Lignin-derived compounds as efficient laccase mediators for decolorization of different types of recalcitrant dyes. Appl Environ Microbiol 71:1775–1784

Cañas A, Alcalde M, Plou FJ, Martínez MJ, Martínez AT, Camarero S (2007) Transformation of polycyclic aromatic hydrocarbons by laccase is strongly enhanced by phenolic compounds present in soil. Environ Sci Technol 41:2964–2971

Capotorti G, Digianvincenzo P, Cesti P, Bernardi A, Guglielmetti G (2004) Pyrene and benzo(a)pyrene metabolism by an *Aspergillus terreus* strain isolated from a polycyclic aromatic hydrocarbons polluted soil. Biodegradation 15:79–85

Chatterjee S, Chattopadhyay P, Roy S, Sen SK (2008) Bioremediation: a tool for cleaning polluted environments. J Appl Biosci 11:594–601

De Groot RC, Woodward BM (1999) Using copper-tolerant fungi to biodegrade wood treated with copper-based preservatives. Int Biodeterior Biodegrad 44:17–27

Jeon JR, Murugesan K, Kim YM, Kim EJ, Chang YS (2008) Synergistic effect of laccase mediators on pentachlorophenol removal by *Ganoderma lucidum* laccase. Appl Microbiol Biotechnol 81:783–790

Johannes C, Majcherczyk A (2000) Natural mediators in the oxidation of polycyclic aromatic hydrocarbons by laccase mediator systems. Appl Environ Microbiol 66:524–528

Juhasz AL, Naidu R (2000) Bioremediation of high molecular weight polycyclic aromatic hydrocarbons: a review of the microbial degradation of benzo[a]pyrene. Int Biodeterior Biodegrad 45:57–88

Kang KH, Dec J, Park H, Bollag JM (2002) Transformation of the fungicide cyprodinil by a laccase of *Trametes villosa* in the presence of phenolic mediators and humic acid. Water Res 36:4907–4915

Koeleman M, Vd Laak JW, Ietswaart H (1999) Dispersion of PAH and heavy metals along motorways in the Netherlands-an overview. Sci Total Environ 235:347–349

Li Z, Sjodin A, Porter EN, Patterson DGJ, Needham LL, Lee S, Russell AG, Mulholland JA (2009) Characterization of $PM_{2.5}$-bound polycyclic aromatic hydrocarbons in Atlanta. Atmos Environ 43:1043–1050

Lo SC, Ho YS, Buswell JA (2001) Effect of phenolic monomers on the production of laccases by the edible mushroom *Pleurotus sajor-caju*, and partial characterization of a major laccase component. Mycologia 93:413–421

Moldes D, Diaz M, Tzanov T, Vidal T (2008) Comparative study of the efficiency of synthetic and natural mediators in laccase-assisted bleaching of eucalyptus kraft pulp. Bioresour Technol 99:7959–7965

Morozova OV, Shumakovich GP, Shleev SV, Yaropolov YI (2007) Laccase-mediator systems and their applications: a review. Appl Biochem Microbiol 43:523–535

Palmer AE, Randall DW, Xu F, Solomon EI (1999) Spectroscopic studies and electronic structure description of the high potential type 1 copper site in fungal laccase: insight into the effect of the axial ligand. J Am Chem Soc 121:7138–7149

Palmieri G, Giardina P, Bianco C, Scaloni A, Capasso A, Sannia G (1997) A novel white laccase from *Pleurotus ostreatus*. J Biol Chem 272:31301–31307

Palmieri G, Giardina P, Bianco C, Fontanella B, Sannia G (2000) Copper induction of laccase isoenzymes in the ligninolytic fungus *Pleurotus ostreatus*. Appl Environ Microbiol 66: 920–924

Park JW, Dec J, Kim JE, Bollag JM (1999) Effect of humic constituents on the transformation of chlorinated phenols and anilines in the presence of oxidoreductive enzymes or birnessite. Environ Sci Technol 33:2028–2034

Pointing SB (2001) Feasibility of bioremediation by white-rot fungi. Appl Microbiol Biotechnol 57:20–33

Rentz JA, Alvarez PJJ, Schnoor JL (2008) Benzo[*a*]pyrene degradation by *Sphingomonas yanoikuyae* JAR02. Environ Pollut 151:669–677

Roy S, Labelle S, Mehta P, Mihoc A, Fortin N, Masson C, Leblanc R, Châteauneuf G, Sura C, Gallipeau C, Olsen C, Delisle S, Labrecque M, Greer CW (2005) Phytoremediation of heavy metal and PAH contaminated brownfield sites. Plant Soil 272:277–290

Sasek V, Cajthaml T, Bhatt M (2003) Use of fungal technology in soil remediation: a case study. Water Air Soil Pollut Focus 3:5–14

Sulistyaningdyah W, Ogawa J, Tanaka H, Maeda C, Shimizu S (2004) Characterization of alkaliphilic laccase activity in the culture supernatant of *Myrothecium verrucaria* 24G–4 in comparison with bilirubin oxidase. FEMS Microbiol Lett 230:209–215

Sun F, Wen D, Kuang Y, Li J, Li J, Zuo W (2010) Concentrations of heavy metals and polycyclic aromatic hydrocarbons in needles of Masson pine *(Pinus massoniana* L.) growing nearby different industrial sources. J Environ Sci 22:1006–1013

Tychanowicz GK, de Souza DF, Souza CGM, Kadowaki MK, Peralta RM (2006) Copper improves the production of laccase by the white-rot fungus *Pleurotus pulmonarius* in solid state fermentation. Braz Arch Biol Technol 49:699–704

4

Isolation and characterization of plant and human pathogenic bacteria from green pepper (*Capsicum annum* L.) in Riyadh, Saudi Arabia

Samiah H. S. Al-Mijalli

Abstract Forty-three bacterial isolates in five genera were recovered from naturally infected green pepper fruits (38 samples) showing dark brown, irregular-shaped splotches. The pathogenicity test was performed on healthy green pepper fruits and red colonies were from inoculated fruits showing the same symptoms and the infected area developed into soft rot. Their identification was based on phenotypic characters and sequence of the gene fragment coding 16S rRNA. Of 43 isolates, 10 showing splotches on green pepper fruits belonged to genus *Serratia* on the basis of phenotypic characters. One representative isolate of the genus *Serratia* has been identified by partial 16S rRNA gene sequencing and phylogenetic analysis as belonging to the *Serratia rubidaea* and has the potential to cause spot on green pepper. Eleven phytopathogenic bacterial isolates were also obtained at the same time but did not induce any splotch symptoms on artificially infected green pepper. Five out of 11 bacterial isolates were identified as *Ralstonia* on the basis of biochemical tests. Partial sequencing of 16S ribosomal gene of representative isolate revealed that the isolate is *Ralstonia solanacearum*. The six remaining isolates were related to *Xanthomonas vesicatoria* on the basis of biochemical tests. Twenty-two of opportunistic human pathogens were isolated at the same time and related to *Proteus* and *Klebsiella*. Opportunistic human pathogens did not produce any symptoms on artificially infected green pepper. One representative isolate for each genus was identified as *Klebsiella oxytoca* and *Proteus mirabilis* based on their partial 16S rRNA gene sequences. The virulence of the *S. rubidaea*, the causal agent of green

pepper fruits splotches was attributed to the production and secretion of a large variety of enzymes capable of degrading the complex polysaccharides of the plant cell wall and membrane constituents.

Keywords Green pepper · Pathogenic bacteria · *Klebsiella oxytoca* · *Proteus mirabilis*

Introduction

The role of fresh fruits and vegetables in nutrition and healthy diet is well recognized and in recent years, many countries have undertaken various initiatives to encourage consumers to eat more of these products. Fruits and vegetables supply much needed vitamins, minerals, and fibers. These play an important role in health through the prevention of heart disease, cancer, and diabetes. The health aspect together with increasing consumer demands for variety and availability, and the changing structure of global trade has led to an increase in trade of fruits and vegetables (Abd-Alla et al. 2011).

Agricultural products can be exposed to microbial contamination through a variety of sources. Although vegetables are good examples of minimally processed foods, there is a high risk of contamination. Since fruits and vegetables are produced in a natural environment, they are vulnerable to contamination by human pathogens. The increased consumption of fruits and vegetables may have unintended consequences with an increase in number of outbreaks. The majority of diseases associated with fresh fruits and vegetables are primarily those transmitted by the fecal–oral route, and therefore, are a result of contamination at some point in the process (De Roever 1998). Therefore good hygienic measures have to be taken during

S. H. S. Al-Mijalli (✉)
Biology Department, College of Sciences, Nora Bent
AbdulRahman University, Riyadh, Saudi Arabia
e-mail: dr.samiah10@hotmail.com; SHSM2000@live.com

the production from farm to table. The world has seen significant changes in eating habits and consumption of fresh products is increasingly becoming important in the diet of many people, especially reflected in the increased demand for organically produced food. In the production and processing of fresh produce quality and hygiene are the most important criteria for the consumers. Such food products are often eaten raw and, if contaminated with pathogenic bacteria may represent a health hazard to consumers (Bruhn 1995).

Bacterial soft rot is a leading cause of postharvest losses of potatoes (Cappellini et al. 1984), tomatoes (Ceponis et al. 1986), peppers (Ceponis et al. 1987), lettuce (Ceponis et al. 1985), and other fresh fruits and vegetables in the marketplace. It is caused by a group of plant pathogens, harmless to humans, that includes *Erwinia carotovora*, pectolytic *Pseudomonas fluorescens* and *P. viridiflava* (Lund 1983). Pectolytic breakdown of affected tissues results in softening, liquefaction, and exudates that can spread bacteria over commodities in bulk storage or display, contaminate food-handling equipment, and protect bacteria from the environment (Snowdon 1990; Wei et al. 1995).

Serratia species were frequently found associated with plants (Grimont et al. 1981). *Serratia marcescens, S. liquefaciens*, and *S. rubidaea* were found in 29, 28, and 11 %, respectively, of vegetable salads served in a hospital in Pittsburgh (Wright et al. 1976). *S. rubidaea*, first described by (Stapp 1940), is an epiphyte on plants. *S. rubidaea* has been isolated from coconuts of Ivory Coast bought in both France and California (Grimont et al. 1981). The relative frequency of *S. rubidaea* in clinical specimens is rare, and there are no data to suggest that the organism is of clinical significance, but clinical significance cannot be totally excluded because of its occurrence in clinical specimens (Farmer et al. 1985).

Sequences of the 16S rRNA gene are generally used as a framework for bacterial classification. Therefore, sequencing of this gene was used as a first identification tool (Garcıa-Martınez et al. 2001). Extracellular enzymes have a number of potential roles in plant disease, including overcoming host defense responses, mobilization of plant cell walls for nutritional purposes, facilitation of movement of bacteria into and between vascular elements, and promotion of bacterial survival on plant material in the soil. Successful management of plant diseases relies on correct diagnosis. Therefore, investigations on pathogenic bacteria and bacterial diseases on the fresh vegetable and fruit plants are economically important. The overall objective of this study was to determine and characterize the bacterial isolates recovered from green pepper fruits spot collected from different vegetable markets and their ability to produce an array of extracellular enzymes capable of degrading the complex polysaccharides of the plant cell wall and membrane constituents.

Materials and methods

Sample collection

Naturally infected green pepper *Capsicum annum* L. spots were collected from different vegetable markets in Riyadh, Saudi Arabia. The collected samples (38) were kept in refrigerator until the time of isolation. They were analyzed within 20 h of sampling.

Bacteriological analysis

Green pepper fruits with spot symptoms were washed with sterilized distilled water, and then treated with 0.5 % solution of hypochlorite (bleach) (Cotter et al. 1985) for 1–2 min to remove the contaminants, rinsed with sterile distilled water and cut into small bits with sterile scalpel. These pieces were immersed in sterilized saline buffer and vortexed strongly. A tenfold dilution series was prepared and 100 µl each of diluted and the undiluted extract was spread (with three replications of each dilution) on yeast dextrose chalk (YDC) medium (Schaad 1988) consisting of 20.0 g/l dextrose, 10.0 g/l yeast extract, and 20.0 g/l $CaCO_3$ with 15.0 g/l of agar in 1 l of distilled water. YDC medium was autoclaved for 15 min at 121 °C.

Cultures were incubated for 3–5 days at 30 °C. Discrete colonies were re-streaked onto YDC plates for pure culture isolation. One colony of the purified presumptive pathogen from each sample was selected and maintained on YDC slants at 4 °C for further tests.

Pathogenicity test

Green pepper fruits were swabbed with 70 % ethanol and washed in sterile water and stabbed with sterile syringe needles at three sites. Inoculations were made by deposition of 5 µl of a bacterial suspension on the upper surface of green pepper fruits. Three fruits were used for each isolate. Inocula were prepared from 48-h-old cultures on nutrient broth medium incubated at 30 °C. Bacterial cells were collected in saline phosphate buffer (pH 7) and adjusted to 10^7 cfu/ml by turbidity measurement (A600). Inoculated green pepper fruits were kept in closed transparent boxes lined with moist blotting and incubated at 30 °C. All fruits were assessed daily for 7 days to record disease symptoms. The causal agent was recovered from green pepper showing the same symptoms on YDC. All bacterial isolates were tested for their ability to produce any symptoms.

Biochemical characterization of bacterial isolates

Bacteriological characteristics of the isolates were examined by using the methods to include: Gram stain, Ryu's test, colony color, oxidase reaction, arginine dihydrolase, nitrate reduction, utilization of carbohydrates, Levin formation, catalase test, gelatin hydrolysis, starch hydrolysis, esculin and Tween 20, as previously described by Bergey's Manual (Brenner et al. 2005).

Sequencing of 16S rRNA gene

DNA was extracted by boiling a small amount of a pure culture plate colony in DNAse- and RNAse-free water (Invitrogen, Carlsbad, CA). For the sequencing of the partial 16S rRNA gene fragments polymerase chain reaction amplifications were performed with universal bacterial primers (Lane 1991) [27f (9–27) GAGTTTGATCM TGGCTCAG and 1492r (1492–1510) ACGGYTACCTT GTTACGACTT] in a total volume of 25 µl. The reaction mixture contained 2.5 µl PCR-buffer, 2.0 µl $MgCl_2$ (25 mM), 2.5 µl dNTPs (2 mM), 0.5 µl of each primer, 15.7 ml RNAse- and DNAse-free water, 0.2 µl BSA (20 mg/ml), 0.1 µl Taq polymerase (5 U $µl^{-1}$) (all MBI Fermentas, St. Leon Rot), and 1 µl of the DNA extract. The reaction was performed using an Eppendorf Ag 22331 Authorized Thermal Cycler (Hamburg, Germany) with an initial denaturation step at 95 °C for 3 min, followed by 29 cycles of denaturation (95 °C for 45 s), annealing (57.3 °C for 20 s) and extension (72 °C for 2 min). The PCR was completed with a terminal extension step for 4 min at 72 °C. PCR products were purified (PCR Purification Kit250, QiaQuicks, Qiagen, Hilden) and quantified photometrically (Ultrospec 4000, Amersham Biosciences, Freiburg). Purified PCR products were cycle sequenced in both directions with the same forward and reverser primers using an Applied Biosystems 3730X-1 DNA Analyzer (Fast Smack Inc. Division DNA synthesis, Kanagawa, Japan). The sequence reads were edited and assembled using BioEdit version 7.0.4 (http://www.mbio.ncsu.edu/ BioEdit/bioedit.html) and clustalW version 1.83 (http:// clustalw.ddbj.nig.ac.jp/top-e.html). BLAST searches were done using the NCBI server at http://www.ncbi.nlm.nih. gov/blast/Blast.cgi. Phylogenetic tree derived from 16S rRNA gene sequence was generated in comparison to 16S rRNA gene sequences from seven different standard bacterial strains obtained from Genbank (Lane 1991). Gene sequences that had been determined were phylogenetically analyzed using the ARB software package (Sanger et al. 1977). New sequences not included in the used ARB database were added from public databases (http://www. ncbi.nlm.nih.gov/BLAST) using BLASTN search to assign the closed relatives. The ARB_Edit tool was used for

automatic sequence alignment and checking and correcting the alignment afterwards. Neighbor-joining algorithms (Drancourt et al. 2000) were used for calculating the trees.

Enzymes' production by *Serratia rubidaea*

Serratia rubidaea isolate was screened for its ability to elaborate hydrolytic enzymes such as lipase enzyme, protease enzyme, polygalacturonase enzyme and alkaline phosphatase enzyme. Bacteria were grown in 50 ml of liquid medium in an Erlenmeyer flask (250 ml) containing (g/l): $MgSO_4·7H_2O$ 0.2, K_2HPO_4 2.0, KH_2PO_4 2 and casein 10 (pH 8) (Chakraborty and Srinivasan 1993). The basal medium for lipase production consisted of (g/l): bacteriological peptone 15.0, yeast extract 5.0, NaCl 2.0, $MgSO_4$ 0.4, K_2HPO_4 0.3, KH_2PO_4 0.3, and olive oil 10.0 ml for lipase induction (Baharum et al. 2003). The basal medium for pectinase production consisted of (g/l): pectin 4, yeast extract 2, NH_4Cl 1, $MgSO_4$ 0.5 (Gomes et al. 1992). Cultures were incubated in an orbital shaking incubator for 36 h at 150 rpm and 37 °C. The culture broth was then centrifuged at 8,000 rpm [may be better in *g* (relative centrifuge force)] to remove cells. The clear supernatant was collected for enzymes assay.

Enzymes assay

Protease activity was assayed by a modified method of Ohara-Nemoto et al. (1994). The reaction was initiated by addition of 1 ml supernatant to 2 ml of reaction mixture containing 2.7 mg of casein per ml in 50 mmol Tris–HCl (pH 8.0) which had been prewarmed at 37 °C. After incubation at 37 °C for 1 hour, the reaction was stopped by addition of 0.5 ml of 15 % ice-cold trichloroacetic acid. The reaction mixture was held on ice for 15 min and then centrifuged; 5 ml of sodium carbonate solution (500 mmol) was added to the reaction mixture followed by 1 ml of Folin and Ciocalteu's phenol reagent (dilute 10 ml of Folin and Ciocalteu's phenol reagent, to 40 ml with distilled water). The soluble peptide in the supernatant fraction was measured with tyrosine as the reference compound. The absorbance at 660 nm of the sample was measured using a spectrophotometer (UNICO UV-2100, USA). One unit of enzyme activity was defined as the amount of enzyme that releases 1 µg of tyrosine per min under the assay conditions. Controls containing autoclaved enzymes instead of active enzymes were used. Lipase activity was measured by universal titrimetric method (Fadıloğlu and SÖylemez 1997).

Polygalacturonase was determined according to the method of Gomes et al. (1992). One unit of activity is defined as that amount of enzyme which catalyzes the release of 1 μmol of reducing groups per min and expressed as U/ml.

Alkaline phosphatase of bacterial culture supernatant was determined by standard assay procedure using alkaline phosphatase kit (ELI Tech, SEPPIM S.A.S.-Zone industrielle-61500 SEES France). One alkaline phosphatase unit was defined as the amount of enzyme which liberates 1 μmol of p-nitrophenol as a result of hydrolysis of p-nitrophenylphosphate (pNPP) in 1 min (Abd-Alla 1994).

Effects of temperature on enzyme production and stability

The effect of temperature on lipase, protease, polygalacturonase, and alkaline phosphatase enzymes production was determined by incubating the culture flasks with different temperature regimes (10, 20, 30, and 40 °C). For determining thermal stability, the enzyme was pre-incubated at different times ranging from 10 to 70 min at maximum temperature (45 °C) and residual activity was measured under standard assay conditions.

Results

Bacterial isolates

A total of 38 samples were randomly taken from different vegetable markets of Riyadh. Forty-three bacterial isolates were obtained from naturally green pepper spot fruits. Four different bacterial colonies were observed on YDC medium. These colonies were distinguished into one type of yellow, two types of white mucoid and one type red.

Pathogenicity test

The pathogenicity test was performed on healthy green pepper fruits and red colonies were re-isolated from inoculated fruits showing splotches. Symptoms appeared as dark brown, irregular-shaped splotches and the infected area developed into soft rot. The other bacterial isolates of yellow and white colonies did not show any spot symptoms on green pepper fruits.

Phenotypic and genotypic characterization of the bacterial isolates

Ten out of 43 bacterial isolates showing spot on green pepper fruits were further characterized using comprehensive range of phenotypic test (Table 1). The isolates might be belonging to the genus of Serratia on the basis of their phenotypic characterization. One representative isolate of the genus Serratia was subjected to the partial 16S rRNA

gene sequences of 612 base pairs. The selected isolates had 16S rRNA gene sequence with 99 % similarity to the closet sequence of S. rubidaea AB004751 in GenBank. Eleven phytopathogenic bacterial isolates were also recovered at the same time. Five out of eleven bacterial isolates were identified as Ralstonia on the basis of biochemical tests. One representative isolate was chosen for further identification using phylogenetic analysis of 16S rRNA gene sequences as the gold standard. The partial 16S rRNA gene sequence of 630 base pairs of the representative isolate had a sequence with 99 % similarity to Ralstonia solanacearum U28224. A phylogenetic tree was constructed from a multiple sequences alignment of 16S rRNA gene sequences (Fig. 1). The other six isolates were related to Xanthomonas vesicatoria on the basis of phenotypic characters.

Other 22 non-phytopathogenic bacterial isolates were recovered at the same time from green pepper fruits. Ten out of 22 belonged to the genus Klebsiella on the basis of biochemical activities. The 16S rRNA gene sequence of 571 bp of the representative isolate was aligned with other 16S rRNA gene sequence using ARB software package to demonstrate the relatedness of the isolate to other major groups. Sequence from the analyzed isolate shared 98 % similarity to known strain of Klebsiella oxytoca AB353048 in GenBank database as supported by phylogenetic tree. The twelve remaining isolates were subjected to varieties of biochemical tests to determine the phenotypic traits and enable their identification. Data presented in Table 1 demonstrated that the isolates clearly belong to the genus of Proteus. The phenotype-based identification was confirmed by phylogenetic analysis. Comparison between 16S rRNA gene sequence of the chosen isolates of genus Proteus and 16S rRNA gene sequences of GenBank database was made by using BLASTN search analysis. Sequencing of 16S rRNA genes of the chosen isolate had 16S rRNA gene with 99 % nucleotides identity to that of Proteus mirabilis EF626945 available in GenBank database. The phylogenetic tree was inferred from 16S rRNA sequence data by the neighbor-joining method (Fig. 1). The tested analyzed isolate was identified as P. mirabilis belonging to the family Enterobacteriaceae.

Effect of temperature on enzymes production

Serratia rubidaea has the ability to produce the extracellular enzyme which was thought to play an important role in host infection. Lipase, protease, polygalacturonase and alkaline phosphatase enzymes were detected in S. rubidaea isolates. The optimum temperature for lipase production by S. rubidaea was 25 °C (Table 2; Fig. 2). It is obvious from the results in Fig. 2 that 30 °C temperature was generally more favorable for protease production by S. rubidaea. The optimum temperature for polygalacturonase production by

Table 1 Biochemical characterization test of bacterial genera isolated from green pepper fruits

Characteristics	Group 1	Group 2	Group 3	Group 4	Group 5
Phenotypic classification	*Serratia*	*Ralstonia*	*Xanthomonas*	*Klebsiella*	*Proteus*
Fluorescent on king's B	−	−	−	−	−
Growth on cetrimide	−	−	−	−	−
Gram's staining	−	−	−	−	−
KOH solubility	+	+	+	+	+
Cytochrome C oxidase	−	+	−	−	−
Nitrate reductase	+	+	−	+	+
Catalase test	+	+	+	+	+
Gelatin hydrolysis	+	+	+	−	−
Casein hydrolysis	+	+	+	−	−
Starch hydrolysis	−	−	−	+	+
Arginine dehydrogenase	−	−	−	+	−
Urease test	−	+	−	−	+
H$_2$S production	−	−	−	+	+
Esculin test	+	+	+	+	−
Voges–Proskauer	−	−	−	−	−
Carbon source utilization					
L-Arabinose	−	−	+	+	−
D-Cellobiose	+	−	+	+	−
D-Fructose	−	+	+	+	−
Citrate	+	−	−	+	−
D-Alanine	+	−	−	+	+
D-Sorbitol	−	+	−	+	+
D-Galactose	+	−	−	+	−
Glycerol	+	+	+	−	+
Glucose	+	−	+	+	+
Lactose	+	−	−	+	+
Maltose	+	+	+	+	−
Mannitol	+	+	+	+	+
Growth at 4 °C	−	−	−	−	−
Growth at 37 °C	+	+	+	+	+
Growth at 41 °C	+	−	+	+	+

Symbols − and + meaning negative and positive

S. rubidaea was 40 °C. The results presented in Fig. 2 revealed that the highest production of alkaline phosphatase by *S. rubidaea* was achieved at 37 °C (Fig. 2; Table 2).

Thermal stability of produced enzymes

Thermal stability was investigated by incubating the enzymes produced by *S. rubidaea*. Lipase enzyme of *S. rubidaea* exhibits the highest thermal stability with 87 % of maximum activity remaining after 10 min, but 50 % of the initial activity retained at 45 °C for 40 min (Fig. 3). Protease enzyme of *S. rubidaea* exhibits the highest thermal stability with 94 % of maximum activity remaining after 10 min, but 87 % of the initial activity retained at 45 °C for 40 min (Fig. 3). The optimum temperature for polygalacturonase produced by *S. rubidaea* was 40 °C. At 45 °C, polygalacturonase enzyme of *S. rubidaea* exhibits the highest thermal stability with 88 % of maximum activity remaining after 10 min, however, 69 % of the initial activity retained for 40 min. Alkaline phosphatase enzyme of *S. rubidaea* exhibits the highest thermal stability with 93 % of maximum activity remaining after 10 min, but 65 % of the initial activity retained at 45 °C for 40 min.

Discussion

Phenotypic and genotypic techniques were used for identification and allowed us to infer the phylogeny of 43

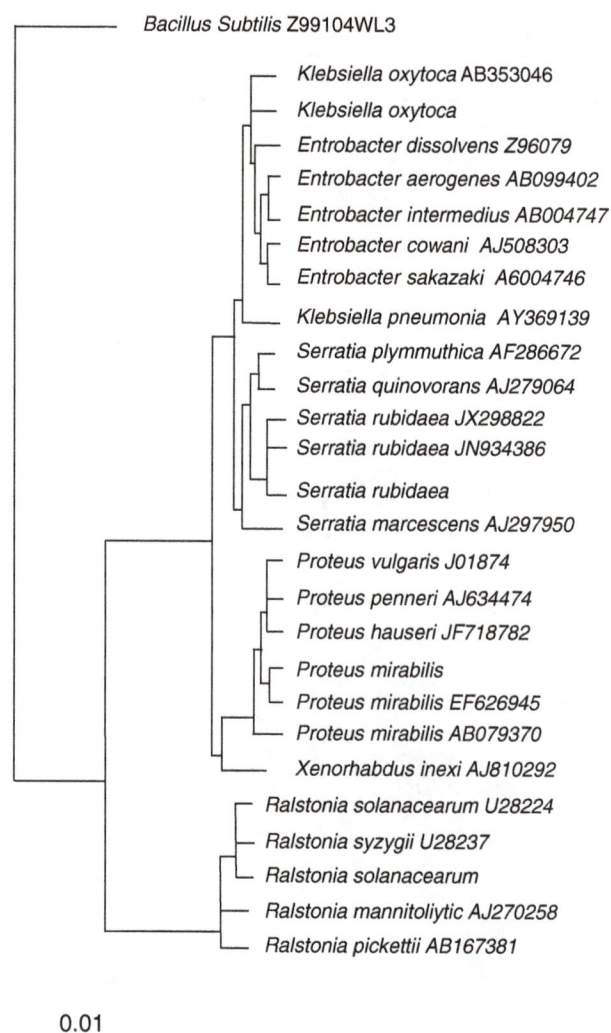

0.01

Fig. 1 *Phylogenetic tree* indicates the phylogenetic relationship of the isolated strains. Isolates are indicated in *bold*. A neighbor-joining tree was calculated using partial 16S rRNA gene sequences (1,362 bp) and a frequency filter included in the ARB software package. *Bacillus subtilis* was used as out-group. The *scale bar* indicates 10 % estimated sequence difference. Accession numbers of the National Centre for Biotechnology Information (NCBI) database of each strain are given in *brackets*

bacterial isolates recovered from green pepper fruits, distributed along five genera (*S. rubidaea, X. vesicatoria, R. solanacearum, K. oxytoca* and *P. mirabilis*) and comprising

three bacterial families. Most pathogenic microbes must access the plant interior, either by penetrating the leaf or root surface directly or by entering through wounds or natural openings, such as stomata or leaf hydathodes (Kroupitski et al. 2009).

Plant pathogens may grow briefly on or in wounded tissue before advancing into healthy tissue. Injection of *Salmonella* into the tomato stem may introduce the pathogen into xylem, which has the principal role of transporting water and nutrients from the root to the extremities of the plant. Additionally, in the secondary xylem, the axial and ray parenchyma store nutrients and water (Blostein 1991) which sustain viability of plants and, possibly, promote survival of pathogenic bacteria. The presence of epiphyseal flora within tissue of fruits and vegetables through various pathways was reported by (Samish et al. 1962). By examining eight internal locations of green pepper fruits, they observed that bacteria are unevenly distributed in the fruit, and entry may be from the stem scar tissue through the core and into the endocarp. This study suggested that some epiphyseal flora might reach internal tissue of tomatoes through natural apertures because of their small size and motility. It may be that bacteria enter fruit tissue more readily in the early stages of fruit development, at a time when various channels are not yet covered by corky or waxy materials (Samish et al. 1962). Broken trichomes on young fruits represent another site of entry of microorganisms. Guo et al. 2001 reported that green pepper fruits, stem and flowers are possible sites at which *Salmonella* may attach and remain viable during fruit development, thus serving as routes or reservoirs for contaminating ripened fruit.

Klebsiella oxytoca is an opportunistic pathogen involved in nosocomial infections and antibiotic-associated diarrhea and hemorrhagic colitis (Högenauer et al. 2006; Gorkiewicz 2009). *K. oxytoca* can cause serious infections, bacteremia, and septic shocks in immunocompromised individuals (Al-Anazi et al. 2008). *P. mirabilis* is a common cause of urinary tract infection (Zunino et al. 2000). Recent studies have shown that enteric bacteria can colonize the interiors of plants (Tyler and Triplett 2008; Berg et al. 2005). Endophytic colonization was shown to result from root infection or contamination of seeds (Tyler and Triplett 2008). The extent of endophytic colonization is

Table 2 Effect of temperature on enzymes production of *Serratia rubidaea* lipase enzyme, protease enzyme, polygalacturonase enzyme, and alkaline phosphatase enzyme

Temperature	Lipase	Protease	Polygalacturonase	Phosphatase
10	550	0.026	750	0.15
25	3,320	0.187	2,500	1.8
30	1,600	0.23	2,500	4.2
37	1,265	0.15	3,900	2.5
40	600	0.075	4,500	0.8
45	550	0.055	1,500	0.32

Data expressed as U/mg protein

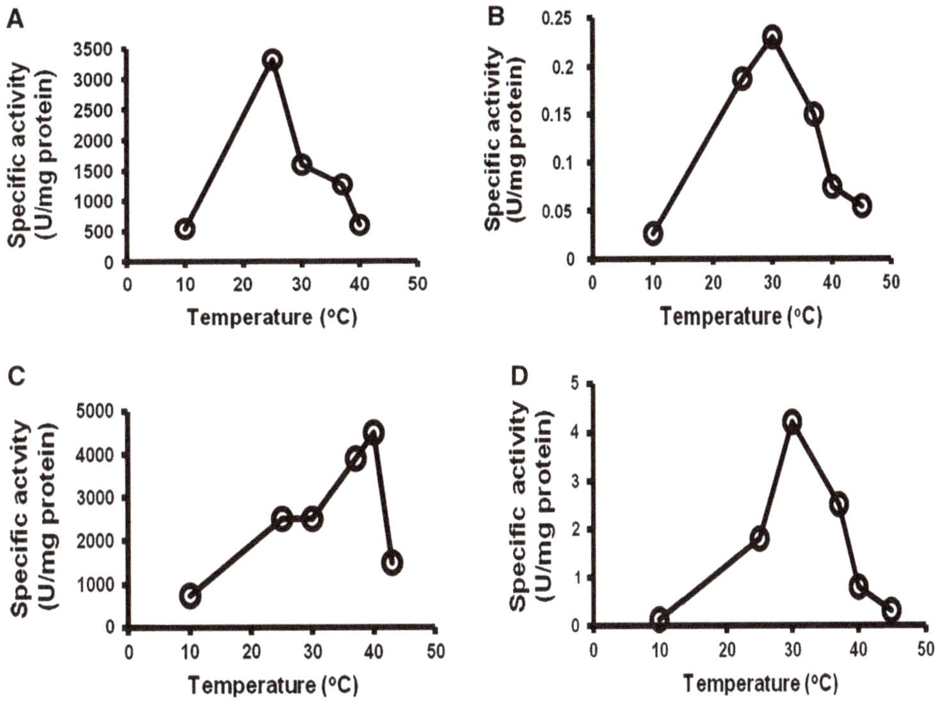

Fig. 2 Effect of temperature on enzymes production of *Serratia rubidaea*. **a** Lipase enzyme, **b** protease enzyme, **c** polygalacturonase enzyme, **d** alkaline phosphatase enzyme

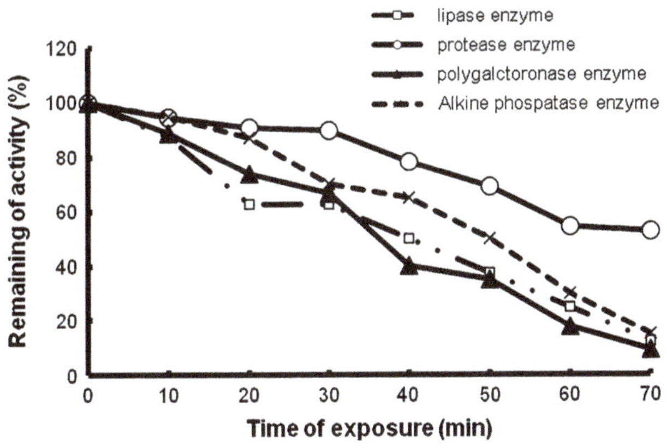

Fig. 3 Thermal stability of enzymes produced by *Serratia rubidaea* at 45 °C

determined by the genetic background of both the microbe and the host plant (Tyler and Triplett 2008).

Serratia rubidaea was the most prevalent bacteria recovered from naturally infected green pepper fruits in this study, representing 85 % of the total. The pathogenicity test was performed on healthy green pepper fruits and red colonies were reisolated from inoculated fruits showing splotches.

The habitats of *S. rubidaea* are not perfectly known. *S. rubidaea* has been repeatedly isolated from coconuts bought in France (originating mostly from Ivory Coast) and in California (Grimont et al. 1981). It has been isolated from coconuts and from vegetable salads, but it has not been reported from water, insects, small mammals, or animal territories (Grimont and Grimont 1991).

Serratia species are important in plant and food microbiology because not only are they involved in food spoilage but also they are opportunistic pathogens that can cause various diseases in humans, animals, and plants. They have been isolated from coconuts and vegetable salads, but not reported from water, insects, small mammals, or animal territories (Grimont and Grimont 1991).

The virulence of the plant pathogen *S. rubidaea* is dependent on the production and secretion of a large variety of plant cell wall-degrading enzymes and membrane constituents, including polygalacturonase, lipase, protease and alkaline phosphatase. The activity of the type of induced enzyme may be influenced by environmental factors. Their activity may be significantly diminished or destroyed by a variety of physical or chemical agents resulting in a loss of the functions performed by the enzymes. A characteristic feature of many phytopathogenic organisms is their ability to produce an array of enzymes capable of degrading the complex polysaccharides of the plant cell wall and membrane constituents. These enzymes are usually produced inductively and are extracellular, highly stable and present in infected host tissue (Bateman and Basham 1976). A major virulence factor in onion pathogenicity is the presence of a polygalacturonase enzyme involved in tissue maceration that is encoded by the plasmid-borne pehA gene (Gonzalez et al. 1997). In conclusion, these bacteria isolated in current study may be pathogenic for humans and could be a threat to human health in food.

Acknowledgments I thank the Nora Bent AbdulRahman University for sustained support and encouragement.

Conflict of interest The author declares that they have no conflict of interest.

References

Abd-Alla MH (1994) Phosphatases and the utilization of organic phosphorus by *Rhizobium leguminosarum* biovar *viceae* phosphatases. Lett Appl Microbiol 18:294–296

Abd-Alla MH, Bashandy SR, Schnell S, Ratering S (2011) Isolation and characterization of *Serratia rubidaea* from dark brown spots of tomato fruits. Phytoparasitica 39(2):175–183

Al-Anazi KA, Al-Jasser AM, Al-Zahrani HA, Chaudhri N, Al-Mohareb FI (2008) *Klebsiella oxytoca* bacteremia causing septic shock in recipients of hematopoietic stem cell transplant: two case reports. Cases J 1:160–164

Baharum SN, Salleh AB, Razak CNA, Basri M et al (2003) Organic solvent tolerant lipase by *Pseudomonas* sp. strain S5: stability of enzyme in organic solvent and physical factors affecting its production. Ann Microbiol 53:75–83

Bateman DF, Basham HG (1976) Degradation of plant cell walls and membranes by microbial enzymes. Physiol Plant Pathol 4:316–335

Berg G, Eberl L, Hartmann A (2005) The rhizosphere as a reservoir for opportunistic human pathogenic bacteria. Environ Microbiol 7:1673–1685

Blostein J (1991) An outbreak of *Salmonella javiana* associated with consumption of watermelon. J Environ Health 56:29–31

Brenner DJ, Krieg NR, Staley JT, Garrity GM (2005) Bergey's manual of systematic bacteriology (The Proteobacteria) part B (The Gammaproteobacteria), vol 2. Springer, New York, pp 633–638

Bruhn CM (1995) Consumer attitudes and market response to irradiated food. J Food Prot 58:175–181

Cappellini RA, Ceponis MJ, Wells JM, Lightner GW (1984) Disorders in potato shipments to the New York market, 1972–1980. Plant Dis 68:1018–1020

Ceponis MJ, Cappellini RA, Lightner GW (1985) Disorders in crisphead lettuce shipments to the New York market, 1972–1984. Plant Dis 69:1016–1020

Ceponis MJ, Cappellini PA, Lightner GW (1986) Disorders in tomato shipments to the New York market, 1972–1984. Plant Dis 70:261–265

Ceponis MJ, Cappellini RA, Lightner GW (1987) Disorders in fresh pepper shipments to the New York market, 1972–1984. Plant Dis 71:380–382

Chakraborty R, Srinivasan M (1993) Production of a thermostable alkaline protease by a new *Pseudomonas* sp. by solid substrate fermentation. Microbiol Biotechnol 8:7–16

Cotter JL, Fader RC, Lilley C, Herndon DN (1985) Chemical parameters, antimicrobial activities, and tissue toxicity of 0.1 and 0.5% sodium hypochlorite solutions. Am Soc Microbiol 28:118–122

Drancourt M, Bollet C, Carlioz A, Martelin R, Gayral JP, Raoult D (2000) 16S Ribosomal DNA sequence Analysis of a large collection of environmental and clinical unidentifiable bacterial isolates. Clin Microbiol 38:3623–3630

Fadıloğlu S, SÖylemez Z (1997) Kinetics of lipase-catalyzed hydrolysis of olive oil. Food Res Int 30:171–175

Farmer JJ III, Davis BR, Hickman-Brenner FW, McWhorter A, Huntley GP et al (1985) Biochemical identification of new species and biogroups of Enterobacteriaceae isolated from clinical specimens. Clin Microbiol 21:46–76

Garcıa-Martınez J, Bescos I, Rodrı′guez-Sala JJ, Valera FR (2001) RISSC: a novel database for ribosomal 16S–23S RNA gene spacer regions. Nucleic Acids Res 29:178–180

Gomes I, Saha RK, Mohiuddin G, Hoq MM (1992) Isolation and characterization of cellulase-free pectinolytic and hemicellulolytic thermophilie fungus. World J Microbiol Biotechnol 8:589–592

Gonzalez CF, Pettit EA, Valadez VA, Provin E (1997) Mobilization, cloning, and sequence determination of a plasmid-encoded polygalacturonase from a phytopathogenic *Burkholderia (Pseudomonas) cepacia*. Mol Plant-Microbe Interact 10:840–851

Gorkiewicz G (2009) Nosocomial and antibiotic-associated diarrhoea caused by organisms other than *Clostridium difficile*. Int J Antimicrob Agents 33:37–41

Grimont F, Grimont PAD (1991) The genus *Serratia*. In: Balows A, Trüper HG, Dworkin M, Harder W, Schleifer KH (eds) The prokaryotes, Springer, Verlag pp. 2822–2848

Grimont PAD, Grimont F, Starr MP (1981) *Serratia* species isolated from plants. Curr Microbiol 5:317–322

Guo X, Chen J, Brackett RE, Beuchat LR (2001) Survival of *Salmonellae* on and in tomato plants from the time of inoculation at flowering and early stages of fruit development through fruit ripening. Appl Environ Microbiol 67:4760–4764

Högenauer C, Langner C, Beubler E, Lippe IT, Schicho R et al (2006) *Klebsiella oxytoca* as a causative organism of antibiotic-associated hemorrhagic colitis. N Engl J Med 355:2418–2426

Kroupitski Y, Golberg D, Belausov E, Pinto R, Swartzberg D et al (2009) Internalization of *Salmonella enterica* in leaves is induced by light and involves chemotaxis and penetration through open stomata. Appl Environ Microbiol 75:6076–6086

Lane DJ (1991) 16S/23S rRNA sequencing, Nucleic acid techniques in bacteria systematics. In: Stackebrandt E, Goodfellow M (eds) John Wiley, New York pp 115–175

Lund BM (1983) Bacterial spoilage. In: Dennis C (ed) Postharvest pathology of fruits and vegetables. Academic Press, London, pp 219–257

Ohara-Nemoto Y, Sasaki M, Kaneko M, Nemoto T, Ota M (1994) Cysteine protease activity of streptococcal pyrogenic exotoxin B. J Clin Microbiol 40:930–936

Roever De (1998) Microbiological safety evaluations and recommendations on fresh produce. Food Control 6:321–347

Samish Z, Etinger-Tulczynska R, Bick M (1962) The microflora within the tissue of fruits and vegetables. Food Sci 28:259–266

Sanger F, Nicklen S, Coulson A (1977) DNA sequencing with chain-terminating inhibitors. Proc Natl Acad Sci USA 74:5463–5467

Schaad NW (1988) Laboratory guide for identification of plant pathogenic bacteria. APS Press, St. Paul 1–15

Snowdon AL (ed) (1990) A color atlas of post- harvest diseases of fruits and vegetables. Vol 2 Vegetables. CRC Press, Boca Raton

Stapp C (1940) Bacterium rubidaeum nov. spec. Zentratbl Bakteriol Abt 102:251–261

Tyler HL, Triplett EW (2008) Plants as a habitat for beneficial and/or human pathogenic bacteria. Annu Rev Phytopathol 46:53–73

Wei CI, Huang TS, Kim JM, Lin WF et al (1995) Growth and survival of *Salmonella* Montevideo on tomatoes and disinfection with chlorinated water. Food Prot 58:829–836

Wright C, Kominos SD, Yee RB (1976) Enterobacteriaceae and *Pseudomonas aeruginosa* recovered from vegetable salads. Appl Environ Microbiol 31:453–454

Zunino P, Geymonat L, Allen AG, Legnani-Fajardo C, Maskell DJ (2000) Virulence of a *Proteus mirabilis* ATF isogenic mutant is not impaired in a mouse model of ascending urinary tract infection. FEMS Immunol Med Microbiol 29:137–143

5

Micropropagation of *Ficus religiosa* L. via leaf explants and comparative evaluation of acetylcholinesterase inhibitory activity in the micropropagated and conventionally grown plants

Priyanka Siwach · Anita Rani Gill

Abstract A high-frequency, season-independent, in vitro regeneration of *Ficus religiosa* was developed, followed by comparative acetylcholinesterase inhibitory (AChEI) activity assay of the in vitro raised and conventionally grown plants. The use of AChEI activity is the most accepted strategy for the treatment of Alzheimer disease. Fully expanded, mature leaves were cut into different segments to initiate the cultures. The middle section of the leaf in vertical orientation with cut portion inserted inside the medium was found most suitable for direct shoot regeneration. Leaf explants responded with nearly consistent frequency (60–66.67 %) throughout the year. To obtain high frequency response with enhanced shoot multiplication rate, 32 plant growth regulator regimes were screened amongst which benzylaminopurine at 5.0 mg/l was found most suitable, yielding 100 % response and maximum number of shoots per explant (7.93); same concentration was also most supportive for repeated multiplication (6.53 shoots). The quality of the shoots and multiplication rate could be significantly enhanced (24.35 shoots) when adenine sulphate, glutamine and phloroglucinol, in an optimised concentration, were additionally supplemented. The clonal nature of the micropropagated plants was confirmed by random amplified polymorphic DNA analysis. A comparative analysis of AChEI activity was carried out amongst the methanolic extracts of stem segments of the mother plant, randomly selected seedlings of different age (4 and 6 months old) of the same mother plant and randomly selected micropropagated plants of different age (3 and 6 months age). The mother plant sample showed effective AChEI activity, with IC_{50} of 66.46 µg/ml while seedlings, of different age groups, performed poorly (6-month-old seedlings, Se-1_{6M}, yielded IC_{50} of 20,538.46 µg/ml, while two randomly selected 4 months' aged seedlings, Se-2_{4M} and Se-3_{4M} exhibited IC_{50} of 19,341.03 and 24,281.70 µg/ml). On the other hand, various micropropagated plants, 2 of 3 months (MiP-1_{3M}, MiP-2_{3M}) and 2 of 6 months (MiP-3_{6M} and MiP-4_{6M}) age behaved like the mother plant, exhibiting IC_{50} values of 71.87, 72.91, 67.65 and 69.65 µg/ml, respectively.

Keywords In vitro regeneration · Shoot multiplication · Clonal nature · RAPD markers · Acetylcholinesterase inhibitory activity · Alzheimer disease

Abbreviations

BAP	Benzylaminopurine
IAA	Indole acetic acid
TDZ	Thidiazuron
ADS	Adenine sulphate
PGR	Plant growth regulator
AChE	Acetylcholinesterase
AD	Alzheimer disease
MiP	Micropropagated plant
MP	Mother plant
Se	Seedling
RAPD	Randomly amplified polymorphic DNA

P. Siwach (✉) · A. R. Gill
Department of Biotechnology, Chaudhary Devi Lal University, Sirsa, Haryana, India
e-mail: psiwach29@gmail.com

Introduction

Ficus religiosa, the Sacred Fig or Bo-Tree, belonging to family Moraceae, is a large heavily branched tree with long petiolated, heart-shaped leaves. The tree is native to India and Nepal where it has great ethano-medicinal and religious importance since times immemorial. Different parts of the tree render applications for more than 50 disorders, which are well documented in Ayurveda, the indigenous Indian medicine system (Singh et al. 2011). Many of these have been confirmed by various pharmacological studies (Kirana et al. 2009; Pandit et al. 2010). One of the most affective and popular use of *F. religiosa* is in the treatment of cognitive decline, improving memory and related central nervous system disorders (Singh and Goel 2009). A number of traditional practitioners in north India prepare specific herbal formulations from stem bark of *F. religiosa* for treatment of memory loss and various neuro-degenerative disorders (Personal communication). The scientific basis for this was revealed by Vinutha et al. (2007), who while analysing the methanolic stem bark extract of *F. religiosa*, found potent acetylcholinesterase inhibitory (AChEI) activity associated with it. The latter is the most accepted strategy for the treatment of Alzheimer disease (AD) and other related diseases (Bertaccini 1982); the inhibitors prolong the half-life of acetylcholine through inhibition of acetylcholinesterase (AChE) (Darvesh et al. 2003). Since, the present day drugs for AD (tacrine, donepezil, rivastigime and galanthamine) are suffering with short-half-lives and/or unfavourable side effects such as hepatotoxicity, low bioavailability, adverse cholinergic side effects and a narrow therapeutic window (Sancheti et al. 2009), *F. religiosa* is being looked upon as a potential source of a new drug for AD. The phytochemical analysis of bark and stem of *F. religiosa* has led to the isolation of phytosterols, amino acids, furanocoumarins, phenolic components, flavonoids, saponins and acid detergent fibres (Ambike and Rao 1967; Swami et al. 1989; Swami and Bisht 1996); however, those responsible for AChEI activity are yet to be identified. There is a need to characterise and exploit the AChEI activity found in the stem bark of *F. religiosa* at the commercial level. Conventionally, the tree is propagated by seeds, which remain viable for a few months and the plants produced are not true to types. The vegetative propagation by cutting is not efficient under varied climatic conditions. The nature and amount of secondary metabolites in different parts of the plants is greatly affected by the environmental condition and so naturally propagated plants are unable to yield consistent yield of secondary metabolites throughout the year (Gurel et al. 2011; Tamara et al. 2011). Micropropagation is a proven method for large-scale production of true-to-type medicinal plants capable of yielding desired plant-derived pharmaceuticals with consistent quality and amount (Pattnaik and Chand 1996; Jiang et al. 2012).

In vitro propagation methods also offer the opportunity to correlate the secondary metabolite production with several parameters like nutritional and hormonal composition of nutrient medium, growth conditions, duration of culture, etc., and so are better for production of plants for commercial, pharmaceutical applications (Gurel et al. 2011).

Some previous work on in vitro propagation of *F. religiosa* L. using nodal segments as explants has been carried out by Jaiswal and Narayan (1985), Deshpande et al. (1998), and Hassan et al. (2009). In an attempt to further improve the micropropagation protocol (Siwach and Gill 2011), it was observed that response of nodal segments as well as frequency of contamination under in vitro conditions was strongly influenced by the season of explants collection, restricting the culture initiation experiment to a particular time-period of the year (Siwach et al. 2011). Similar observations were also observed with apical shoot explants (data not shown). The seasonal influence on the establishment and response of in vitro cultures of perennial trees is due to their periodic development and is one of the major hurdles towards commercialisation of micropropagation for these plants as it cannot be overcome by environmental or nutritional manipulations (Siril and Dhar 1997; McCown 2000).

Therefore, a need was felt to formulate a commercially viable micropropagation protocol for *F. religiosa* L., which would be beyond the seasonal constraints. Axillary and apical meristematic cells (on the nodal and apical explants respectively), of the perennial trees, are genetically directed to divide actively during the active growth season, and this periodic growth direction may not be there with the differentiated tissues like leaves. There exists one study reporting the use of leaves as explants for in vitro propagation of *F. religiosa* (Narayan and Jaiswal 1986). However, the study pertains to indirect shoot organogenesis from leaves through the callus phase and does not discuss about the seasonal influence on the in vitro response of leaves' explants. So during the present study, experiments were initiated with fully expanded mature leaf segments to optimise season independent protocol for *F. religiosa*. It was planned to go specifically for direct shoot regeneration followed by confirming the clonal nature of plants by randomly amplified polymorphic DNA (RAPD) based molecular analysis. The latter has been reported as a reliable method for monitoring the genetic stability of micropropagated plants in many species (Khan and Spoor 2001). To explore the pharmaceutical potential of micropropagated plants, present study was extended to carry out comparative AChE inhibitory activity assay among the stem tissue of randomly selected micropropagated plants, randomly selected seedlings of the same mother plant and mother plant itself.

Table 1 Effect of different orientations of various cut leaf sections of *Ficus religiosa* L. on shoot organogenesis (cultured on MS medium having 1.0 mg/l BAP)

Cut leaf section	Orientation	Response (%)	No. of shoot buds[A]	No. of shoots[B]	Length of shoots[B]
Proximal section with petiole	Cut portion inserted vertically in medium	$40.00^{abc} \pm 9.09$	$3.33^b \pm 0.15$	$3.40^b \pm 0.18$	$1.97^{bc} \pm 0.18$
	Adaxial side touching the medium	$36.67^{abc} \pm 8.95$	$2.80^c \pm 0.11$	$2.53^c \pm 0.11$	$2.13^{bc} \pm 0.19$
	Abaxial side touching the medium	$33.33^{bc} \pm 8.75$	$1.33^e \pm 0.11$	$1.47^d \pm 0.13$	$1.67^c \pm 0.09$
Middle section	Cut portion inserted vertically in medium	$63.33^a \pm 8.95$	$4.83^a \pm 0.18$	$4.57^a \pm 0.13$	$2.67^a \pm 0.19$
	Adaxial side touching the medium	$60.00^{ab} \pm 9.09$	$3.47^b \pm 0.21$	$3.37^b \pm 0.19$	$2.13^{bc} \pm 0.12$
	Abaxial side touching the medium	$56.67^{abc} \pm 9.20$	$3.23^{bc} \pm 0.20$	$2.33^c \pm 0.11$	$2.03^{bc} \pm 00.11$
Distal section with leaf tip	Cut portion inserted vertically in medium	$36.67^{abc} \pm 8.95$	$2.23^d \pm 0.18$	$2.17^c \pm 0.14$	$2.47^{ab} \pm 0.17$
	Adaxial side touching the medium	$33.33^{bc} \pm 8.75$	$1.53^e \pm 0.21$	$1.17^d \pm 0.14$	$2.17^{bc} \pm 0.18$
	Abaxial side touching the medium	$30.00^c \pm 8.50$	$1.33^e \pm 0.09$	$1.17^d \pm 0.08$	$2.13^{bc} \pm 0.17$

Data are means from 10 replicates $\pm$ SE and those representing similar letter in the appropriate column are not significantly different (ANOVA, $P \leq 0.05$), (groupings applying to whole table)

[A] Observed after 4 weeks of culture

[B] Observed after 60 days of initial culture

Materials and methods

Plant materials

Fully grown, healthy looking leaves were excised from third to sixth node from the tip of a healthy branch of a 45–50-year-old tree of *F. religiosa* L., selected as the mother plant (MP) for the present study, growing near the campus area of Chaudhary Devi Lal University, Sirsa, Haryana, India. The leaves were kept for 15 min under running tap water to remove the traces of dust and were surface sterilised as described earlier (Siwach and Gill 2011). The surface sterilised leaves were used as explants for direct shoot regeneration in the present study.

Seeds of the same mother plant were also sown in the month of March (seeds are available during months of March to June in North India) under greenhouse conditions and were properly taken care of. Among various seedlings so obtained, three seedlings of different ages were randomly selected. One seedling was of 6 months age (named as Se-1_{6m}) while other two were of 4 months age (named as Se-2_{4M} and Se-3_{4M}). These three seedlings were used for comparison with micropropagated plants in different analysis, reported below.

Effect of explant orientation on shoot organogenesis

The surface sterilised leaves, as obtained above, were cut into three sections—proximal section with petiole, middle

section and distal section with leaf tip. Each section was further cultured in three orientations-cut portion inserted in medium, adaxial surface touching the medium and abaxial surface touching the medium. Murashige and Skoog's medium (1962), (MS) supplemented with 1.0 mg/l benzylaminopurine (BAP) was used for this experiment. The number of explants initiating shoot buds (percentage of response) and the average number of shoot buds per explant was recorded after 28 days of culture. The explants with sprouted shoot buds/shoots were shifted to the same medium as such within an interval of 30 days so as to score the number of shoots and the length of shoots after 60 days of initial culture (Table 1). Every treatment contained ten replicates and the experiment was repeated thrice. These experiments were carried out during January–February, 2009.

Using the most suitable orientation of the leaf segments as obtained from above experiment, following two studies were carried out in parallel.

Effect of season of explant collection on percentage of response of explants

The cut leaf segments, in the most suitable orientation as deduced from above experiment, were cultured every month from March 2009–February 2010. The MS basal medium supplemented with 1.0 mg/l BAP was employed for this study. There were 20 replicates for each month and the observations were recorded after 4 weeks of culture.

Table 2 Effect of various concentrations and combinations of BAP, TDZ and IAA on shoot organogenesis from leaf explants

PGR (mg/l)			Response (%)	No. of shoot buds	No. of shoots[B]	Length of shoots (cm)
BAP	TDZ	IAA				
1.0	–	–	$63.33^{bcdefg} \pm 8.95$	$4.80^{fg} \pm 0.23$	$4.53^{d} \pm 0.16$	$2.67^{abcde} \pm 0.16$
1.5	–	–	$60.0^{cdefgh} \pm 9.09$	$3.93^{hijk} \pm 0.33$	$4.57^{d} \pm 0.18$	$2.47^{defg} \pm 0.09$
2.0	–	–	$66.67^{bcdef} \pm 8.75$	$3.93^{hijk} \pm 0.28$	$4.33^{de} \pm 0.24$	$2.23^{efgh} \pm 0.12$
2.5	–	–	$66.67^{bcdef} \pm 8.75$	$3.20^{kl} \pm 0.25$	$4.07^{de} \pm 0.29$	$2.07^{ghi} \pm 0.14$
3.0	–	–	$73.3^{bcde} \pm 8.21$	$3.73^{ijkl} \pm 0.35$	$2.53^{jk} \pm 0.22$	$2.50^{cdefg} \pm 0.14$
3.5	–	–	$76.67^{abcd} \pm 7.85$	$3.93^{hijk} \pm 0.41$	$3.13^{hi} \pm 0.21$	$2.53^{bcdef} \pm 0.11$
4.0	–	–	$80.0^{abc} \pm 7.42$	$4.23^{ghi} \pm 0.31$	$4.20^{de} \pm 0.29$	$2.33^{defgh} \pm 0.18$
4.5	–	–	$83.33^{abc} \pm 6.92$	$4.17^{ghij} \pm 0.19$	$4.47^{d} \pm 0.19$	$2.93^{abc} \pm 0.15$
5.0	–	–	$100.0^{a} \pm 0$	$5.83^{e} \pm 0.19$	$7.93^{a} \pm 0.20$	$2.97^{ab} \pm 0.13$
5.5	–	–	$90.0^{ab} \pm 5.57$	$4.13^{ghij} \pm 0.31$	$3.23^{ghi} \pm 0.08$	$2.20^{fgh} \pm 0.10$
6.0	–	–	$63.33^{bcdefg} \pm 8.95$	$2.13^{n} \pm 0.08$	$1.17^{m} \pm 0.07$	$2.17^{fgh} \pm 0.09$
1.0	–	0.5	$43.33^{fghij} \pm 9.20$	$3.70^{ijkl} \pm 0.22$	$7.27^{b} \pm 0.23$	$2.77^{abcd} \pm 0.17$
1.5	–	0.5	$40.0^{fghij} \pm 9.09$	$3.23^{kl} \pm 0.10$	$3.47^{fgh} \pm 0.19$	$3.07^{a} \pm 0.15$
2.0	–	0.5	$36.67^{ghij} \pm 8.95$	$3.43^{ijkl} \pm 0.12$	$3.97^{def} \pm 0.21$	$2.43^{defg} \pm 0.11$
2.5	–	0.5	$36.67^{ghij} \pm 8.95$	$3.07^{lm} \pm 0.18$	$3.33^{ghi} \pm 0.22$	$1.67^{hijk} \pm 0.15$
3.0	–	0.5	$40.0^{fghij} \pm 9.09$	$3.33^{jkl} \pm 0.23$	$3.77^{efg} \pm 0.21$	$1.57^{ijk} \pm 0.18$
3.5	–	0.5	$36.67^{ghij} \pm 8.95$	$3.33^{jkl} \pm 0.22$	$2.77^{ij} \pm 0.27$	$1.43^{jkl} \pm 0.16$
4.0	–	0.5	$33.33^{hij} \pm 8.75$	$3.93^{hijk} \pm 0.12$	$2.10^{k} \pm 0.19$	$2.57^{bcdef} \pm 0.18$
4.5	–	0.5	$46.67^{efghij} \pm 9.26$	$3.78^{ijkl} \pm 0.20$	$4.30^{de} \pm 0.26$	$2.97^{ab} \pm 0.20$
5.0	–	0.5	$63.33^{bcdefg} \pm 8.95$	$4.67^{gh} \pm 0.15$	$5.53^{c} \pm 0.32$	$2.67^{abcde} \pm 0.27$
5.5	–	0.5	$50.0^{defghi} \pm 9.28$	$2.23^{n} \pm 0.09$	$3.13^{hi} \pm 0.18$	$2.13^{fgh} \pm 0.13$
6.0	–	0.5	$36.67^{ghij} \pm 8.95$	$2.13^{n} \pm 0.08$	$1.07^{m} \pm 0.08$	$1.97^{hij} \pm 0.20$
–	1.0	–	$63.33^{bcdefg} \pm 8.95$	$5.50^{ef} \pm 0.16$	$2.77^{ij} \pm 0.23$	$1.07^{kl} \pm 0.05$
–	1.5	–	$60.0^{cdefgh} \pm 9.09$	$4.83^{fg} \pm 0.25$	$2.33^{jk} \pm 0.22$	$1.37^{jkl} \pm 0.16$
–	2.0	–	$43.33^{fghij} \pm 9.20$	$2.43^{mn} \pm 0.41^{A}$	$2.10^{k} \pm 0.13$	$1.33^{jkl} \pm 0.14$
–	2.5	–	$33.33^{hij} \pm 8.75$	$2.33^{mn} \pm 0.22^{A}$	$2.03^{kl} \pm 0.19$	$1.57^{ijk} \pm 0.14$
–	3.0	–	$30.33^{ij} \pm 8.51$	$1.97^{n} \pm 0.22^{A}$	$1.83^{l} \pm 0.19$	$1.47^{jkl} \pm 0.14$
–	1.0	0.5	$26.67^{ij} \pm 8.21$	$10.17^{b} \pm 0.39^{A}$	0	–
–	1.5	0.5	$33.33^{hij} \pm 8.75$	$12.33^{a} \pm 0.36^{A}$	0	–
–	2.0	0.5	$30.0^{ij} \pm 8.51$	$9.30^{c} \pm 0.36^{A}$	0	–
–	2.5	0.5	$23.33^{ij} \pm 7.85$	$8.87^{c} \pm 0.27^{A}$	0	–
–	3.0	0.5	$20.0^{ij} \pm 7.43$	$7.07^{d} \pm 0.17^{A}$	0	–

Data are means from 30 replicates $\pm$ SE and those representing similar letter in the appropriate column are not significantly different (ANOVA, $P \leq 0.05$), (groupings applying to whole table)
[A] Extensive callus formation, [B]observed after 60 days of culture

Effect of plant growth regulators on direct shoot regeneration

It was carried out during the period of March–July, 2009. The cut leaf sections in most suitable orientation as obtained above were inoculated on MS medium supplemented with various concentrations and combinations of BAP, thidiazuron (TDZ) and indole acetic acid (IAA). The observations were made for shoot organogenesis as reported above (Table 2). For each of these treatments, 20 explants were used and the experiment was repeated thrice.

Shoot multiplication during sub-culturing

The shoots regenerated from leaf explant were excised from the explant surface and were cultured in cluster of 2–3 shoots on six different medium. MS medium having most suitable PGR regime deduced from above study was taken as control. Two more concentrations of the selected PGR, lower than that in control, were also screened. These three media were further modified with specific combination of three additives—glutamine (200 mg/l), adenine sulphate (ADS) (150 mg/l) and phloroglucinol (100 mg/l), which

Table 3 Effect of different concentrations of BAP, with or without optimised additives, on shoot multiplication as well as quality of shoots (observations recorded after 5 weeks of culture)

BAP (mg/l)	Medium additives	No. of shoots	Length of shoots (cm)	Comments[A]
5.0 (control)	No additive	$6.53^d \pm 0.14$	$2.97^{ab} \pm 0.13$	$--$
2.5	No additive	$4.07^e \pm 0.29$	$2.07^c \pm 0.14$	$--$
1.0	No additive	$4.53^e \pm 0.16$	$2.67^b \pm 0.16$	$--$
5.0	With optimised additives[B]	$24.37^a \pm 0.34$	$3.17^a \pm 0.15$	$+++$
2.5	With optimised additives[B]	$17.17^b \pm 0.09$	$2.23^c \pm 0.13$	$+++$
1.0	With optimised additives[B]	$15.03^c \pm 0.21$	$2.17^c \pm 0.08$	$+++$

Data are means from 30 replicates $\pm$ SE and those representing similar letter in the appropriate columns are not significantly different (ANOVA, $P \leq 0.05$) (groupings applying to whole table)

[A] $--$(Chlorosis, leaf fall), $+++$ (no chlorosis, no leaf fall, increased vigour of shoots)

[B] 200 mg/l glutamine+ 150 mg/l adenine sulphate+ 100 mg/l phloroglucinol

was earlier reported to support the shoot proliferation during repeated sub-culturing (Siwach and Gill 2011). Observations were recorded after 35 days of culture (Table 3). For each of the above treatment, 20 replicates were used and the experiment was repeated thrice.

In vitro rooting, acclimatisation and transplantation

Individual shoots were subjected to in vitro rooting, acclimatisation and transplantation as reported previously (Siwach and Gill 2011).

Culture conditions

MS medium, modified with different growth regulators as per requirement as reported above, was supplemented with 3 % (w/v) sucrose and solidified with 0.8 % (w/v) agar. The pH of medium was adjusted to 5.8 using 0.1 N NaOH or 0.1 N HCl, before autoclaving at 15 kg/cm^2 and 121 °C for 20 min. The cultures were maintained at 25 ± 2 °C, at a photoperiod of 16 h (80 µmol m^{-2} s^{-1}) in a culture room.

Checking the clonal fidelity of micropropagated plants

Six micropropagated plants, of different age (1, 2, 4, 5, 6 and 7 months aged, after transplantation) were randomly selected for this. Three seedlings-Se-1$_{6M}$, Se-2$_{4M}$ and Se-3$_{4M}$, discussed above in the "Plant material" sub-section, were also selected to check their clonal fidelity. Mother plant sample was taken as reference sample. Young leaves were excised from the above listed ten samples and were subjected to DNA isolation using Cetyl trimethyl ammonium bromide (CTAB) method (Sanghai-Maroof et al. 1984). Qualitative and quantitative assessment of total genomic DNA was carried out by spectrophotometer as well as agarose gel electrophoresis, followed by DNA purification.

Twenty RAPD primers, procured from Bangalore Genie Pvt. Ltd., were used for amplification (Table 4). Polymerase Chain Reaction (PCR) was carried out in a total volume of 15 µl with each reaction tube comprising of 8.1 µl of PCR water, 1.5 µl of PCR buffer containing MgCl$_2$ (10X), 2.4 µl of dNTPs mixture (5 mM), 1.2 µl of primers (10 µM), 1.5 µl (10 ng/µl) of template DNA, and 0.3 µl of Taq DNA polymerase. Amplification was carried out using Bio-Rad thermal-cycler and was run for 35 cycles; each cycle consisting of a denaturation step at 94 °C (1 min), a primer annealing step at 36 °C (1 min) followed by amplification at 72 °C (3 min). Amplified products were loaded on 1.5 % agarose gel along with 100 bp ladder (GE Healthcare Life Sciences) and electrophoresis was carried out at 100 V. Gel was stained with 25 µg/ml ethidium bromide and photographed on a Gel Documentation Polaroid system (Bio-Rad).

Sample preparation for AChE inhibitory activity assay

Mother plant sample (MP) was taken as reference for this. Four micropropagated plants of different age (two plants after 3 months of transplantation—MiP-1$_{3M}$ and MiP-2$_{3M}$, and two plants after 6 months of transplantation—MiP-3$_{6M}$ and MiP-4$_{6M}$) were randomly selected. Three seedlings—Se-1$_{6M}$, Se-2$_{4M}$ and Se-3$_{4M}$ were also taken for AChEI activity assay. The MP selected was of 45–50 years old, the age when higher secondary metabolites are reported to be present in stem bark. On the other hand, the other plants selected were of 3–6 months age as reported above. So to avoid the huge difference in the development stage of these plants, young, fresh, thin branches of mother plant was taken as sample (Fig. 1a) instead of stem bark or older branches. Likewise for seedlings and micropropagated plants also, young branches were excised (Fig. 1b). After excision, young branches were defoliated and cut into small-sized pieces followed by thorough washing with

Table 4 Amplification profile obtained with selected 20 RAPD primers, during the present study

Sr no.	Primer sequence	Band size of amplified products (bp)				
		Mother plant	Micropropagated plants	Se-1[6M]	Se-2[4M]	Se-3[4M]
1	CAGGCCCTTC	500	500	500, 670[A]	500	500, 600[A]
2	TGCCGAGCTG	1,200, 700	1,200, 700	700, 840[A]	700	700
3	AGTCAGCCAC	500	500	500	500	500
4	AATCGGGCTG	200, 300, 400, 500	200, 300, 400, 500	200, 300, 400, 500	200, 300, 400, 500	200, 300, 400, 500
5	AGGGGTCTTG	NA	NA	NA	NA	NA
6	GGTCCCTGAC	NA	NA	NA	NA	NA
7	GAAACGGGTG	NA	NA	NA	NA	NA
8	GTGACGTAGG	NRA	NRA	NRA	NRA	NRA
9	GGGTTTCGCC	240, 340, 440	240, 340, 440	300[A], 590[A], 610[A]	240, 340, 440, 590[A], 610[A]	240, 340, 440, 590[A], 610[A]
10	GTGATCGCAG	NRA	NRA	NRA	NRA	NRA
11	TAATCGGCGT	260, 350, 410, 600, 700	260, 350, 410, 600, 700	150[A], 260, 350, 410, 600, 700, 900[A]	350, 410, 600, 700	150[A], 260, 350, 410, 600, 700
12	TCGGCGATAG	280, 370, 480	280, 370, 480	280, 370, 480	280, 370, 480	280, 370, 480
13	CAGCACCCAC	280, 340, 500, 520, 600	280, 340, 500, 520, 600	280, 340, 500, 520, 600	280, 340, 500, 520, 600	280, 340, 500, 520, 600
14	TCTGTGCTGG	170, 380, 480, 690, 790	170, 380, 480, 690, 790	170, 380, 480, 690, 790	170, 380, 480, 690, 790	170, 380, 480, 690, 790
15	TTCCAAACCC	NA	NA	NA	NA	NA
16	TGCCATCGAA	NA	NA	NA	NA	NA
17	GGCGGCTTGT	300, 400	300, 400	300, 400	300, 400	300, 400
18	ATTGACCGT	550	550	550	550	550
19	CAAACGTCGG	300, 550, 700	300, 550, 700	300, 550, 700	300, 550, 700	300, 550, 700
20	GTTGCGATCC	420, 580, 640	420, 580, 640	420, 580, 640	420, 580, 640	420, 580, 640

NA No amplification, *N-RA* non-reproducible band pattern

[A] Polymorphic bands

sterilised double distilled water. These were then dried in an oven at 32 °C. The dried pieces were finely ground to a fine powder using mixer grinder. The 15 g of dried fine powder of each sample was macerated separately with HPLC (high performance liquid chromatography) grade methanol (75 ml) at room temperature for 24 h. The extracts were obtained by filtration using Whatman No. 1 filter paper and concentrated to dryness. Working concentrations of methanolic extracts of each of the eight samples were prepared by completely dissolving the dried extract in methanol in ratio 1:1 to have stock solution of strength 1 mg/ml.

AChE inhibitory activity assay

For each of the eight samples, five dilutions of 20, 40, 60, 80 and 100 µg/ml concentration were prepared. The assay for measuring AChEI activity was modified from the method described by Ellman et al. (1961) and Ingkaninan et al.

(2000). Briefly, 4 µl of 3 mM dithiobis nitrobenzoic acid (DTNB), 20 µl of 15 mM acetylthiocholine iodide (ATCI), 130 µl Tris–HCl and 20 µl of each diluted sample was taken and added to the wells of microplate, followed by 50 µl of 0.28 U/ml AChE enzyme. The microplate was then read at 405 nm every 5 s for 2 min by a CERES microplate reader (Spectra Max Plus 384, Molecular Devices, SoftMax Pro 5 S·No. SMP500-16135-DPVW). Mean absorbance per minute (A) was calculated for each dilution of different samples. Each plate had one blank well and one control well also. Percentage of inhibition of enzyme activity by a given concentration of sample was calculated by using the formula:

$$\text{Per cent inhibition} = [(A_{\text{control}} - A_{\text{extract}})/A_{\text{control}}] \times 100$$

The percentage of inhibition obtained above was plotted vs. corresponding concentration of the extract of each sample. There were three replicates for each diluted concentration. The experiment was repeated for two more times.

Fig. 1 Source of stem tissue (for methanolic extract preparation). **a** Young branches of mother plant (of 45–50 year age), *bar* 10 cm; **b** young branches of seedlings/ micropropagated plants (of age 3–6 months), *bar* 2.5 cm

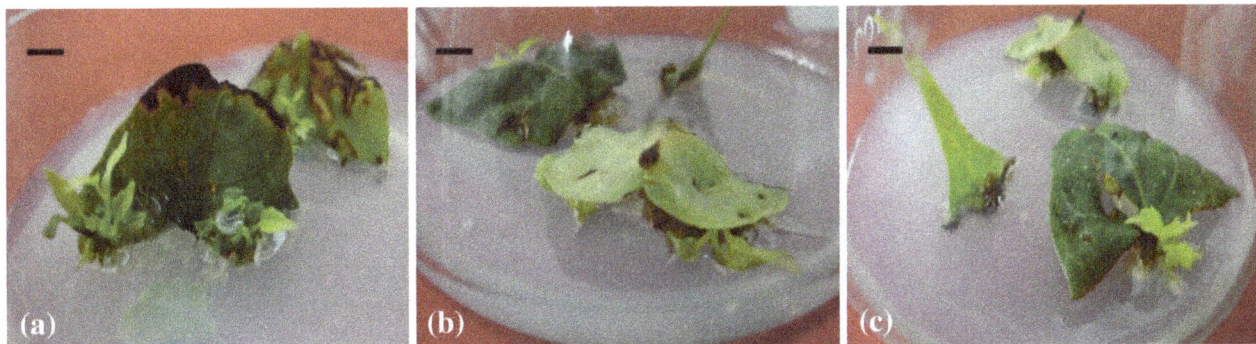

Fig. 2 Induction of shoot buds/shoots from middle section of leaf cultured in; **a** vertical orientation with one cut end inserted inside the medium, *bar* 0.4 cm; **b** horizontal orientation with adaxial surface touching medium, *bar* 0.7 cm; **c** horizontal orientation with abaxial surface touching the medium, *bar* 0.7 cm

Calculating the IC_{50} value of each sample

The obtained graph of each sample fitted perfectly in Logarithmic curve shape and was subjected to regression analysis using logarithmic equation. The IC_{50} value (i.e. the concentration at which the enzyme activity is inhibited by 50 %) was calculated with the obtained equation.

Statistical analysis

The experiments were set up in completely randomised design (CRD). The data were analysed by Analysis of Variance (ANOVA) followed by Duncan multiple range test (DMRT). Data analysis was carried out by using SPSS version 18.

Results

Effect of explant orientation on shoot organogenesis

The leaf sections, in all the orientations, exhibited shoot-buds formation after 17–20 days of culture. Of the three types of cut leaf sections, middle section of the leaf performed better than the proximal or distal sections in each orientation (Table 1). Middle section responded with maximum frequency (63.33 %) in vertical orientation with cut portion inserted inside the medium, which was not significantly different to that obtained with horizontal orientation with adaxial surface (60.00 %) or abaxial surface (56.67 %) touching the medium (Table 1). Significantly higher number of shoots buds (4.83) which proliferated into equally higher number of shoots (4.57) with considerably more lengths (2.67 cm) was obtained with middle section of leaf in vertical orientation (Table 1). Distal sections and proximal sections responded poorly (40 % or less than it) in all the three orientations. Regardless of the nature and orientation, shoots originated from the cut ends of leaf segments having direct contact with the medium (Fig. 2a–c).

Middle section of the leaves with cut portion inside the medium was selected for the subsequent studies and here onwards it will be referred as explant/leaf explant.

Effect of season on shoot organogenesis from leaf explant

With the same medium condition, percentage of response of leaf explant varied from a maximum of 66.67 % (during the months of April, May, June, October and December) to

(a)

(b)

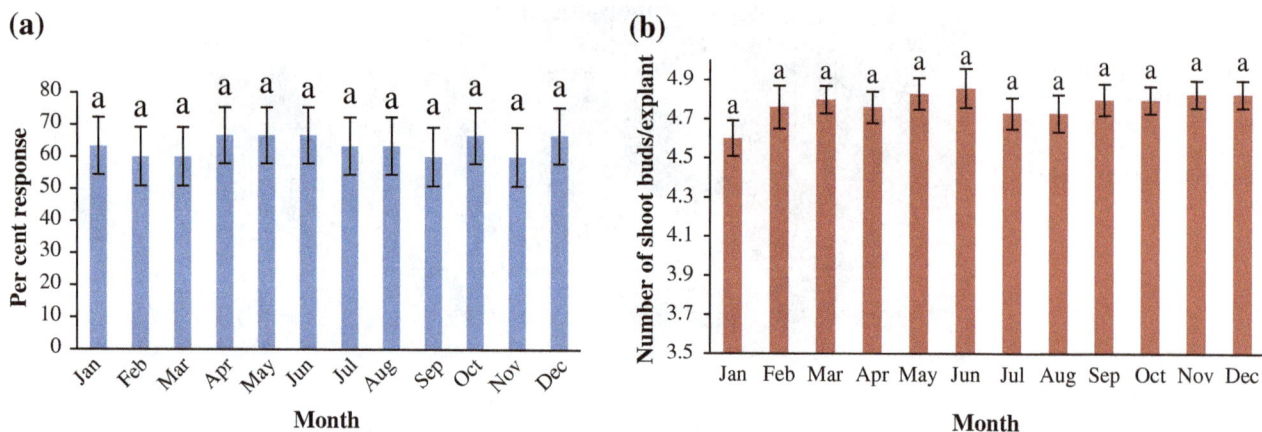

Fig. 3 Effect of month of explant collection on; **a** percent response of explant; **b** average number of shoot buds per leaf explant (observed after 28 days of culture)

Fig. 4 Shoot organogenesis from middle section of the leaf in vertical orientation, when cultured on MS medium having 5.0 mg/l BAP; **a** shoot buds as observed after 28 days of culture, *bar* 0.6 cm; **b** shoots as observed after 60 days of initial culture, *bar* 0.5 cm

a minimum of 60 % (during the months of February, March, September, November), the difference being statistically non-significant (Fig. 3a). Further, the maximum number of shoot buds per explant (4.86) was obtained for the month of June and this was not significantly different to that obtained during rest of the months (Fig. 3b).

Effect of plant growth regulators on direct shoot regeneration

Leaf explants responded differently to different concentrations of BAP and TDZ. With increase in concentrations of BAP from 1.0 up to 5.0 mg/l, a continuous increase in percentage of response was observed, maximum being on 5.0 mg/l (100 %), afterwards the frequency decreased. On the other hand, with increase in concentration of TDZ from 1.0 to 3.0 mg/l, percentage of response continuously decreased; maximum was observed on 1.0 mg/l (63.33 %) (Table 2). Further, beyond 1.5 mg/l of TDZ, shoot organogenesis was accompanied with extensive callus formation at the explant surface.

Addition of IAA (0.5 mg/l) in the medium already containing BAP or TDZ, making different combinations,

significantly lowered the percentage of response (Table 2). A decrease in the number of shoot buds/shoots per explant, after addition of IAA to different concentrations of BAP, was also observed except for the combination of 0.5 mg/l IAA with 1.0 mg/l BAP which induced considerably higher number of shoots (7.27) per explant, though the percentage of response was low (43.33 %). Addition of IAA to different concentrations of TDZ induced a peculiar type of response, considerably higher number of shoot buds (7.07–12.23) were observed per explant but these shoot buds could not differentiate into full grown shoots.

Of the thirty-two PGR regimes screened, 5.0 mg/l BAP, besides supporting maximum percent response (100 %), induced proliferation of maximum number of shoot buds (5.83) (as observed after 4 weeks of culture) (Fig. 4a) which later on turned into maximum number of shoots (7.93) with higher lengths (2.97 cm, on an average) (as observed after 60 days of initial culture) (Fig. 4b) (Table 2). Comparable number of shoots (7.27) was obtained on combination of 0.5 mg/l IAA with 1.0 mg/l BAP, but as percentage of response on this particular combination was very low (43.33 %), this was not selected for further studies.

Fig. 5 Shoots multiplication during sub-culturing on; **a** MS medium having 5.0 mg/l BAP, *bar* 0.7 cm; **b** MS medium having 5.0 mg/l BAP with additional supplementation of 200 mg/l glutamine, 150 mg/l adenine sulphate and 100 mg/l phloroglucinol, *bar* 0.5 cm

Shoot multiplication during sub-culturing

The shoots so obtained, when excised from the explant and cultured in a cluster of 2–3 shoots on the most suitable medium obtained above i.e. MS medium having 5.0 mg/l BAP (control), mild chlorosis of leaves and stem as well as curling and browning of leaves edges and early leaf fall (at low frequency) was observed (Fig. 5a). Further, the number of shoots obtained on this medium, at the end of fifth week, was also less (6.53) (Table 3). Lowering of BAP concentration to 2.5 and 1.0 mg/l further lowered the multiplication rate while health of shoots remained poor (Table 3). To improve quality of shoots as well as to enhance shoot multiplication rate, three additives were added in optimal concentration (200 mg/l glutamine+ 150 mg/l ADS+ 100 mg/l phloroglucinol); optimisation of these additives being discussed in our previous report (Siwach and Gill 2011). Supplementation of these additives to each of the three concentration of BAP significantly increased the shoot multiplication rate as well as improved the quality of the shoots (Table 3). Of the three concentrations of BAP, significantly higher number of shoots (24.37) was obtained when this combination of glutamine, ADS and phloroglucinol was supplemented to 5.0 mg/l BAP (Fig. 5b) indicating the suitability of higher cytokinin concentration even during repeated sub-culturing. The shoots so obtained were isolated after 5 weeks of culture and were sub-cultured on this medium and the process was repeated 4–5 times, before subjecting the individual shoot to rooting.

In vitro rooting, acclimatisation and transplantation

The efficient in vitro rooting of the shoots, regenerated above, was obtained on MS medium supplemented with 2.0 mg/l indole butyric acid (IBA) along with 0.1 mg/l IAA, with a frequency of 95 %, as discussed in earlier report (Siwach and Gill 2011) (Fig. 6). The plantlets were successfully acclimatised (Fig. 7) and transferred to field

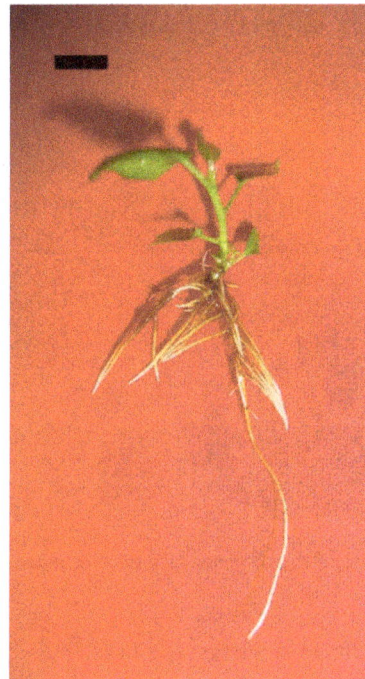

Fig. 6 In vitro rooting of micro-shoots obtained on semi-solid MS medium supplemented with 2.0 mg/l IBA and 0.1 mg/l IAA, *bar* 1.0 cm

conditions (Fig. 8) with a survival rate of 90 % on conditions reported earlier (Siwach and Gill 2011).

Checking the clonal fidelity of micropropagated plants

Out of 20 RAPD primers, amplification with reproducible band pattern was obtained with 13 (primers 1, 2, 3, 4, 9, 11, 12, 13, 14, 17, 18, 19, 20); rest seven primers either did not yield amplification or gave poor non-reproducible band pattern (Table 4). With these thirteen primers monomorphic band pattern was observed in the six randomly selected micropropagated plants and the MP sample, which confirmed the clonal nature of micropropagated plants (Table 4).

Fig. 7 Acclimatisation of transplanted plants under culture room conditions, *bar* 5 cm

Fig. 8 Successful acclimatised, potted plants of different age, *bar* 10 cm

The selected three seedlings exhibited monomorphic band profile among themselves as well as the MP and selected micropropagated plants, when amplified with nine primers-3, 4, 12, 13, 14, 17, 18, 19 and 20 (Table 4). With rest four primers (1, 2, 9, 11), the seedlings yielded different band amplification profile. With primer 1, Se-2_{4M} yielded monomorphic band pattern with MP (single band of 500 bp), while one polymorphic band of 670 and 600 bp was observed for Se-1_{6M} and Se-3_{4M} respectively. With primer 2, a monomorphic band of 700 bp was observed in all the three seedlings (monomorphic with MP and microproagated plants also), while polymorphic band of 840 bp was observed for Se-1_{6M}. The MP and micropropagate plants also yielded a band of 1,200 bp with primer 2 and this band was absent in all the three seedlings. The amplification with primer 9 yielded 3 monomorphic bands of 240, 340, 440 bp size in MP, micropropagated plants, Se-2_{4M}, and Se-3_{4M}, while two polymorphic bands of 590 and 610 bp size were found only in the three seedlings, while a polymorphic band of 300 bp was observed only in Se-1_{6M} (Fig. 9). Likewise, with primer

11, monorphic bands of sizes 350, 410, 600, 700 bp were found in all the ten samples, polymorphic band of 150 bp was found in Se-1_{6M} and Se-3_{4M} and polymorphic band of 900 was observed only in Se-1_{6M} (Fig. 9).

AChE inhibitory assay

Different concentration of the extracts, of the selected eight samples (MP, Se-1_{6M}, Se-2_{4M}, Se-3_{4M}, MiP-1_{3M}, MiP-2_{3M}, MiP-3_{6M} and MiP-4_{6M}) inhibited the AChE enzyme by different degree, as revealed by AChE inhibitory assay (Fig. 10). The inhibitions obtained with different samples could be categorised in three groups: low (5–20 %), moderate (20–50 %) and good (50–100 %) inhibition. MP exhibited low inhibition at low concentrations of 20 µg/ml (15.35 %), moderate inhibition at 40 and 60 µg/ml (35.75 and 46.48 %, respectively), while at concentration of 80 and 100 µg/ml, good inhibition of 55.80 and 61.58 %, respectively, was observed (Fig. 10a). The IC_{50} value obtained with the mother plant sample was 66.49 µg/ml (Table 5). The 6 months' aged seedlings (Se-1_{6M}) exhibited very poor ability (<20 %) to inhibit AChE enzyme at all the concentrations; the highest concentration of 100 µg/ml, could inhibit the AChE enzyme by only 16.86 % (Fig. 10b). The IC_{50} value obtained for Se was very high (20,538.30 µg/ml) (Table 5). Similar observations were made for two other seedlings of 4 months age; IC_{50} for Se-2_{4M} and Se-3_{4M} being 19,341.03 and 24,281.70 µg/ml, respectively. Contrary to the seedlings, the plants obtained by micropropagation, showed very effective AChE inhibitory activity. At each concentration, the methanolic extracts of all the four selected micropropagated plants exhibited AChE inhibitory activity quite similar to that obtained with MP; good inhibition (more than 50 %) was observed at concentrations of 80 and 100 µg/

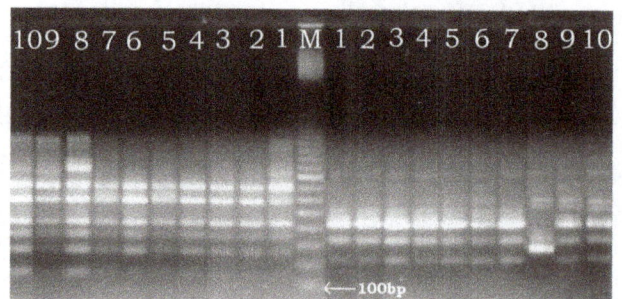

Fig. 9 A representative RAPD reaction for checking the clonal fidelity of micropropagated plants and seedling. M: 100 bp ladder (GE Healthcare Life Sciences), Samples on the right of the ladder were amplified with primer 9 (GGGTTTCGCC) and samples on the left of ladder were amplified using primer 11 (TATTCGGCGT). *Lane 1* mother plant (MP); *lane 2–7* micropropagated plants of 1, 2, 4, 5, 6, 7 months age respectively; *lane 8* seedling of 6 months age (Se-1_{6M}); *lane 9* seedling of 4 months age (Se-2_{4M}); *lane 10* another seedling of 4 months age (Se-3_{4M})

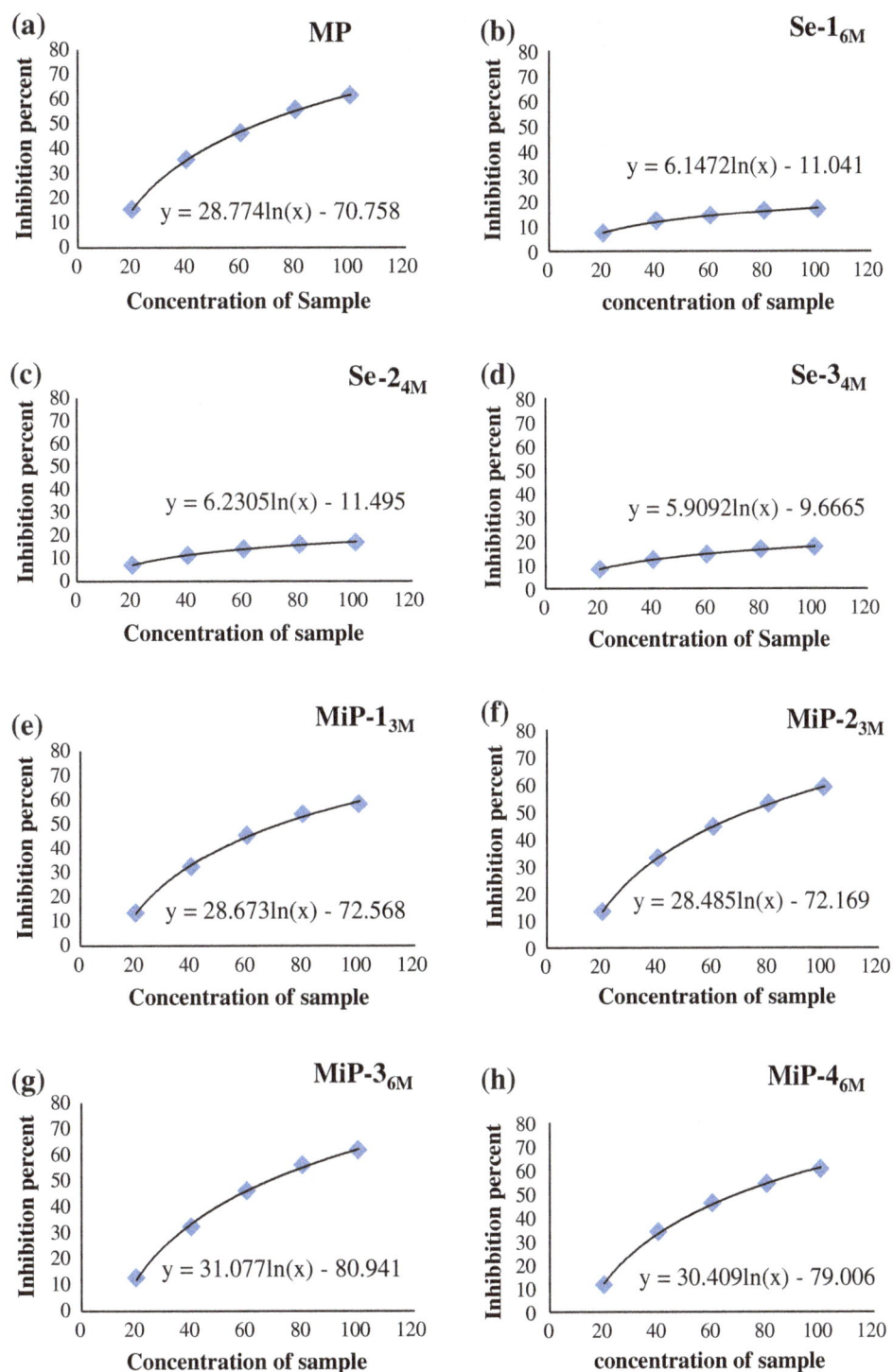

Fig. 10 Logarithmic curves representing the percentage of inhibition of AChE enzyme by respective concentrations of different samples viz. **a** MP, **b** Se-1_{6M}, **c** Se-2_{4M}, **d** Se-3_{4M}, **e** MiP-1_{3M}, **f** MiP-2_{3M}, **g** MiP-3_{6M}, **h** MiP-4_{6M}

ml (Fig. 10c–f). In fact, the four selected micropropagated plants, MiP-1_{3M}, MiP-2_{3M}, MiP-3_{6M} and MiP-4_{6M} exhibited IC$_{50}$ values of 71.87, 72.91, 67.65 and 69.65 µg/ml, which was quite similar to that obtained with MP (66.49 µg/ml) (Table 5).

Discussions

A highly efficient, season-independent, micropropagation protocol has been developed for *F. religiosa* L. using leaf explants, followed by comparative AChE inhibitory

Table 5 IC_{50} values of mother plant (MP), three randomly selected seedlings (Se) and four randomly selected micropropagated plants (MiP) of *F. religiosa*, as calculated by the acetylcholinesterase inhibitory (AChEI) activity assay

	Sample	IC_{50} value* (µg/ml)
1	MP (mother plant)	66.49 ± 01.39
2	Se-1_{6M} (of 6 months age)	20,538.46 ± 1.22
3	Se-2_{4M} (of 4 months age)	19,341.03 ± 0.47
4	Se-3_{4M} (of 4 months age)	24,281.70 ± 0.32
5	MiP-1_{3M} (of 3 months age)	71.87 ± 01.28
6	MiP-2_{3M} (of 3 months age)	72.91 ± 00.12
7	MiP-3_{6M} (of 6 months age)	67.65 ± 00.41
8	MiP-4_{6M} (of 6 months age)	69.65 ± 00.42

*Values are means from 3 repeat experiments (each experiment further having three replicates) ± SE

activity assay of the in vitro raised and conventionally grown plants, for the first time during the present study.

A significant effect of nature and orientation of leaf segments was observed on direct shoot organogenesis; middle section of leaves with cut end inserted inside the medium was found most suitable (Table 1). The orientation of leaf explant has also been reported to play an important role in the micropropagation of some other plants like Marion Blackberry (Meng et al. 2004). In vitro response of explants from woody perennials is reported to be strongly influenced by season of explant collection (Debergh and Maene 1984; Debergh and Read 1991). Such an observation has also been reported for the nodal explants of *F. religiosa* (Siwach et al. 2011). However, during the present study, the percentage of response as well as average number of shoots buds per leaf explant was found nearly consistent (with non-significant differences) during all the 12 months of the year (Fig. 3). Such report, mentioning leaves as explants with no seasonal constraints, did not exist previously for any other plant species. One of the possible explanations is that the axillary and apical meristematic cells, of the perennial trees, are genetically directed to divide actively during the active growth season, and this periodic growth direction may not be there with the differentiated tissues like leaves, though to conclude this fact, more studies are needed.

The optimal PGR requirement for direct shoot regeneration from nodal segments of *F. religiosa* has been found different by different groups. Deshpande et al. (1998) reported 5 mg/l BAP as suitable for initial bud break and 1.0 mg/l BAP as for further shoot proliferation, Hassan et al. (2009) found 0.5 mg/l BAP in combination with 0.1 mg/l IAA as most suitable while Siwach and Gill (2011) found combination of 1.0 mg/l BAP and 0.5 mg/l IAA best for shoot induction and multiplication. During the present study, with leaves' segments as explants, a

different PGR requirement was observed for direct shoot regeneration. Of the two cytokinins, BAP was found better than TDZ for shoot induction and proliferation (Table 2). Amongst various concentrations of BAP, alone as well as in combination with 0.5 mg/l IAA, used during the present study, BAP at 5.0 mg/l was found most suitable for percentage of response (100 %) as well as for inducing maximum number of shoots per explant (7.93) (Table 2). The suitability of higher level of BAP for shoot induction and proliferation from the leaf explant has also been documented for *Piper nigrum* (Ahmad et al. 2010) and *Brassica rapa* (Abbasi et al. 2011). Though combination of 1.0 mg/l BAP and 0.5 mg/l IAA also induced considerably high number of shoots (7.27), it could not be considered as suitable medium because of poor response (43.33 %). Induction of high number of shoots on this particular medium could be attributed to low cytokinin to auxin ratio but the reason for poor response on it could not be explained with present data and literature available. The low cytokinin to auxin ratio has also been reported earlier as the most suitable ratio for efficient shoot induction from the nodal explants of *F. religiosa* (Hassan et al. 2009; Siwach and Gill 2011).

Addition of IAA in the medium having BAP or TDZ was found inhibitory for percentage of response. Similar findings of inhibitory effect of combination of auxin and cytokinin for shoot organogenesis frequency have also been reported for some other plant species (Ahmad et al. 2010; Abbasi et al. 2011). Combination of IAA and TDZ was found very efficient in initial shoot bud induction, however, these buds could not grow into shoots and remained stunted. Similar observation was observed with TDZ alone in the medium, during the direct shoot organogenesis from nodal explants of *F. religiosa* (Siwach and Gill 2011). TDZ is a substituted phenyl urea (*N*-phenyl-1,2,3-thidiazol-5-yl urea) that has immense potential as a cytokinin in shoot organogenesis in a large number of plant systems, especially in the woody species (Mansouri and Preece 2009). On the other hand, some studies have reported the negative impact of TDZ on shoot proliferation (Feng et al. 2010), however how this effect was further augmented by addition of 0.5 mg/l IAA during the present study could not be explained by the existing literature.

The shoots when excised from the explant and subcultured on same medium (BAP 5.0 mg/l) exhibited chlorosis of stem and leaves, curling and browning of leaves edges as well as early leaf fall (Fig. 5a). Such observations were also made during the shoot proliferation from nodal segments of *F. religiosa* (Siwach and Gill 2011) as well as for some other plant species (Husain et al. 2008). The suitability of a particular combination of three additives (200 mg/l glutamine+ 150 mg/l ADS+ 100 mg/l phloroglucinol) for overcoming these problems has been reported and discussed well earlier (Siwach and Gill 2011). During present study, higher concentration of BAP (5.0 mg/l) was also found supportive for shoot multiplication during repeated sub-culturing, as compared to lower one (Table 3). This was contrary to a general observation where cytokinin, at high concentration, favours shoot induction while lowering of concentration is optimal for subsequent shoot proliferation (Deshpande et al. 1998).

The protocol so developed as discussed above is different and more efficient from the existing protocols for micropropagation of *F. religiosa*. Jaiswal and Narayan (1985) have reported shoot regeneration via indirect organogenesis through callus phase initiated from nodal explants of *F. religiosa*, maximum number of 6–10 shoots being obtained from a callus piece. Likewise Narayan and Jaiswal (1986) have reported differentiation of plantlets, 5–10 in number, from leaf callus of *F. religiosa*. Micropropagation via the callus phase has chances of somaclonal variations and so cannot be said to be clonal propagation. Direct shoot regeneration in *F. religiosa* L. has been reported by Deshpande et al. (1998) and in their study, maximum of 5–6 shoots were obtained from a single explant which may be considered as low regeneration rate for a commercially efficient and economically viable micropropagation system. Further no observation has been made for shoot multiplication during repeated sub-culturing, no mention of any observation/problem common to tissue culture of higher age woody perennials, has been made. Hassan et al. (2009) have reported a comparative high rate of multiplication (10–15 shoots per explant) from nodal explants of *F. religiosa* but still the study lacked the observations regarding health of the shoots during continuous sub-culturing process. An efficient micropropagation protocol for *F. religiosa*, using nodal segments as explants (giving 35 shoots per explant) was reported by us earlier (Siwach and Gill 2011) but culture initiation part had to be restricted to a particular time-period of the year as the response of nodal segments was found to be strongly affected by the season of explant collection (Siwach et al. 2011). The present study observed no effect of season on the in vitro response of mature leaves segments of *F. religiosa* and hence leaves were concluded as suitable explants for obtaining season independent micropropagation protocol. Leaves have been taken as explants for micropropagation of many other woody plant species like *Prunus serotina* (Liu and Pijut 2008), *Elaeocarpus robustus* (Rahman et al. 2003) but weather the response is affected by the season of leaf explant collection has not been studied. Such a finding will be of great importance for developing commercial micropropagation protocol of woody perennials.

The RAPD molecular marker analysis was found successful for checking the genetic fidelity of plants, during the present study. Out of twenty primers, five primers (primers 5, 6, 7, 15, 16) resulted in no amplification in all the samples probably because of lack of formation of stable primer template structure during PCR reaction. Two primers (8 and 10) did not yield stable and consistent amplification profile for any of the selected samples indicating towards the redundancy of complementary sites for these two primers. The 13 primers (primers 1, 2, 3, 4, 9, 11, 12, 13, 14, 17, 18, 19, 20), which resulted in stable amplification in all the samples, exhibited monomorphism among the selected micropropagated plants and MP confirming the clonal nature of these plants. Out of these 13, amplification with four primers (1, 2, 9, 11), exhibited polymorphism among seedlings and MP. Such an observation supports the occurrence of genetic changes during gamete formation and so confirms the propagation via seeds as non-clonal way of propagation. RAPD markers have also been used successfully to assess genetic stability among micropropagated plants of a number of species earlier e.g. *Ribes nigrum* L. (Khan and Spoor 2001).

The 45–50 years old *F. religiosa* tree, used as mother plant during the present study was found to possess higher AChE inhibitory activity (IC_{50} of 66.49 µg/ml) than that reported for *F. religiosa* (IC_{50} of 73.69 µg/ml) by Vinutha et al. (2007). It can be attributed to the difference in the genotype, physiological state as well as environmental conditions between the samples of the two studies.

The micropropagated plants of 3 and 6 months age were found to exhibit AChE inhibitory activity quite similar to that of mother plant of 45–50 years age (Fig. 10). Contrarily, seedling of the 6 months age and 4 months age exhibited very poor ability to inhibit AChE enzyme. The normal course of development observed with perennial trees include the nil or poor production of secondary metabolites during primary years of growth and development and generally after 20–25 years of growth, secondary metabolites synthesis is accelerated leading to higher level of these metabolites in different parts of the tree (Kulkarni 2000). This could explain the huge difference in the AChE inhibitory activity of the higher age MP (IC_{50}-66.49 µg/ml), 6 months old seedling Se-1_{6m} (20,538.30 µg/ml) and 4 months old seedlings—Se-2_{4M} (19,341.03 µg/ml), Se-3_{4M} (24,281.70 µg/ml). The explants, when isolated from the higher age plants, have secondary metabolic pathways genes in the active mode (i.e. switched on state) and these genes generally remain in same state in the plants, regenerated from these explants (Kulkarni 2000). So, in vitro raised plants (from explants of higher age plant) start producing secondary metabolites much earlier than the plants obtained by seed germination and are better source for pharmaceutical application (Gurel et al. 2011; Rao et al. 2011; Jiang et al. 2012). This is the possible explanation of obtaining higher AChE inhibitory activity in the methanolic extracts of 3 and 6 months aged micropropagated plants, compared to that in the 4 and 6 months old seedling. Such observation is of great significance for exploiting the pharmaceutical applications of *F. religiosa* towards AD treatment as it rules out the need of waiting for so many years to let the plant start secondary metabolites synthesis and hence avoids the issues of social, environmental and religious concerns and gives an alternative for large scale commercial production.

Conclusion

In conclusion, our studies demonstrated that the leaves explant can be a better alternative to apical or nodal explants for developing a season independent regeneration protocol. Higher concentration of BAP (5.0 mg/l) was most suitable for direct shoot regeneration as well as shoot multiplication. During repeated sub-culturing, the medium needed to be additionally supplemented with ADS (150 mg/l), glutamine (200 mg/l) and phloroglucinol (100 mg/l) to overcome the problems of chlorosis and leaf fall. The micropropagated plants were genetically stable as was revealed by RAPD analysis of randomly selected plants of different age. Acetylcholinesterase inhibitory (AChEI) activity of micropropagated plants was quite effective, similar to that of mother plant while 6 months old seedling performed very poorly towards enzyme inhibition. The findings of present study give an alternative source for exploiting AChEI activity at large scale and will be of great use in developing new solutions to Alzheimer disease.

Acknowledgments The award of junior research fellowship to Anita Rani Gill under the 'Rajiv Gandhi National Fellowship Scheme' by the 'University Grants Commission' New Delhi, India is acknowledged. We also acknowledge the financial assistance provided by Ch. Devi Lal University, Sirsa, Haryana, India, for all the laboratory requirements.

Conflict of interest It is declared that no conflict of interest is associated with the publication of this paper.

References

Abbasi BH, Khan M, Guo B, Bokhari SA, Khan MA (2011) Efficient regeneration and antioxidative enzyme activities in *Brassica rapa* var. turnip. Plant Cell Tiss Org Cult 105:337–344

Ahmad N, Fazal H, Abbasi BH, Rashid M, Mahmood T, Fatima N (2010) Efficient regeneration and antioxidant potential in regenerated tissues of *Piper nigrum* L. Plant Cell Tiss Org Cult 102(1):129–134

Ambike SH, Rao MRR (1967) Studies on a phytosterolin from the bark of *Ficus religiosa* Part-I. Ind J Pharm 29:91–92

Bertaccini G (1982) Substance P. In: Handbook of Experimental Pharmacology, vol 59, no. 2, Springer, Berlin, pp 85–105

Darvesh S, Walsh R, Kumar R, Caines A, Roberts S, Magee D, Rockwood K, Martin E (2003) Inhibition of human cholinesterases by drugs used to treat Alzheimer disease. Alzheimer Dis Assoc Disord 17:117–126

Debergh P, Maene L (1984) Pathological and physiological problems related to the in vitro culture of plants. Parasitica 40:69–75

Debergh PC, Read PE (1991) Micropropagation. In: Debergh PC, Zimmerman RH (eds) Micropropagation, technology and application. Kluwer Academic Publishers, The Netherlands, pp 1–13

Deshpande SR, Josekutty PC, Prathapasenam G (1998) Plant regeneration from axillary buds of a mature tree of *Ficus religiosa* L. Plant Cell Rep 17(6–7):571–573

Ellman GL, Courtney KD, Andres V, Featherstone RM (1961) A new and rapid colorimetric determination of acetylcholinesterase activity. Biochem Pharmacol 7:88–95

Feng JCYXM, Shang XL, Li JD, Wu YX (2010) Factors influencing efficiency of shoot regeneration in *Ziziphus jujube* Mill 'Huizao'. Plant Cell Tiss Org Cult 101:111–117

Gurel E, Yucesan B, Aglic E, Gurel S, Verma S, Sokmen M, Sokmen A (2011) Regeneration and cardiotonic glycoside production in *Digitalis Davisiana* Heywood (Alanya Foxglove). Plant Cell Tiss Org Cult 104:217–255

Hassan AKMS, Afroz F, Jahan MAA, Khatun R (2009) In vitro regeneration through apical and axillary shoot proliferation of *Ficus religiosa* L.—A multipurpose woody medicinal plant. Plant Tiss Cult Biotechnol 19(1):71–78

Husain MK, Anis M, Shahzad A (2008) In vitro propagation of a multipurpose leguminous tree (*Pterocarpus marsupium* Roxb.) using nodal explants. Acta Physiol Plant 30:353–359

Ingkaninan K, Best D, Heijden VD, Hofte AJP, Karabatak B, Irth H, Tjaden UR, Greef VD, Verpoorte R (2000) High-performance liquid chromatography with on-line coupled UV, mass spectrometric and biochemical detection for identification of acetylcholinesterase inhibitors from natural products. J Chromatrogr A 872:61–73

Jaiswal VS, Narayan P (1985) Regeneration of plantlets from the callus of stem segments of adult plants of *Ficus religiosa* L. Plant Cell Rep 4:256–258

Jiang W, Chen L, Pan Q, Qiu Y, Shen Y, Fu C (2012) An efficient regeneration system via direct and indirect organogenesis for the medicinal plant *Dysosma versipellis* (Hance) M. Cheng and its potential as a podophyllotoxin source. Acta Physiol Plant 34:631–639

Khan S, Spoor W (2001) Evaluation of genetic stability in the Blackcurrent plants regenerated via micropropagation using RAPD-PCR technique. Pak J Bot 33(4):411–417

Kirana H, Aggrawal SS, Srinivasan BP (2009) Aqueous extract of *Ficus religiosa* Linn: reduces oxidative stress in experimentally induced type 2 diabetic rats. Indian J Exp Biol 47:822–826

Kulkarni AA (2000) Micropropagation and secondary metabolite studies in Taxus spp. and *Withania somnifera* (L) Dunal. Ph.D. Thesis submitted to the University of Pune

Liu X, Pijut M (2008) Plant regeneration from in vitro leaves of mature black cherry (*Prunus serotina*). Plant Cell Tiss Org Cult 94:113–123

Mansouri K, Preece JE (2009) The influence of plant growth regulators on explant performance bud break and shoot growth from large stem segments of *Acer saccharinum* L. Plant Cell Tiss Org Cult 99:313–318

McCown BH (2000) Recalcitrance of woody and herbaceous perennials plants: dealing with genetic predetermination. In Vitro Cell Dev Biol Plant 36:149–154

Meng R, Chen THH, Fin CE, Li Y (2004) Improving in vitro plant regeneration from leaf and petiole explants of 'Marion' blackberry. HortScience 39(2):316–320

Murashige T, Skoog F (1962) A revised medium for rapid growth and bioassays with tobacco tissue cultures. Plant Physiol 15:473–497

Narayan P, Jaiswal VS (1986) Differentiation of plantlets from leaf callus of *F. religiosa* L. Indian J Exp Biol 24:193–194

Pandit R, Phadke A, Jagtap A (2010) Antidiabetic effect of *Ficus religiosa* extract in streptozotocin-induced diabetic rats. J Ethnopharmacol 128:462–466

Pattnaik SK, Chand PK (1996) In vitro propagation of the medicinal herbs *Ocimum americanum* L O canum Sims (hoary basil) *Ocimum sanctum* L (holy basil). Plant Cell Rep 15:846–850

Rahman MM, Amin MN, Azad MA, Begum F, Karim MK (2003) In vitro rapid regeneraion of plantlets from leaf explant of native-olive (*Elaeocarpus robustus* Roxb.). J Biol Sci 3(8):718–725

Rao K, Chodisetti B, Gandi S, Mangamoori LN, Giri A (2011) Direct and indirect organogenesis of *alpinia galanga* and the phytochemical analysis. Appl Biochem Biotechnol 165:1366–1378

Sancheti S, Sancheti S, Um BH, Seo SY (2009) 1, 2, 3, 4, 6-penta-*O*galloyl-D-glucose: a cholinesterase inhibitor from *Terminalia chebula*. S Afr J Bot

Sanghai-Maroof MA, Soliman K, Jorgensen RA, Allard RW (1984) Ribosomal DNA spacer length polymorphisms in barley: mendelian inheritance, chromosomal location and population dynamics. Proced Natl Acad Sci 81:8014–8018

Singh D, Goel RK (2009) Anticonvulsant effect of *Ficus religiosa*: role of serotonergic pathways. J Ethnopharmacol 123(2): 330–334

Singh D, Singh B, Goel RK (2011) Traditional uses, phytochemistry and pharmacology of *Ficus religiosa*: a review. J Ethnopharmacol 134(3):565–583

Siril EA, Dhar U (1997) Micropropagation of mature Chinese tallow tree (*Sapium sebiferum* Roxb). Plant Cell Rep 16:637–640

Siwach P, Gill AR (2011) Enhanced shoot multiplication in *Ficus religiosa* L. in the presence of adenine sulphate, glutamine and phloroglucinol. Physiol Mol Biol Plants 17(3):271–280

Siwach P, Gill AR, Kumari K (2011) Effect of season explants growth regulators and sugar level on induction and long term maintenance of callus cultures of *Ficus religiosa* L. Afr J Biotechnol 10(24):4879–4886

Swami KD, Bisht NPS (1996) Constituents of *Ficus religiosa* and *Ficus infectoria* and their biological activity. In Chem Soc 73:631

Swami KD, Malik GS, Bisht NPS (1989) Chemical investigation of stem bark of *Ficus religiosa* and *Prosopis spicigera*. J Ind Chem Soc 66:288–289

Tamara S, Al-Qudah, Rida A, Shibli, Feras Q, Alali (2011) In vitro propagation and secondary metabolites production in wild germander (*Teucrium polium* L). In Vitro Cell Dev Boil Plant 47:496–505

Vinutha B, Prashanth D, Salma K, Sreeja SL, Pratiti D, Padmaja R, Radhika S, Amit A, Venkateshwarlu K, Deepak M (2007) Screening of selected Indian medicinal plants for acetylcholinesterase inhibitory activity. J Ethnopharmacol 109(2):359–363

6

Two-stage culture procedure using thidiazuron for efficient micropropagation of *Stevia rebaudiana*, an anti-diabetic medicinal herb

Pallavi Singh · Padmanabh Dwivedi

Abstract *Stevia rebaudiana* Bertoni, member of Asteraceae family, has bio-active compounds stevioside and rebaudioside which taste about 300 times sweeter than sucrose. It regulates blood sugar, prevents hypertension and tooth decay as well as used in treatment of skin disorders having high medicinal values, and hence there is a need for generating the plant on large scale. We have developed an efficient micropropagation protocol on half strength Murashige and Skoog (MS) media, using two-stage culture procedures. Varying concentrations of cytokinins, i.e., benzylaminopurine, kinetin and thidiazuron (TDZ) were supplemented in the nutrient media to observe their effects on shoot development. All the cytokinins promoted shoot formation, however, best response was observed in the TDZ (0.5 mg/l). The shoots from selected induction medium were sub-cultured on the multiplication media. The media containing 0.01 mg/l TDZ produced maximum number of shoot (11.00 ± 0.40) with longer shoots (7.17 ± 0.16) and highest number of leaves (61.00 ± 1.29). Rooting response was best observed in one-fourth strength on MS media supplemented with indole-3-butyric acid (1.0 mg/l) and activated charcoal (50 mg/l) with (11.00 ± 0.40) number of roots. The plantlets thus obtained were hardened and transferred to the pots with soil and sand mixture, where the survival rate was 80 % after 2 months. Quantitative analysis of stevioside content in leaves of in vivo mother plant and in vitro plantlets was carried out by high performance liquid chromatography. A remarkable increase in stevioside content was noticed in the in vitro-raised plants as compared to in vivo grown plants. The protocol reported here might be useful in genetic improvement and high stevioside production.

Keywords Micropropagation · Murashige and Skoog medium · *Stevia rebaudiana* · Stevioside · Thidiazuron

Abbreviations

MS	Murashige and Skoog
BAP	Benzylaminopurine
Kn	Kinetin
TDZ	Thidiazuron
IBA	Indole-3-butyricacid
HPLC	High performance liquid chromatography

Introduction

Stevia rebaudiana Bertoni, a member of Asteraceae family, native to certain regions of South America-Brazil and Paraguay (Alhady 2011), is one of the important anti-diabetic medicinal herbs. It is indigenous to the Rio Monday Valley of the Amambay mountain region at altitudes between 200 and 500 m (Pande and Gupta 2013). The compounds in its leaves, stevioside and rebaudioside taste about 300 times sweeter than sucrose (Geuns 2003). It is used as sweetening agent and has enormous commercial importance. Its other medicinal uses include regulating blood sugar, preventing hypertension and tooth decay, and treatment of skin disorders (Singh and Rao 2005). *Stevia* also has healing effect on blemishes, wound cuts and scratches, besides being helpful in weight and blood pressure management. Conventional propagation in

P. Singh · P. Dwivedi (✉)
Laboratory of Plant Tissue Culture and Stress Physiology, Department of Plant Physiology, Institute of Agricultural Sciences, Banaras Hindu University, Varanasi, India
e-mail: pdwivedi25@rediffmail.com

this plant is restricted due to the poor seed viability coupled with very low germination rate. Role of vegetative propagation method is also limited as specific habitat conditions are mandatory to grow the plants in addition to low acclimatization rate in soil. A suitable alternative method to prepare sufficient amount of plants within short time duration is the use of in vitro cultures. There are reports of in vitro clonal propagation of *Stevia* using nodal segments (Sivaram and Mukundan 2003; Mitra and Pal 2007; Singh et al. 2012). In vitro clonal propagation of *Stevia* has been carried out using nodal, inter-nodal segment (Uddin et al. 2006; Ahmed et al. 2007; Sairkar et al. 2009; Thiyagarajan and Venkatachalam 2012), leaf (Ali et al. 2010) and shoot-tip explants (Anbazhagan et al. 2010; Das et al. 2011). The present study was undertaken to evaluate the effectiveness of two-stage culture procedures as means of micropropagation of *S. rebaudiana* in half strength MS media, which has not been attempted before, using various cytokinins—benzylaminopurine (BAP), kinetin (Kn) and thidiazuron (TDZ). The study also compares the stevioside content in the in vivo and in vitro leaves, supporting the effectiveness of micropropagation protocol generated.

Materials and methods

Explant source

Healthy plants of *S. rebaudiana* were collected from local herbal nurseries and established in horticulture garden of BHU campus. Nodal explants (1.5–2.0 cm long and 0.2–0.4 cm thick) were washed thoroughly for 15 min under running tap water, treated with Tween 80 (2–3 drops in 100 ml) for 7 min followed by treatment with 0.002 % (w/v) bavistin for 2–3 min. These were surface sterilized with 0.1 % (w/v) $HgCl_2$ for 2–3 min and washed 4–5 times with sterile double distilled water.

Shoot induction media

Murashige and Skoog medium (1962) (half strength) was used supplemented with 3 % (w/v) sucrose, ascorbic acid (50 mg/l), gibberellic acid (1 mg/l), solidified with 0.8 % (w/v) agar; pH was adjusted to 5.8 prior to autoclaving at 121 °C for 20 min. The nodal explants were inoculated with different concentrations of TDZ (0.01, 0.03, 0.05, 0.1, 0.2, 0.5 mg/l), BAP (0.2, 0.5, 1.0 mg/l) and Kn (0.2, 0.5, 1.0 mg/l) to observe their effect on shoot development and multiplication. The cultures were kept under cool, fluorescent light (16 h photoperiod) at 25 ± 2 °C in the culture room. Data were noted after 3 weeks of inoculation.

Multiplication media

Shoots (3-week-old) obtained from the shoot induction media containing 0.2 mg/l BAP (B), 0.2 mg/l Kn (E), 0.2 mg/l each of BAP and Kn (H), 0.5 mg/l TDZ (P), as shown in Table 1, were sub-cultured on multiplication media. The multiplication medium consisted of half MS supplemented with TDZ (0.01 mg/l); half MS devoid of any growth regulator and original shoot induction medium (B, E, H and P), as shown in Table 2; cultures without growth regulators served as control. Total number of shoots, length of shoots and number of leaves were observed after 4 weeks of culture.

In vitro root induction

The elongated shoots with length more than 5.0 cm were excised from the culture flasks and transferred to the rooting media amended with 0.2, 0.5 and 1.0 mg/l IBA under aseptic condition. The growth regulator was added separately to half strength and one-fourth strength MS media containing 3 % sucrose to determine the effect of MS salt concentration on root induction (Table 3). The media was additionally supplemented with activated charcoal (50 mg/l), ascorbic acid (50 mg/l), polyvinyl-polypyrrolidone (100 mg/l) and gibberellic acid (0.5 mg/l). Total number of roots per shoot as well as length of the roots was measured after 4 weeks of culture.

Hardening

Rooted plants were carefully removed from the culture flasks, washed with sterile water to remove agar media, placed in the plastic cups filled with sterilized perlite. The plants were covered with polythene bags to maintain high humidity. These plants were maintained in the culture room for 3 weeks with the following atmospheric conditions: temperature, 25 ± 2 °C; light, 16 h photoperiod. Afterwards these were transferred to pots containing sterilized garden soil and sand (1:1).

Observation recorded and statistical analysis

Observations were recorded and different parameters (number of shoots, length of shoots and number of leaves) were examined using 8–10 replicates. Data were subjected to Duncan's multiple range test (Duncan 1955).

Stevioside analysis

Biochemical analysis of the in vivo and in vitro-raised plants was performed using more sensitive and rapid method of HPLC (Shimadzu, Japan) analysis to confirm the presence of stevioside in leaves. Stevioside standard was obtained from Sigma Ltd., USA (95 % purity).

Conditions for HPLC study: column used C-18, mobile phase used methanol:water (80:20) with the flow rate of 1.5 ml/min (injection 10 μl). Peak detection was made at 210 nm at the room temp of 25 °C.

Standard preparation: 1 mg of stevioside sample was dissolved in 10 ml of methanol. Then 10 μl was applied to HPLC chromatogram.

Sample preparation from *Stevia* leaves: *Stevia* leaves were dried at 50 °C for 24 h in dark and pulverized to uniform size. This material (2 g) was extracted in boiling water (2 × 50 ml) and further boiled for 30 s. Cooled extract was first filtered through the filter paper and then microfiltered (0.45 μm) before it was ready for analysis.

Calculation of percentage of stevioside (X) in the sample was done as per the formula*:

$$\%X = [WS/W] \times [fX \times AX/AS] \times 100$$

where, WS is the amount (mg) of stevioside in the standard solution W is the amount (mg) of sample in the sample solution AS is the peak area for stevioside from the standard solution AX is the peak area of X for the sample solution fX is the ratio of the formula weight of X to the formula weight of stevioside: 1.00 (stevioside).

Prepared at the 68th Joint FAO/WHO Expert Committee on Food Additives JECFA (2007).

Results and discussion

Shoot induction

The nodal explants were inoculated in half strength MS media with various concentration of BAP (0.2, 0.5, 1.0 mg/l). Almost at every concentration, bud break was seen during 3–7 days after inoculation. The parameters recorded (number of shoots, length of shoot and number of leaves) have shown differences at various concentrations in the induction media. 0.2 mg/l BAP showed better growth in terms of number of shoots (2.25 ± 0.25), shoot length (1.20 ± 0.27 cm), number of leaves (15.75 ± 1.84). Multiple shoot observed from 1.0 to 2.0 mg/l BAP (MS media) and maximum response as well as healthy shoot was noticed at 1 mg/l BAP (Ranganathan 2012). When kinetin was used, 0.2 mg/l showed better response in terms of all the parameters; number of shoots (2.00 ± 0.00), shoot length (1.95 ± 0.35 cm), though media with 0.5 mg/l of kinetin has shown higher number of leaves (20.00 ± 0.81). BAP and Kinetin at different concentrations (0.2, 0.5, 1.0 mg/l of each) when applied together, produced the best response at lower concentration (0.2 mg/l), in terms of number of shoots (2.00 ± 0.00), shoot length (3.10 ± 0.96 cm) and number of leaves (11.50 ± 0.50). Mehta et al. (2012) reported that best shooting response was observed on MS media containing 0.5 mg/l BAP + 2.0 mg/l Kn (average number of

Table 1 Effect of various cytokinins on in vitro shoot induction in *Stevia rebaudiana*

	Plant growth regulator (mg/l)			Number of shoots	Shoot length (cm)	Number of leaves
	BAP	Kinetin	TDZ			
A	0.0	0.0	0.0	1.50 ± 0.28[a]	1.75 ± 0.14[bcd]	14.50 ± 1.50[bc]
B	0.2	–	–	2.25 ± 0.25[a]	1.20 ± 0.27[abc]	15.75 ± 1.84[bc]
C	0.5	–	–	1.75 ± 0.25[a]	0.80 ± 0.08[ab]	15.25 ± 1.79[bc]
D	1.0	–	–	2.00 ± 0.00[a]	0.45 ± 0.09[a]	12.00 ± 1.63[ab]
E	–	0.2	–	2.00 ± 0.00[a]	1.95 ± 0.35[cde]	12.25 ± 0.25[ab]
F	–	0.5	–	1.50 ± 0.28[a]	1.00 ± 0.08[ab]	20.00 ± 0.81[de]
G	–	1.0	–	1.75 ± 0.25[a]	1.15 ± 0.05[abc]	17.50 ± 0.50[cd]
H	0.2	0.2	–	2.00 ± 0.00[a]	3.10 ± 0.96[f]	11.50 ± 0.50[ab]
I	0.5	0.5	–	2.25 ± 0.25[a]	0.82 ± 0.19[ab]	9.50 ± 0.95[a]
J	1.0	1.0	–	1.75 ± 0.25[a]	0.95 ± 0.12[ab]	8.00 ± 1.63[a]
K	–	–	0.01	2.00 ± 0.00[a]	2.82 ± 0.11[ef]	22.75 ± 0.94[ef]
L	–	–	0.03	2.00 ± 0.00[a]	2.60 ± 0.21[def]	23.50 ± 1.25[ef]
M	–	–	0.05	2.00 ± 0.00[a]	2.77 ± 0.13[ef]	24.25 ± 0.47[f]
N	–	–	0.1	2.00 ± 0.00[a]	2.65 ± 0.12[def]	24.75 ± 0.94[fg]
O	–	–	0.2	2.00 ± 0.00[a]	2.30 ± 0.13[def]	28.50 ± 1.25[g]
P	–	–	0.5	3.00 ± 0.57[b]	2.20 ± 0.11[def]	33.00 ± 2.88[h]

Parameters have been recorded after 3 weeks of culture. Data are in the form of mean ± SEM, and means followed by the same letter within the columns are not significantly different ($P = 0.05$) using Duncan's multiple range test

Table 2 Effect of cytokinins on shoot multiplication of *Stevia rebaudiana*

Induction medium	Multiplication medium	Number of shoots	Shoot length (cm)	Number of leaves
BAP (0.2 mg/l)	B	5.75 ± 0.25^{bc}	4.32 ± 0.48^{a}	46.25 ± 0.62^{cd}
	½ MS	3.25 ± 0.62^{a}	4.20 ± 0.16^{a}	37.50 ± 1.70^{b}
	½ MS + TDZ (0.01)	9.25 ± 0.25^{d}	6.67 ± 0.26^{d}	50.50 ± 0.95^{ef}
Kn (0.2 mg/l)	E	4.75 ± 0.25^{b}	5.05 ± 0.09^{b}	44.75 ± 0.48^{c}
	½ MS	2.50 ± 0.28^{a}	4.82 ± 0.16^{ab}	30.50 ± 0.95^{a}
	½ MS + TDZ (0.01)	8.50 ± 0.28^{d}	6.95 ± 0.95^{d}	50.00 ± 0.81^{ef}
BAP and Kn (0.2 mg/l each)	H	5.25 ± 0.25^{bc}	5.20 ± 0.14^{b}	40.75 ± 0.75^{b}
	½ MS	3.00 ± 0.40^{a}	5.90 ± 0.12^{c}	30.50 ± 1.70^{a}
	½ MS + TDZ (0.01)	8.75 ± 0.25^{d}	8.05 ± 0.22^{e}	49.00 ± 1.29^{de}
TDZ (0.5 mg/l)	P	6.00 ± 0.40^{c}	4.85 ± 0.12^{ab}	52.50 ± 1.25^{f}
	½ MS	4.75 ± 0.25^{b}	5.02 ± 0.16^{b}	40.00 ± 0.81^{b}
	½ MS + TDZ (0.01)	11.00 ± 0.40^{e}	7.17 ± 0.16^{d}	61.00 ± 1.29^{g}

Parameters have been recorded after 4 weeks of transfer in multiplication media. Data are in the form of mean ± SEM, and means followed by the same letter within the columns are not significantly different ($P = 0.05$) using Duncan's multiple range test

Table 3 Effect of MS salt concentration and IBA on in vitro root induction

Rooting media	IBA (mg/l)	Number of roots	Root length (cm)
½ MS media	0.2	1.75 ± 0.25^{a}	1.85 ± 0.09^{a}
	0.5	3.25 ± 0.25^{b}	1.87 ± 0.11^{a}
	1.0	6.00 ± 0.40^{c}	2.20 ± 0.18^{a}
¼ MS media	0.2	2.75 ± 0.25^{ab}	2.12 ± 0.11^{a}
	0.5	6.25 ± 0.85^{c}	2.80 ± 0.18^{b}
	1.0	11.00 ± 0.40^{d}	4.62 ± 0.19^{c}

Parameters have been recorded after 4 weeks of transfer in rooting media. Data are in the form of mean ± SEM, and means followed by the same letter within the columns are not significantly different ($P = 0.05$) using Duncan's multiple range test

shoots 3.42 ± 0.39) and 0.5 mg/l BAP + 0.5 mg/l Kn (average shoot length 7.54 ± 0.31 cm). BAP (0.2 mg/l alone) and BAP/Kn (0.5 mg/l each) showed same result in terms of number of shoots, but considering all the parameters together, BAP (0.2 mg/l) produced the best response in shoot induction media (Table 1).

Nodal explants inoculated on half strength MS media having concentration of TDZ (0.01, 0.03, 0.05, 0.1, 0.2, 0.5 mg/l) showed shoot development at every concentration (Fig. 1a); 0.5 mg/l was found to be the most effective, i.e., number of shoots (3.00 ± 0.57), shoot length (2.20 ± 0.11 cm) and number of leaves (33.00 ± 2.88) (Table 1). It was noted that at lower concentration of TDZ, number of shoots was less and shoot length high, and on increasing the concentration, multiple shooting increases, but with decreased shoot length. Mithila et al. (2003) reported that low concentration of TDZ induced shoot organogenesis of African violet explants, whereas at higher doses (5–10 μM) somatic embryos were formed. Among

other agents with cytokinin activity, comparatively low amount of TDZ promotes shoot multiplication in several plants (Guo et al. 2011). It was also reported that TDZ induced better response than BA (6-benzyl adenine) in shoot regeneration in peanut (Victor et al. 1999; Gairi and Rashid, 2004). Role of TDZ in shoot induction and multiple shooting has been reported in other plants as well; the highest rate of shoot regeneration from *Echinacea purpurea* leaf explants cultured on medium with TDZ at 2.5 μM or higher was reported (Jones et al. 2007). TDZ, a phenylurea type plant growth regulator, was earlier used as a cotton defoliant (Arndt et al. 1976). Later, it was believed to exhibit strong cytokinin-like activity almost similar to that of *N*6-substituted adenine derivatives (Mok et al. 1982; Gyulai et al. 1995). In the present study, TDZ appears to mimic cytokinin-like activity causing the release of lateral buds (Wang et al. 1986) and showed better response in terms of shoot regeneration efficiency, compared to other cytokinins, similar to other findings where TDZ produced shoots comparable to or greater than that of other cytokinins (Fiola et al. 1990; Malik and Saxena 1992).

At initial stage, best response of shoot induction, considering all the parameters together was observed at BAP (0.2 mg/l), Kn (0.2 mg/l), BAP/Kn (0.2 mg/l each) and TDZ (0.5 mg/l) (Table 1). These best concentrations, respectively, denoted as B, E, H and P were selected and transferred to different shoot multiplication media (Table 2).

Shoot multiplication

Shoots cultured in the same induction medium B, E, H, P and half MS medium devoid of growth regulator (Fig. 1b) produced minimal response of shoot multiplication;

Fig. 1 In vitro shoot multiplication and rooting of *Stevia rebaudiana*. **a** Shoot formation in ½ MS supplemented with 0.01 mg/l TDZ. **b** Cultures in induction media TDZ (0.5 mg/l) transferred to ½ MS multiplication media without hormone. **c** Cultures in induction media TDZ (0.5 mg/l) transferred to ½ MS multiplication media supplemented with TDZ (0.01 mg/l). **d** In vitro rooting in ¼ MS supplemented with IBA. **e** Hardened plants in perlite. **f** In vitro-raised *Stevia rebaudiana* in garden soil and sand mixture

however, those sub-cultured in the half MS multiplication medium containing TDZ produced significantly higher number of shoots (Fig. 1c), the best response being observed in TDZ (0.01 mg/l). Shoots obtained from induction medium P containing TDZ (0.5 mg/l) produced highest number of shoots (11.00 ± 0.40) in half MS multiplication media containing 0.01 mg/l TDZ (Table 2). However, shoot obtained from the induction medium B containing BAP (0.2 mg/l), E containing Kn (0.2 mg/l) and H containing BAP/Kn (0.2 mg/l each), produced (9.25 ± 0.25), (8.50 ± 0.28), (8.75 ± 0.25) shoots, respectively, when transferred separately in MS half containing 0.01 mg/l TDZ. The number of shoots obtained is lesser when compared with those obtained from medium P containing 0.5 mg/l TDZ (Table 2).

Shoot elongation

Shoot length increased in all the media supplemented with TDZ (Fig. 1c). Shoot obtained from induction medium H

(BAP/Kn 0.2 mg/l each) showed highest shoot length (8.05 ± 0.22) when transferred in half MS multiplication medium supplemented with 0.01 mg/l TDZ (Table 2). It is to be noticed that there is a minute difference in shoot length of cultures obtained from induction medium H (BAP/Kn 0.2 mg/l each) and P (TDZ 0.5 mg/l), when they were transferred separately in half MS medium supplemented with 0.01 mg/l TDZ. The shoots sub-cultured in the same induction medium (B, E, H, P) and half MS medium devoid of growth regulator (Fig. 1b) did not significantly increase the length of the shoots (Table 2). Also, it is observed that shoot cultured in the medium containing TDZ produced significantly more number of leaves (Fig. 1c). Highest number of leaves (61.00 ± 1.29) was observed in the multiplication media with ½ MS + TDZ (0.01 mg/l), the shoots when cultured in the same induction medium (B, E, H, P) and half MS medium devoid of growth hormone showed comparatively lesser number of leaves, i.e., in the range of 30–52 leaves (Table 2).

Root induction and hardening

Elongated shoots were separated from the shoot multiplication media and transferred to the rooting media. Root induction was observed in all cultures supplemented with different concentrations of IBA (Fig. 1d). IBA added to one-fourth MS medium produced a better rooting response as compared to half MS (Table 3). Highest number of roots (11.00) with longer roots (4.62 cm) was obtained on one-fourth MS medium supplemented with 1.0 mg/l IBA (Table 3). Use of activated charcoal in the rooting media facilitated rooting, as also reported in other studies (Komalivalli and Rao 2000; Priyadarshini et al. 2007). Rooted plants were hardened in perlite (Fig. 1e) where all of them were healthy and viable for 3 weeks in the culture room. The survival rate of the transferred rooted plants into the pots containing garden soil and sand (1:1) was 80 %, after 2 months (Fig. 1f). Two-stage culture procedure seems to be a versatile protocol for efficient induction and multiple shoot formation, in *Stevia* similar to that noticed in *Cassia angustifolia* (Iram and Anis 2007) and *Pterocarpus marsupium* (Husain et al. 2007). A two-stage culture procedure has been developed for highly efficient shoot regeneration from leaf and internode explants of *Bacopa monnieri* (Ceasar et al. 2010).

Stevioside content

The presence of the active principles was confirmed in both the in vivo- and in vitro-derived leaves of *Stevia*. The identification and quantification of stevioside content in the samples were done by comparing the retention time and peak area of sample with that of the standard. In samples of in vitro plants, HPLC analysis revealed that stevioside content was higher than those of in vivo plants. Initial study showed that stevioside production was tissue and age dependent (data not shown). The percentage stevioside content of the samples, in the in vivo and in vitro leaves of *Stevia* was found to be 7.017 ± 0.058 and 9.236 ± 0.046, respectively (Table 4); the in vitro-raised plants had higher stevioside content.

In conclusion, the present study demonstrates an efficient micropropagation protocol of *S. rebaudiana* using two-stage culture procedures. Furthermore, the micropropagation protocol generated did not affect the content of stevioside in *Stevia* leaves as shown by its higher content in the in vitro-raised leaves. The protocol might be useful in germplasm conservation and stevioside production.

Acknowledgments The authors are thankful to the University Grants Commission (UGC), New Delhi for the financial assistance in the form of a major research project.

Conflict of interest We declare that we do not have any conflict of interest in the publication. Further, we state that this manuscript is original, unpublished and not under simultaneous consideration by another journal.

Table 4 Stevioside content in in vivo and in vitro leaf samples

S. no.	Samples	Medium	Retention time (mm:ss)	Peak area (mAs)	Stevioside content (%)
1	Standard	–	2:22	596.1	–
2	In vivo plants	–	2:21	8,421.6	7.017 ± 0.058
3	In vitro plants (3-week-old)	Half strength MS	2:24	11,040.7	9.236 ± 0.046
4	In vitro-raised plants (8-week-old)	Half strength MS	2:24	11,040.7	9.236 ± 0.046

Two replicates of each sample were used for HPLC analysis (mean value calculated)

References

Ahmed MB, Salahin M, Karim R, Razvy MA, Hannan MM, Sultana R, Hossain M, Islam R (2007) An efficient method for in vitro clonal propagation of newly introduced sweetener from plant *Stevia rebaudiana* Bertoni in Bangladesh. Am-Eurasian J Sci Res 2(2):121–125

Alhady MRAA (2011) Micropropagation of *Stevia rebaudiana* Bertoni—a new sweetening crop in Egypt. Glob J Biotechnol Biochem 6(4):178–182

Ali A, Gull I, Naz S, Afghan S (2010) Biochemical investigation during different stages of in vitro propagation of *Stevia rebaudiana*. Pak J Bot 42(4):2827–2837

Anbazhagan M, Kalpana M, Rajendran R, Natarajan V, Dhanavel D (2010) In vitro production of *Stevia rebaudiana* Bertoni. Emir J Food Agric 22(3):216–222

Arndt F, Rusch R, Stillfried HV (1976) SN 49537, a new cotton defoliant. Plant Physiol 57:99

Ceasar SA, Maxwell SL, Prasad KB, Karthigan M, Ignacimuthu S (2010) Highly efficient shoot regeneration of *Bacopa monnieri* (L.) using a two-stage culture procedure and assessment of genetic integrity of micropropagated plants by RAPD. Acta Physiol Plant 32:443–452

Das A, Gantait S, Mandal N (2011) Micropropagation of an elite medicinal plant: *Stevia rebaudiana* Bert. Int J Agric Res 6:40–48

Duncan DB (1955) Multiple range and multiple *F*-tests. Biometrics 11:1–42

Fiola JA, Hassan MA, Swartz HJ, Bors RH, McNicols R (1990) Effect of thidiazuron, light fluence rates and kanamycin on in vitro shoot organogenesis from excised *Rubus* cotyledons and leaves. Plant Cell Tissue Org Cult 20:223–228

Gairi A, Rashid A (2004) Direct differentiation of somatic embryos on different regions of intact seedlings of *Azadirachta* in response to thidiazuron. J Plant Physiol 161:1073–1077

Geuns JMC (2003) Molecules of interest: Stevioside. Phytochemistry 64:913–921

Guo B, Abbasi BH, Zeb A, Xu LL, Wei YH (2011) Thidiazuron: a multi-dimensional plant growth regulator. Afr J Biotechnol 10(45):8984–9000

Gyulai G, Jekkel Z, Kiss J, Heszky LE (1995) A selective auxin and cytokinin bioassay based on root and shoot formation in vitro. J Plant Physiol 145:379–382

Husain MK, Avis M, Shahzad A (2007) In vitro propagation of Indian kino (*Pterocarpus marsupium* Roxb.) using thidiazuron. In Vitro Cell Dev Biol Plant 43:59–64

Iram S, Anis M (2007) In vitro shoot multiplication and plantlet regeneration from nodal explants of *Cassia angustifolia* Vahl. a medicinal plant. Acta Phys Plant 29:233–238

JECFA (2007) Prepared at the 68th JECFA (2007) and published in FAO JECFA Monographs 4 (2007), superseding tentative specifications prepared at the 63rd JECFA (2004), in the combined compendium of food additive specifications, FAO JECFA Monographs 1 (2005)

Jones MP, Yi Z, Murch SJ, Saxena PK (2007) Thidiazuron-induced regeneration of *Echinacea purpurea* L.: micropogation in solid and liquid culture systems. Plant Cell Rep 26:13–19

Komalivalli N, Rao MV (2000) In vitro micropropagation of *G. sylvestre*—a multipurpose medicinal plant. Plant Cell Tissue Org Cult 61:97–105

Malik KA, Saxena PK (1992) Regeneration of *Phaseolus vulgaris* L. High-frequency induction of direct shoot formation in intact seedlings by N6-benzylaminopurine and thidiazuron. Planta 186:384–389

Mehta J, Sain M, Sharma DR, Gehlot P, Sharma P, Dhaker JK (2012) Micropropagation of an anti diabetic plant—*Stevia rebaudiana* Bertoni (natural sweetener) in Hadoti region of south-east Rajasthan, India. ISCA J Biol Sci 1(3):37–42

Mithila J, Hall JC, Victor JMR, Saxena PK (2003) Thidiazuron induces shoot organogenesis at low concentrations and somatic embryogenesis at high concentrations on leaf and petiole explants of African violet (*Saintpaulia ionantha* Wendl.). Plant Cell Rep 21(5):408–414

Mitra A, Pal A (2007) In vitro regeneration of *Stevia rebaudiana* Bert. from the nodal explants. J Plant Biochem Biotechnol 16:59–62

Mok MC, Mok DWS, Armstrong DJ, Shudo K, Isogai Y, Okamoto T (1982) Cytokinin activity of N-phenyl-N-1,2,3-thiadiazol-5-ylurea (thidiazuron). Phytochemistry 21:1509–1511

Murashige T, Skoog F (1962) A revised medium for rapid growth and bioassays with tobacco tissue cultures. Physiol Plant 15:473–497

Pande SS, Gupta P (2013) Plant tissue culture of *Stevia rebaudiana* (Bertoni): a review. J Pharmacogn Phytother 5(1):26–33

Priyadarshini GR, Kumar A, Janifer X (2007) Micropropagation studies in *Stevia rebaudiana* Bertoni. In: Kukreja AK, Mathur AK, Banerjee S, Mathur A, Sharma A, Khanuja SPS (eds) Proceedings of national symposium on plant biotechnology: new frontiers. CIMAP, Lucknow, pp 121–127

Ranganathan J (2012) Studies on micropropagation of *Stevia rebaudiana*. Int J Pharmacol Biol Arch 3(2):315–320

Sairkar P, Chandravanshi MK, Shukla NK, Mehrotra MN (2009) Mass propagation of an economically important medicinal plant *Stevia rebaudiana* using in vitro propagation technique. J Med Plants Res 3(4):266–270

Singh SD, Rao GP (2005) *Stevia*: the herbal sugar of 21st century. Sugar Technol 7:17–24

Singh P, Dwivedi P, Atri N (2012) In vitro shoot regeneration of *Stevia rebaudiana* through callus and nodal segments. Int J Agric Environ Biotechnol 5(2):101–108

Sivaram L, Mukundan U (2003) In vitro culture studies on *Stevia rebaudiana*. In Vitro Cell Dev Biol Plant 39:520–523

Thiyagarajan M, Venkatachalam P (2012) Large scale in vitro propagation of *Stevia rebaudiana* Bert. for commercial application: pharmaceutically important and antidiabetic medicinal herb. Ind Crops Prod 37(1):111–117

Uddin MS, Chowdhury MS, Khan MM, Uddin MB, Ahmed R, Betan MA (2006) In vitro propagation of *Stevia rebaudiana* Bert in Bangladesh. Afr J Biotechnol 5:1238–1240

Victor JMR, Murthy BNS, Murch SJ, KrishnaRaj S, Saxena PK (1999) Role of endogenous purine metabolism in thidiazuron-induced somatic embryogenesis of peanut (*Arachis hypogaea*). Plant Growth Regul 28:41–47

Wang SY, Steffens GL, Faust M (1986) Breaking bud dormancy in apple with a plant bioregulator, thidiazuron. Phytochemistry 25:311–317

7

Identification of relevant non-target organisms exposed to weevil-resistant Bt sweetpotato in Uganda

R. J. Rukarwa · S. B. Mukasa · B. Odongo ·
G. Ssemakula · M. Ghislain

Abstract Assessment of the impact of transgenic crops on non-target organisms (NTO) is a prerequisite to their release into the target environment for commercial use. Transgenic sweetpotato varieties expressing Cry proteins (Bt sweetpotato) are under development to provide effective protection against sweetpotato weevils (Coleoptera) which cause severe economic losses in sub-Saharan Africa. Like any other pest control technologies, genetically engineered crops expressing insecticidal proteins need to be evaluated to assess potential negative effects on non-target organisms that provide important services to the ecosystem. Beneficial arthropods in sweetpotato production systems can include pollinators, decomposers, and predators and parasitoids of the target insect pest(s). Non-target arthropod species commonly found in sweetpotato fields that are related taxonomically to the target pests were identified through expert consultation and literature review in Uganda where Bt sweetpotato is expected to be initially evaluated. Results indicate the presence of few relevant non-target Coleopterans that could be affected by Coleopteran Bt sweetpotato varieties: ground, rove and ladybird beetles. These insects are important predators in sweetpotato fields. Additionally, honeybee (hymenoptera) is the main pollinator of sweetpotato and used for honey production. Numerous studies have shown that honeybees are unaffected by the Cry proteins currently deployed which are homologous to those of the weevil-resistant Bt sweetpotato. However, because of their feeding behaviour, Bt sweetpotato represents an extremely low hazard due to negligible exposure. Hence, we conclude that there is good evidence from literature and expert opinion that relevant NTOs in sweetpotato fields are unlikely to be affected by the introduction of Bt sweetpotato in Uganda.

Keywords Environmental risk assessment · Cry proteins · Sweetpotato weevil

R. J. Rukarwa (✉) · S. B. Mukasa
School of Agricultural Sciences, Makerere University,
P.O. Box 7062, Kampala, Uganda
e-mail: rrukarwa@yahoo.com

B. Odongo
African Institute for Capacity Development,
P.O. Box 46179, Nairobi GPO 00100, Kenya

G. Ssemakula
National Crop Resources Research Institute (NaCRRI),
P.O. Box 7084, Namulonge, Kampala, Uganda

M. Ghislain
International Potato Center, P.O. Box 25171,
Nairobi 00603, Kenya

Weevil-resistant (Bt) sweetpotato for Uganda

Sweetpotato (*Ipomoea batatas* (L) Lam.) is an important crop in all tropical areas of the world. In Uganda, sweetpotato is grown as a staple food in low-input farming systems (Smit 1997). For some farmers, the crop also supplements family income. Strategies to reduce losses due to pests would impact directly on livelihood of millions of rural households by enhancing food security. Sweetpotato weevils, *Cylas puncticollis* Boheman and *C. brunneus* F., are the major production constraints in sub-Saharan Africa (SSA), whereas in the Americas and Asia, *C. formicarius* F. is the major pest (Sorensen 2009). In areas where weevils are endemic, production losses range between 60 and 100 % (Smit 1997; Stathers et al. 2003). In Uganda, a

survey on the socioeconomic impact of sweetpotato weevils indicates an average yield loss of over 28 % between
wet and dry seasons (B. Kiiza, pers. comm, Makerere
University, Kampala, Uganda). Even low levels of *Cylas*
spp. infestation can reduce root quality and marketable
yield because the plants produce unpalatable terpenoids in
response to weevil feeding. In addition, fungal rotting
occurring as a consequence of weevil tunnelling in the
storage roots produces several compounds including
ipomeamarone which is particularly toxic to animals
(Pandey 2008). Hence, control of weevils through host
plant resistance would bring significant benefits to low-
input farmers (Qaim 2001).

Considerable research has been conducted to identify
host plant resistance to *Cylas* spp. in sweetpotato and significant progress has been made to release varieties less
affected by weevils (Stathers et al. 2003; Jackson et al.
2012; Muyinza et al. 2012). However, improved varieties
with high levels of *Cylas* spp. resistance are not yet
available. Progress in breeding weevil-resistant cultivars
has been slow due to the genetic complexity of the crop
(hexaploid and highly heterozygous) and lack of attrac-
tiveness of deep-rooting varieties which is the most
effective breeding target to avoid weevil infestation (Stathers et al. 2003). Another option could be to breed for
enhanced accumulation of the biochemical component of
resistance of the variety New Kawogo, but its inheritance
and impact on nutritional quality of the storage roots
remain to be elucidated (Stevenson et al. 2009). Conversely, genetic transformation for insect resistance is one
of the attractive options to improve sweetpotato production
as has been witnessed in insect-resistant (Bt) varieties of
maize and cotton in sub-Saharan Africa (Thomson 2008).
High levels of resistance have been achieved against
coleopteran pests by expressing toxins derived from
Bacillus thuringiensis in the crop plant (Betz et al. 2000;
Qaim et al. 2008).

Accordingly, in Uganda Cry proteins were tested for
activity against the African sweetpotato weevil resulting
in the identification of three samples of *Bacillus thuringiensis* (Bt) endotoxins, Cry7Aa1, ET33/34 and
Cry3Ca1 which were found to be active against *C.
puncticollis* and *C. brunneus* in artificial diet assays
(Ekobu et al. 2010). Therefore, Bt sweetpotato varieties
expressing these Cry proteins might be protected against
weevils. To that end, the corresponding genes were
introduced into sweetpotato via *Agrobacterium tumefac-
iens* genetic transformation (Ghislain et al. 2013).
Assuming weevil pests will be controlled through the
expression of the Cry proteins in sweetpotato, it is
important to assess the impact of these proteins on other
organisms in the sweetpotato growing environments, like
other insect control technologies.

Framework for non-target organisms' risk assessment of Bt crops

Environmental risk assessment of a Bt crop considers the
impact of the Cry protein on the target pest, but also on
non-target organisms either directly or indirectly (OECD
2007). To identify relevant non-target organisms, it is
important to understand the mode of action of Cry proteins.
All Cry toxins characterised so far bind to specific receptors on the plasma membrane of midgut epithelium cells in
susceptible insects which form oligomeric transmembrane
pores causing osmotic lysis (Aronson and Shai 2001; Bravo
et al. 2007). Some Cry proteins have multiple receptors, or
may bind to multiple sites on a single receptor and it has
been demonstrated that receptor binding is necessary but
not sufficient for toxicity (de Maagd et al. 2001). Experiments using sub-lethal concentrations have also revealed
that there may be other relevant interactions between Cry
proteins and their target insects (Aronson and Shai 2001).
Zhang et al. (2006) also suggested that toxicity could be
related to G-protein-mediated apoptosis following receptor
binding instead of forming oligomers resulting in pore
formation. Preliminary research results indicate that the
Cry proteins used to control weevils in sweetpotato have
similar mode of action as the other Cry proteins (Hernandez-Martinez et al. 2010; B. Escriche, University of
Valencia, Spain, pers. comm.).

Non-target organisms (NTO) are species not targeted for
control using a particular Cry protein expressed in transgenic plants, but may become exposed to it by feeding
directly on plant tissues or indirectly on herbivores or
parasites, or by direct ingestion via the environment, such
as in the soil or water (Groot and Dicke 2002). Although all
organisms of relevance to sweetpotato are arthropods, we
will use NTO when discussing non-target arthropods due to
the global acceptance of NTO. Non-target risk assessment
is a process based on scientific principles that aims at the
evaluation of the potential adverse effects of transgenic
plants on the non-target organisms of environmental relevance (OECD 2007; Romeis et al. 2008).

Problem formulation

The initial step of risk assessment is problem formulation
which is an important step that leads the risk assessment
process to successful risk characterisation (Raybould et al.
2007; Romeis et al. 2011). The Environmental Protection
Agency (EPA) of the USA has elaborated in 1998 guidelines on ecological risk assessment which sets the basis for
NTO risk assessment (EPA 1998). Problem formulation, in
an ideal sense, develops a concise problem statement, a risk
hypothesis, a conceptual model and an analysis plan. The

risk hypothesis represents an assumption regarding the cause–effect relationships among attributes of the risk characterisation, including sources, exposure routes, end points, responses and measures relevant to the risk assessment. The conceptual model describes key relationships between a transgenic plant occurring in the environment and its linkages to an assessment end point (Raybould 2007). It sets the problem in perspective and establishes the proposed relationships that need evaluation, and the analysis plan establishes the appropriate risk formulation to be considered in the risk characterisation.

Assessment end points

A second conceptual element of the NTO risk assessment is the assessment end point which is an explicit expression of the environmental value to be protected (EPA 1998). This necessitates defining species and ecosystem functions that could be adversely affected by the Bt plant and that require protection from harm. Assessment end points are made operational into quantitatively measurable end points. An appropriate measurement end point for NTO testing is relative fitness or some component of relative fitness, which is the relative lifetime survival and reproduction of the exposed versus unexposed non-target species (Andow and Hilbeck 2004). It is therefore required that NTO tests consider both toxic effects (mortality, longevity) and sub-lethal effects. The sub-lethal effects are assessed through growth pattern, development rate, reproduction parameters (number and size of offspring, percentage of eggs hatching, sex ratio of progeny, age of sexual maturity), and, when appropriate, behavioural characteristics (searching efficiency, predation rates, food choice). In field conditions, the abundance and species diversity of certain groups of NTO at a relevant life stage are typical measurement end points. The choice of specific measurement end points shall be done according to the problem formulation on a case-by-case basis (Romeis et al. 2011).

Species selection

Non-target organisms in Bt crop fields feeding directly or indirectly on the crop or residues are exposed to the Cry protein expressed in the pest-resistant plant. Hence, NTO risk assessment has to be done for some of these species when there is a reasonable doubt that they may suffer a negative impact due to a real exposure. For practical reasons, only a small fraction of all possible terrestrial organisms can be considered for regulatory testing. Therefore, to assess the effect of insect-resistant plants on NTO, appropriate species should be selected (Romeis et al.

2008; Garcia-Alonso et al. 2006). It is necessary to select suitable species which can act as surrogates for species that should, but cannot, be tested (Garcia-Alonso et al. 2006; Romeis et al. 2013). The use of appropriate surrogates is a widely accepted concept for scientific experimentation and enables one to design high-quality and repeatable laboratory (and semi-field) studies with clear measurement end points and the ability to extrapolate results to other species. Non-target species subject to the NTO risk assessment should be chosen from different ecological functions such as herbivores, pollinators, predators and parasitoids of pest organisms and decomposers in the soil (Romeis et al. 2006). The NTO risk assessment may also consider species with special aesthetic, economic or cultural value or species of national importance. These species are regionally specific and can be evaluated within the ecological risk assessment independent of their ecological function. To reflect biogeographical variation, it is crucial to determine what relevant species are likely to occur in the cropping systems where the transgenic plant is expected to be grown.

Framework for NTO risk assessment of Bt sweetpotato in Uganda

In this section, we will apply the NTO risk assessment described above to identify relevant NTOs which could be affected by the cultivation of Bt sweetpotato and recommendations.

Problem formulation

In Uganda, one possible concern is that Bt sweetpotato plants may have an adverse effect on biodiversity and its functioning at several levels, through interactions with populations of other species associated with Bt sweetpotato fields. Because the environment is to be protected from harm according to protection goals set out by Ugandan legislation (The National Environment Act-Cap 153 1995), protection of species richness and ecological functions should be considered in this risk assessment. The receiving environment is the Bt sweetpotato cultivated fields and the wider environment (other adjacent Bt or non-Bt cultivations). For the benefit of sustainable production, the interest is to maintain a certain level of biodiversity in sweetpotato fields, providing essential functions such as biological control of pests, decomposition of plant materials and maintenance of soil quality and fertility. Bt sweetpotato varieties need to be evaluated to determine if they are directly and/or indirectly (through food web interactions) potentially harmful to species guilds involved in ecosystem functions. Problem formulation in our case is the identification of potential hazards such as exposure to the Cry proteins through a comparison of the Bt sweetpotato with their conventional counterpart.

Assessment end points

Before commercialisation of Bt sweetpotato, an appropriate assessment end point for initial testing in Bt sweetpotato will be the relative survival and reproduction of NTOs. These parameters are a particularly useful measurable assessment end point in relation to Bt sweetpotato, because they relate directly to risk. Survival experiments should last at least through all relevant developmental stages of the selected NTO, including adult parameters such as age-specific mortality. In principle, the duration of the test should correspond to the time the non-target species would be exposed to the Bt sweetpotato plants or crop residues. In our case, NTO survival experiments would be conducted through all developmental stages, including adult life stage parameters such as age-specific mortality. If the Bt sweetpotato has a negative impact on NTOs in the field, its effect could be observed at any developmental stage during their life cycle. Usually, inappropriate assessment end points may misdirect research or regulatory efforts, and may even lead to the imposition of unnecessary controls to reduce risk.

Species selection

In tropical ecosystems, there is usually a relatively high number of NTO species that may be exposed to Bt crop plants. Considering that not all species can be evaluated, a representative subset of NTO species should be selected for consideration in the risk assessment based on their known ecological functions. A decision on which focal NTO species are to be used is based on the identification of arthropods associated with the crop and then followed by selection of focal test species.

Identification of functional groups of arthropods associated with sweetpotato fields

In Uganda like in the rest of Africa, sweetpotato is not a native species, but has been grown long enough to be considered as a traditional food crop. It is a crop grown typically with very little input: sometimes fertilisers but no insecticides or nematicides in SSA. Over the years, field experiments have been conducted in Uganda by National Agricultural Research Organisation (NARO) scientists in different sweetpotato-growing districts to determine pests and beneficial organisms associated with sweetpotato

(Ames et al. 1996; Smit 1997; Stathers et al. 2005; Sorensen 2009). Insects representing more than 30 species of eight orders and in different developmental stages were found to be prevalent in sweetpotato fields (Table 1). Their levels of abundance differ according to the seasons and agro-ecological zones. Individuals belonging to eight species of six families were noted as major pests of sweetpotato and are those that farmers have to monitor as part of a sustainable integrated pest management (IPM) system in sub-Saharan Africa (Stathers et al. 2005).

Furthermore, individuals of 19 species were minor pests, while 9 species belonging to nine families were represented by beneficial insects. The beneficial insects noted were pollinators, decomposers, predators and/or parasitoids of insect pests. Among the homopterans pests observed, white fly (*Bemisia tabaci*) and aphids (e.g. *Myzus persicae*) are vectors of viral diseases. Most of the minor pests observed are cosmopolitan, polyphagous and are pests of other crops. These include *Phyllophaga* spp., *Hapatesus* spp., *Leucinodes orbonalis*, *Spodoptera* spp., *Omopyge sudanica*, *Macrotermes bellicosus*, *Gryllus* spp., *Zonocerous variegates*, *Attractomorpha psitticina*, *Locusta migratoria*, *Bemisia tabaci*, *Myzus persicae*, *Macrosiphum euphorbiae*, *Leptoglossus gonagra* and *Nezara viridula*. In addition, nine spp. of non-insect organisms (Table 2) were also found to be common in sweetpotato fields. Three of these non-insects were beneficial, while five are pests of sweetpotato. The identification of the arthropod complex in sweetpotato fields helps to ascertain value or risk of each species. Samples of arthropod specimen mentioned in this paper have been preserved in the entomology laboratory at National Crop Resources Research Institute (NaCRRI), Namulonge.

Selection of focal species

Based on the considerations addressed on the identification of the functional groups of arthropods associated with sweetpotato fields and categorisation of NTO, focal species need to be selected from each functional category of NTO group. The functional groups commonly associated with sweetpotato fields are pollinators, decomposers, predators and parasitoids. The following criteria should be considered in choosing the appropriate focal test species.

(a) *The mode of action and specificity of the insecticidal protein and the impact of that protein on non-target species closely related to the target pest* Cry proteins identified as active against *Cylas* spp. (Cry7Aa1, ET33/34, and Cry3Ca1) are typically of the type affecting primarily coleopteran species (Crickmore et al. 2013). Hence, the most likely affected NTO to initially evaluate should be within coleoptera.

Table 1 Insects associated with sweetpotato fields in Uganda (based on insect collection and rearing facility at NaCRRI and literature review)

Order	Family	Species	Common name	Importance	Abundance
Coleoptera	Brentidae	*Cylas puncticolis*	African Sweetpotato weevil	Major Pest	Common
		Cylas brunneus	African Sweetpotato weevil	Major Pest	Common
	Scarabeidae	*Phyllophaga* spp.	White grub	Pest	Common
	Meloidae	*Epicauta* spp.	Blister beetle	Pest	Common during flowering
	Curculionidae	*Peloropus batatae*	Peloropus Weevil	Pest	Fairly common
		Blosyrus obliguatus	Rough Sweetpotato weevil	Major Pest	Fairly common
		Alcidodes dentipes	Striped Sweetpotato weevil	Pest	Fairly Common
	Coccinellidae	*Delphastus catalinae*	Ladybird beetle	Predator	Common
	Chrysomelidae	*Aspidomorpha* spp.	Tortoise shell beetle	Major Pest	Common
	Elateridae	*Hapatesus* spp.	Wireworm	Pest	Common
	Carabidae	*Poecilus chalcites*	Ground beetle	Predator	Common
	Staphylinidae	*Aleochara bilineata*	Rove beetle	Predator	Common
Lepidoptera	Nymphalidae	*Acraea acerata*	Sweetpotato butterfly	Major Pest	Common
		Synanthedon dascyeles	Clearwing Moth	Major Pest	Common
	Crambidae	*Leucinodes orbonalis*	Eggplant fruit borer	Pest	Rare
	Noctuidae	*Agrotis subterranea*	Granulate cutworm	Pest	Fairly common
		Spodoptera spp.	Armyworm	Major Pest	Common
	Sphingidae	*Agrius cingulata*	Sweetpotato hornworm	Major Pest	Fairly Common
		Agrius convolvuli	Sweetpotato moth	Pest	Fairly Common
		Hippotion celerio	Taro hawkmoth	Pest	Fairly Common
Isoptera	Termitidae	*Macrotermes bellicosus*	Termite	Pest	Common
Orthoptera	Gryllidae	*Gryllus* spp.	Field cricket	Pest	Common
	Pyrgomoriphidae	*Zonocerous variegatus*	Elegant grasshopper	Pest	Common
	Acrididae	*Attractomorpha psitticina*	Slant-faced grasshopper	Pest	Rare
		Locusta migratoria	Migratory locust	Pest	Rare
Hemiptera	Aleyrodidae	*Bemisia tabaci*	Sweetpotato whitefly	Pest/vector	Common
	Aphididae	*Myzus persicae*	Aphid	Pest/vector	Fairly common
		Macrosiphum euphorbiae	Potato aphid	Pest	Common
	Coreidae	*Leptoglossus gonagra*	Squash bug	Pest	Common
	Pentatomidae	*Nezara viridula*	Green stink bug	Pest	Common
	Reduviidae	*Sycanus* spp.	Assassin bug	Predator	Common
Dermaptera	Forficuliidae	*Doru taeniatum*	Earwig	Predator	Common
Hymenoptera	Ichneumonidae	*Charops* spp.	Ichneumon wasp	Parasitoid	Common
	Braconidae	*Meteorus autographae*	Braconid wasp	Parasitoid	Common
	Apidae	*Apis mellifera*	Honeybee	Pollinator	Common
Diptera	Tachinidae	*Caricelia normula*	Tachinid fly	Parasitoid	Common

Table 2 Non-insect arthropods associated with sweetpotato fields in Uganda

Order	Family	Species	Common name	Importance	Abundance
Haplotaxida	Lumbricidae	*Eisenia foetida*	Earthworm	Decomposer	Common
Araneae	Oxyopidae	*Oxyopes* spp.	Lynx spider	Predator	Common
	Glycosidase	*Lycos* Spp.	Wolf spider	Predator	Common
Nematoda	Hoplolaimidae	*Rotylenchulus reniformis*	Reniform nematode	Pest	Common
	Meloidogynidae	*Meloidogyne arenaria*	Root knot nematode	Pest	Common
Rodentia	Muridae	*Mus musculus*	Mouse	Pest	Common
		Spalax spp.	Rat	Pest	Common
Diplopoda	Odontopygidae	*Omopyge sudanica*	Millipede	Pest	Common

Therefore, we may consider ground beetles, (Carabidae: *Poecilus chalcites*), rove beetles (Staphylinidae: *Aleochara bilineata*) and ladybird beetles (Coccinellidae: *Delphastus catalinae*) that occur in the same taxonomic order (coleoptera) as the target species. Weevil species such as striped sweetpotato weevil (*Alcidodes dentipes*), rough sweetpotato weevil (*Blosyrus obliguatus*) and peloropus weevil (*Peloropus batatae*) belonging to the same superfamily (Curculionoidea, as the target pests) are either considered as minor pests or not ecologically relevant. There are few examples of cross-order activity for Cry proteins (Tailor et al. 1992; van Frankenhuyzen 2009). However, previous research has shown that that this cross-order activity does not threaten the environmental safety of Bt-based pest control, because Cry proteins tend to be much less toxic to taxa outside of the primary specificity range (van Frankenhuyzen 2009). Furthermore, the large body of published literature provides no indication that the currently grown Bt crops cause direct adverse effects on arthropods that are not closely taxonomically related to the target pest (Romeis et al. 2006; Wolfenbarger et al. 2008; Duan et al. 2010).

(b) *Exposure based on habitat and field abundance* Relevant NTO should represent species that are abundant in the crop and have known relevant routes of exposure to the insecticidal protein (Romeis et al. 2013). Exposure could be direct, from deliberate or incidental feeding on crop tissues or decaying plant material, or indirect, from feeding on herbivores that feed on the crop. For example, testing ground beetles (Carabidae) is relevant for coleopteran insecticidal proteins produced in sweetpotato, but their exposure is low since these insects are primarily predators of organisms unaffected by Cry proteins in sweetpotato fields, living especially at the soil surface or within the soil where the roots are located. The same can be said of the rove beetle (Staphynilidae). Ladybird beetle (Coccinellidae) adults are active fliers and feed on pollen; they are unlikely to be affected because Bt protein expression in the pollen is low and nectar is a plant secretion, not a tissue and has no cellular content (Ferry et al. 2007). The ladybird larvae feed primarily on aphids feeding on sweetpotato leaves and exposure of the larvae to the Bt toxins is considered to be relatively low, as aphids contain no or only trace amounts of the toxins due to the fact that they feed on the phloem sap which does not contain Cry proteins (Raps et al. 2001; Romeis and Meissle 2011). Although no NTO risk assessment for Bt crops have been conducted on *Delphastus catalinae*, other Coccinellidae species have been shown to

be unaffected by coleopteran pest-resistant Bt crops (Duan et al. 2002; Ferry et al. 2007; Li and Romeis 2010; Álvarez-Alfageme et al. 2011). In the case of maize, Cry3Bb1 and Cry34Ab1/Cry35Ab1 proteins' impact on ground, rove and ladybird species was reviewed and essentially found not to persist in the environment due to rapid degradation in the soil (Wolt et al. 2007). These Cry proteins are closely related to those currently used to engineer resistance to sweetpotato weevils (Cry7Aa1 and Cry3Ca1 are close to the Cry3Bb1 and the ET33/ET34 to the Cry34Ab1/Cry35Ab1) (Crickmore et al. 2013). Furthermore, these Cry proteins used to control weevils in sweetpotato have similar mode of action as the other Cry proteins (B. Escriche, University of Valencia, Spain, pers. comm.). Therefore, Bt sweetpotato is unlikely to cause harm to the above-mentioned species.

(c) *Ecological and taxonomic diversity* Relevant NTO may include a broad range of invertebrates, particularly economically or socially beneficial species that represent diverse habitats. In our case; honeybee (*Apis mellifera*) is the main pollinator of sweetpotato, which may forage for sweetpotato pollen and therefore could be exposed to Cry proteins. Earthworms (*Eisenia foetida*) are important decomposers, and sweetpotato butterfly (*Acraea acerata*) feeds on plant canopy and would be surrogate to lepidopteron arthropods which feed on the sweetpotato canopy. In all three cases, coleopteran-specific Cry proteins are unlikely to cause any harm because of their target specificity. Indeed, a meta-analysis of Cry protein impact on honeybees resulted in no harm, as shown by recent studies (Duan et al. 2008; Hendriksma et al. 2012). Similarly, field studies have also shown no significant differences in earthworm populations in fields planted with Cry1Ab1 or Cry3Bb1 proteins (van der Merwe et al. 2012).

(d) *Ability to conservatively estimate field exposure* In the laboratory, the potential level of exposure of test species to insecticidal proteins in the field has to be identified. Farmers rarely cultivate only one landrace in one area; even when they have adopted an improved variety, they will maintain some level of diversity because production does not target a single use. It is unlikely that a Bt sweetpotato improved variety will displace landraces, because these are grown for their culinary or taste properties. Therefore, an accurate estimate of exposure of the relevant NTO will be difficult to agree on. Hence, the concentration of the insecticidal protein in the plant tissue on which the NTO feeds provides a worst-case estimate of the environmental exposure concentration. Such data for the Bt sweetpotato is not yet available, because Bt sweetpotato is still under development in Uganda.

(e) *Whether a suitable test system exists for laboratory analysis* Relevant NTOs adaptable to a laboratory bioassay system and suitable protocols are necessary for testing. When feasible, the organism life stage that is most susceptible to the insecticidal protein should be tested. Protocols typically include information on test end points, positive/negative controls, acceptable control mortality, sample sizes and statistical power analyses. For a number of chosen species on Bt sweetpotato, standard testing protocols are not yet available but a number of protocols are available from tests conducted with other crops and/or related invertebrate species, which could be adapted for testing the effect of insecticidal proteins being expressed in sweetpotato.

In general, non-target organisms that are related taxonomically to the target pests are most likely to be affected similarly by the Bt Cry protein (Romeis et al. 2008). In the case of Bt sweetpotato, the rove, ground and ladybird beetles are the primary relevant NTO. Accepting a much lower probability of impact, the honeybee as the main pollinator of sweetpotato and a charismatic insect may be looked at as an NTO for Bt

sweetpotato. However, numerous impact studies have been published over the last decade and have been subject to meta-analyses drawing very clear conclusions of non-impact of Cry proteins on NTO (Marvier et al. 2007; Wolt et al. 2007; Duan et al. 2008; Wolfenbarger et al. 2008). Species that are not exposed to the Cry proteins or from kingdoms never reported to be affected by other Cry proteins do not need to be tested to draw a negligible-risk conclusion (Peterson et al. 2011; Prischl et al. 2012).

Once the relevant non-target species are selected and their surrogates identified, these would be evaluated moving through the tiered testing procedure that has been recommended and well accepted by regulators and risk assessors (Garcia-Alonso et al. 2006; Romeis et al. 2006, 2008; USEPA 2007). In the case of ground, rove and ladybird beetles, these can be used directly as test species. The procedure starts with laboratory tests (lower tier), followed by semi-field (glasshouse or screenhouse) and field (higher tier) tests if necessary (Fig. 1). However, the tiers should not be just considered as sequential steps in a linear approach, because the response of arthropods between the tiers is necessary during the assessment, to determine whether to stop or proceed to the next tier (Kos et al. 2009). Lower-tier tests serve to identify potential

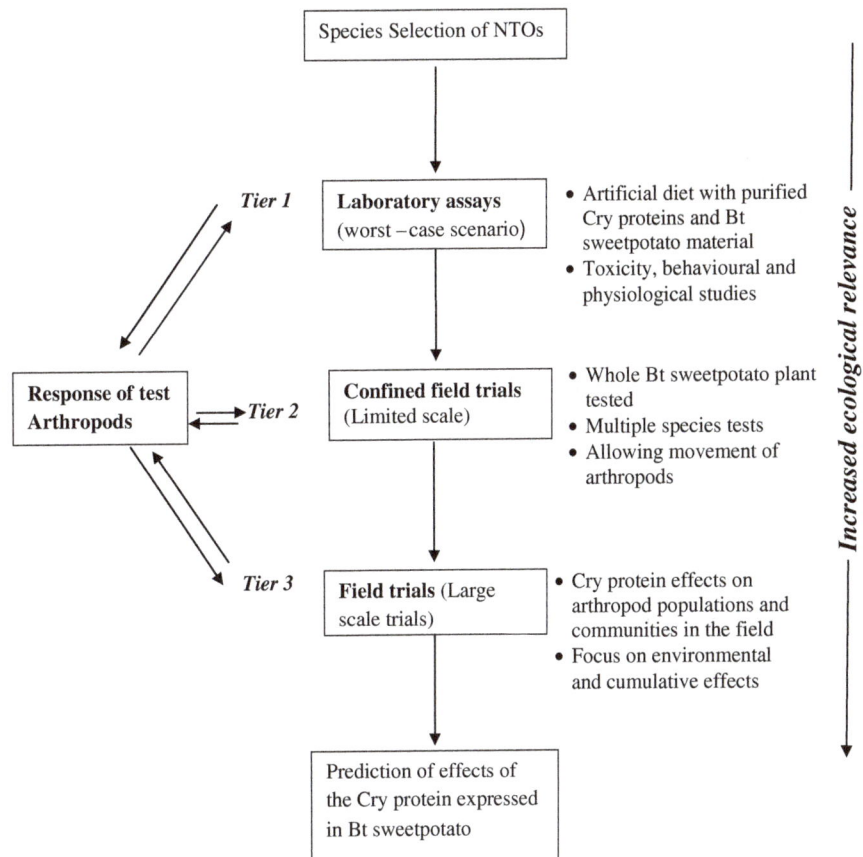

Fig. 1 Tiered approach for testing the effect of Cry proteins non-target organisms found in Bt sweetpotato fields

hazards and are typically conducted in controlled conditions. Lower-tier tests are designed to measure a specific end point under worst-case conditions using protein concentrations that are normally 10–100 times higher than those present in plant tissues (USEPA 2007). Lower-tier studies must be properly designed and executed to maximise the probability that compounds with adverse effects are detected. The confidence in the conclusions drawn from these studies mainly depends on the study's ability to detect potential hazards, if present (Romeis et al. 2008; Duan et al. 2010). The Cry protein level of the Cry7Aa1, ET33/ET34 and Cry3Ca1 will first be determined for the transgenic events causing mortality in both the storage roots and leaves, and then 10–100 times higher than those present in these tissues will be evaluated in artificial diet bioassays. The use of storage root-specific promoters (sporamin and β-amylase) in sweetpotato is expected to reduce the amount of Cry proteins in leaves.

Conclusion

This review provides the scientific rationale for risk assessment of Bt sweetpotato to assist regulatory decision making. The risk hypotheses are developed from current knowledge about the crop, the Cry proteins, the receiving environment and the interactions of the three. It therefore makes maximum use of the existing data and aims to minimise collection of data that are irrelevant to the risk assessment of non-target arthropods. Accordingly, we have identified the ground, rove and ladybird beetles as the primary relevant NTOs which may potentially be affected by cultivation of Bt sweetpotato. Honeybees may be considered as relevant due to their ecological role, but there is solid scientific evidence from literature indicating no harm. Potential hazards are evaluated with representative surrogate/indicator species that are selected case by case for their suitability and amenability to test relevant risk. For effective NTO assessment threshold, values need to be defined that elicit the advance to higher tiers as has been done for environmental risk assessments of conventional pesticides. These values will be available when a transgenic event with efficacy to control weevils will be available. At this point, it is too speculative to estimate what this threshold could be based on solely the LC_{50} which is currently the only toxicity value known. Tissue-specific or enhanced promoters and different Cry protein combinations will influence this threshold value. It is important to note that defining the threshold values is not solely a scientific question, but also depends on whether policy makers are concerned about under- or over-estimating risks considering that sweetpotato is an introduced crop and that Bt sweetpotato will bring food security benefits to vulnerable

populations. The NTO testing approach presented above minimises the likelihood of unexpected negative impact on other organisms and help decision makers to authorise the release of weevil-resistant sweetpotato plants with confidence that it will not have undesirable effects on NTOs.

Acknowledgments We acknowledge the Regional Universities Forum for Capacity Building in Agriculture (RUFORUM) for funding this work.

Conflict of interest The authors declare that they have no conflict of interest.

References

Álvarez-Alfageme F, Bigler F, Romeis J (2011) Laboratory toxicity studies demonstrate no adverse effects of Cry1Ab and Cry3Bb1 to larvae of Adalia bipunctata (Coleoptera: Coccinellidae): the importance of study design. Transgenic Res 20:467–479.

Ames T, Smit NEJM, Braun AR, O'Sullivan JN, Skoglund L (1996) Sweetpotato: major pests diseases and nutritional disorders. International Potato Center (CIP), Lima

Andow DA, Hilbeck A (2004) Science-based risk assessment for non-target effects of transgenic crops. BioSci 54(7):637–649

Aronson AI, Shai Y (2001) Why Bacillus thuringiensis insecticidal toxins are so effective: unique features of their mode of action. FEMS Microbiol Lett 195:1–8

Betz FS, Hammond BG, Fuchs RL (2000) Safety and advantages of Bacillus thuringiensis-protected plants to control insect pests. Regul Toxicol Pharm 32(2):156–173

Bravo A, Gill SS, Soberón M (2007) Mode of action of Bacillus thuringiensis Cry and Cyt toxins and their potential for insect control. Toxicon 49:423–435.

Crickmore N, Baum J, Bravo A, Lereclus D, Narva K, Sampson K, Schnepf E, Sun M, Zeigler DR (2013) Bacillus thuringiensis toxin nomenclature. http://www.btnomenclature.info/. Accessed 20 Mar 2013

de Maagd RA, Bravo A, Crickmore N (2001) How Bacillus thuringiensis has evolved specific toxins to colonize the insect world. Trends Genet 17(4):193–199

Duan JJ, Head G, McKee MJ, Nickson TE, Martin JW, Sayegh FS (2002) Evaluation of dietary effects of transgenic corn pollen expressing Cry3Bb1 protein on a non-target ladybird beetle, Coleomegilla maculata. Entomol Exp Appl 104(2–3):271–280.

Duan JJ, Marvier M, Huesing J, Dively G, Huang ZY (2008) A meta-analysis of effects of Bt crops on honey bees (Hymenoptera: Apidae). PLoS ONE 3(1):e1415.

Duan JJ, Lundgren JG, Naranjo S, Marvier M (2010) Extrapolating non-target risk of Bt crops from laboratory to field. Biol Lett 6:74–77.

Ekobu M, Solera M, Kyamanywa S, Mwanga ROM, Odongo B, Ghislain M, Moar WJ (2010) Toxicity of seven Bacillus thuringiensis Cry proteins against Cylas puncticollis and Cylas *brunneus* (Coleoptera: Brentidae) using a novel artificial diet. J Econ Entomol 103(4):1493–1502

EPA (1998) Guidelines for Ecological Risk Assessment. EPA/630/R-95/002F, April 1998 Final. United States Environmental Protection Agency. Washington, D.C

Ferry N, Mulligan EA, Majerus ME, Gatehouse AM (2007) Bitrophic and tritrophic effects of Bt Cry3A transgenic potato on beneficial, non-target, beetles. Transgenic Res 16(6):795–812.

Garcia-Alonso M, Jacobs E, Raybould A, Nickson TE, Sowig P, Willekens H, van der Kouwe P, Layton R, Amijee F, Fuentes AM, Tencalla F (2006) A tiered system for assessing the risk of genetically modified plants to non-target organisms. Environ Biosaf Res 5:57–65

Ghislain M, Tovar J, Prentice K, Ormachea M, Rivera C, Manrique S, Kreuze J, Rukarwa R, Sefasi A, Mukasa S, Ssemakula G, Wamalwa L, Machuka J (2013) Weevil resistant sweetpotato through biotechnology. Acta Hortic (ISHS) 974:91–98

Groot AT, Dicke M (2002) Insect-resistant transgenic plants in a multi-trophic context. Plant J 31:387–406

Hendriksma HP, Härtel S, Babendreier D, von der Ohe W, Steffan-Dewenter I (2012) Effects of multiple Bt proteins and GNA lectin on in vitro-reared honey bee larvae. Apidologie 1–12.

Hernandez-Martinez P, Moar W, Escriche B (2010) Proteolytic processing of *Bacillus thuringiensis* Cry7Aa toxin and specific binding to brush border membrane vesicles of three sweetpotato weevil species (Coleoptera: Brentidae), abstract presented at the 43th meeting of the Society for Invertebrate Pathology, 11–15 May Trabzon, Turkey

Jackson DM, Harrison HF, Ryan-Bohac JR (2012) Insect resistance in sweetpotato plant introduction accessions. J Econ Entomol 105(2):651–658.

Kos M, van Loon JJ, Dicke M, Vet LE (2009) Transgenic plants as vital components of integrated pest management. Trends Biotechnol 27(11):621–627.

Li Y, Romeis J (2010) *Bt* maize expressing Cry3Bb1 does not harm the spider mite, *Tetranychus urticae*, or its ladybird beetle predator, *Stethorus punctillum*. Biol Control 53(3):337–344.

Marvier M, McCreedy C, Regetz J, Kareiva P (2007) A meta-analysis of effects of Bt cotton and maize on nontarget invertebrates. Science 316(5830):1475–1477.

Muyinza H, Talwana HL, Mwanga RO, Stevenson PC (2012) Sweetpotato weevil (*Cylas* spp.) resistance in African sweetpotato germplasm. Int J Pest Manage 58:73–81

OECD (2007) Consensus document on safety information on transgenic plants expressing *Bacillus thuringiensis*-derived insect control protein. Series on harmonisation of regulatory oversight in biotechnology, vol 42. Organisation for Economic Co-operation and Development, Paris

Pandey G (2008) Acute toxicity of ipomeamarone, a phytotoxin isolated from the injured sweet potato. Pharmacogn Mag 4:89–92

Peterson JA, Lundgren JG, Harwood JD (2011) Interactions of transgenic *Bacillus thuringiensis* insecticidal crops with spiders (Araneae). J Arachnol 39(1):1–21

Prischl M, Hackl E, Pastar M, Pfeiffer S, Sessitsch A (2012) Genetically modified Bt maize lines containing *cry3Bb1*, *cry1A105* or *cry1Ab2* do not affect the structure and functioning of root-associated endophyte communities. Appl Soil Ecol 54:39–48.

Qaim M (2001) A prospective evaluation of biotechnology in semi-subsistence agriculture. Agr Econ 25:165–175

Qaim M, Pray C, Zilberman D (2008) Economic and social considerations in the adoption of Bt Crops. In: Romeis J,

Shelton A, Kennedy G (eds) Integration of insect-resistant genetically modified crops within IPM programs, vol 5. Springer, Netherlands, pp 329–356

Raps A, Kehr J, Gugerli P, Moar WJ, Bigler F, Hilbeck A (2001) Immunological analysis of phloem sap of *Bacillus thuringiensis* corn and of the non-target herbivore *Rhopalosiphum padi* (Homoptera: Aphididae) for the presence of Cry1Ab. Mol Ecol 10:525–533

Raybould A (2007) Environmental risk assessment of genetically modified crops: general principles and risks to non-target organisms. BioAssay 2:8

Raybould A, Stacey D, Vlachos D, Graser G, Li X, Joseph R (2007) Non-target organisms risk assessment of MIR604 maize expressing mCry3A for control of corn rootworms. J Appl Entomol 131:391–399

Romeis J, Meissle M (2011) Non-target risk assessment of Bt crops-cry protein uptake by aphids. J Appl Ent 135:1–6.

Romeis J, Meissle M, Bigler F (2006) Transgenic crops expressing *Bacillus thuringiensis* toxins and biological control. Nature Biotechnol 24:63–71

Romeis J, Barsch D, Bigler F, Candolfi MP, Gielkens MMC, Hartley SE, Hellmich RI, Huesing JE, Jepson PC, Layton R, Quemada H, Raybould A, Rose RI, Schiemann J, Sears MK, Shelton AM, Sweet J, Vaituzis Z, Wolt JD (2008) Assessment of risk of insect-resistant transgenic crops to non target arthropods. Nature Biotechnol 26:203–208

Romeis J, Hellmich RL, Candolfi MP, Carstens K, De Schrijver A, Gatehouse AMR, Herman RA, Huesing JE, McLean MA, Raybould A, Shelton AM, Waggoner A (2011) Recommendations for the design of laboratory studies on non-target arthropods for risk assessment of genetically engineered plants. Transgenic Res 20(1):1–22.

Romeis J, Raybould A, Bigler F, Candolfi MP, Hellmich RL, Huesing JE, Shelton AM (2013) Deriving criteria to select arthropod species for laboratory tests to assess the ecological risks from cultivating arthropod-resistant genetically engineered crops. Chemosphere 90(3):901–909.

Smit NEJM (1997) Integrated Pest Management for sweetpotato in Eastern Africa. PhD Thesis, Agricultural University Wageningen, The Netherlands

Sorensen KA (2009) Sweetpotato insects: identification, biology and management. In: Loebenstein G, Thottappilly G (eds) The Sweetpotato. Springer-Verlag New York Inc, New York, pp 161–188

Stathers TE, Rees D, Kabi S, Mbilinyi L, Smit NEJM, Kiozya H, Jeremiah S, Nyango A, Jeffries D (2003) Sweetpotato infestation by *Cylas* spp. in East Africa: I: Cultivar differences in field infestation and the role of plant factors. Int J Pest Manage 49:131–140

Stathers T, Namanda S, Mwanga ROM, Khisa G, Kapinga R (2005) Manual for sweetpotato integrated production and pest management farmer field schools in sub-Saharan Africa. International Potato Centre, Kampala

Stevenson PC, Muyinza H, Hall DR, Porter EA, Farman D, Talwana H, Mwanga ROM (2009) Chemical basis for resistance in sweetpotato *Ipomoea batatas* to the sweetpotato weevil *Cylas puncticollis*. Pure Appl Chem 81(1):141–151

Tailor R, Teppett J, Gibb G, Stephen P, Derek P, Linda J, Ely S (1992) Identification and characterization of a novel *Bacillus thuringiensis* δ-endotoxin entomocidal to coleopteran and lepidopteran larvae. Mol Microbiol 6:1211–1217

Thomson JA (2008) The role of biotechnology for agricultural sustainability in Africa. Philos Trans R Soc Lond B Biol Sci 363:905–913.

USEPA (2007) White paper on tier-based testing for the effects of proteinaceous insecticidal plant-incorporated protectants on non-target arthropods for regulatory risk assessments. U.S. Environmental Protection Agency (USEPA), Washington D. C. http://www.epa.gov/pesticides/biopesti-cides/pips/non-target-arthropods.pdf. Accessed 20 Jan 2013

van der Merwe F, Bezuidenhout C, van den Berg J, Maboeta M (2012) Effects of Cry1Ab transgenic maize on lifecycle and biomarker responses of the earthworm, *Eisenia Andrei*. Sensors 12(12):17155–17167.

van Frankenhuyzen K (2009) Insecticidal activity of *Bacillus thuringiensis* crystal proteins. J Invertebr Pathol 101:1–16

Wolfenbarger LL, Naranjo SE, Lundgren JG, Bitzer RJ, Watrud LS (2008) Bt crops effects on functional guilds of non-target arthropods: a meta-analysis. PLoS ONE 3(5):e2118.

Wolt JD, Prasifka JR, Hellmich RL (2007) Ecological safety assessment of insecticidal proteins introduced into biotech crops. In: Hammond BG (ed) Food safety of proteins in agricultural biotechnology, vol 172. CRC Press.

Zhang X, Candas M, Griko NB, Taussig R, Bulla LA Jr (2006) A mechanism of cell death involving an adenylyl cyclase/PKA signaling pathway is induced by the Cry1Ab toxin of *Bacillus thuringiensis*. Proc Natl Acad Sci 103(26):9897–9902.

8

Biocontrol potential of three novel *Trichoderma* strains: isolation, evaluation and formulation

A. K. Mukherjee · A. Sampath Kumar ·
S. Kranthi · P. K. Mukherjee

Abstract We have isolated three novel strains of *Trichoderma* (two *T. harzianum* and one *T. atroviride*) from wild mushroom and tree bark, and evaluated their biocontrol potential against *Sclerotium delphinii* infecting cultivated cotton seedlings. *T. harzianum* strain CICR-G, isolated as a natural mycoparasite on a tree-pathogenic *Ganoderma* sp. exhibited the highest disease suppression ability. This isolate was formulated into a talcum-based product and evaluated against the pathogen in non-sterile soil. This isolate conidiated profusely under conditions that are non-conducive for conidiation by three other *Trichoderma* species tested, thus having an added advantage from commercial perspective.

Keywords *Trichoderma* · Biological control · Formulation · *Sclerotium delphinii* · Cotton

Introduction

Trichoderma spp. are widely used as commercial biofungicides all over the world (Harman 2006; Harman et al. 2004; Howell 2006; Lorito et al. 2010; Schuster and Schmoll 2010; Shoresh et al. 2010; Verma et al. 2007). In India alone, more than 250 commercial formulations are available (Singh et al. 2012), but almost all of them are based on

A. K. Mukherjee · A. Sampath Kumar · S. Kranthi ·
P. K. Mukherjee (✉)
Central Institute for Cotton Research, PB 2, Shankar Nagar PO,
Nagpur 440010, Maharashtra, India
e-mail: prasunmala@gmail.com

Present Address:
A. K. Mukherjee
Central Rice Research Institute, Cuttack, Odisha, India

a single strain of *T. viride* (recently reclassified as *T. asperelloides*; Mukherjee et al. 2013b), isolated from rhizosphere (Sankar and Jeyarajan 1996). Soil/rhizosphere has been classically viewed as the main habitat of *Trichoderma*, even though the maximum diversity of this species occurs aboveground e.g., on tree bark and wild mushrooms, and mycotrophy is viewed as the ancestral trait of this genus (Druzhinina et al. 2011). Consequently, only a few strains have been isolated from soil/rhizosphere and used as commercial biopesticides, and the above ground source remained largely unexploited in agriculture, except, perhaps, for a few endophytic strains, such as *T. gamsii* (http://www.clemson.edu/extension/horticulture/fruit_vegetable/peach/diseases/arr_biological.html). In the present study, we have isolated three novel *Trichoderma* strains from wild mushroom and tree bark and evaluated their potential as biocontrol agents. We have also developed a formulation product based on the most effective strain and evaluated this formulation as seed treatment for suppression of seed and root rot of cotton caused by *Sclerotium delphinii*, an emerging pathogen of cultivated cotton (Mukherjee et al. 2013a).

Results and discussion

Continued commercial success of *Trichoderma* would depend on identification of novel strains adapted to local conditions. Since the diversity of *Trichoderma* is profound on the above-ground, the success of novel strains to be developed as biocontrol products would be greater if the newer isolates are obtained that are naturally mycoparasites, as against collecting a large number of typical saprophytes (from soil) and mass screening. The current study has focused on isolation of *Trichoderma* from wild mushroom

Fig. 1 Natural occurrence and cultural characteristics of three isolates of *Trichoderma*. *Top left*: occurrence of *T. atroviride* CICR-A on *Acacia* sp. bark; *Top middle*: occurrence of *T. harzianum* CICR-E on *Eucalyptus* sp. bark; *Top right*: occurrence of *T. harzianum* CICR-G on a basidiocarp of *Ganoderma* sp. *Middle panel*: cultural characteristics on PDA, photographed after 5 days of inoculation. *Lower panel*: cultural characteristics on PDA, photographed after 10 days of inoculation. *Inset*: conidiophore structures observed under microscope

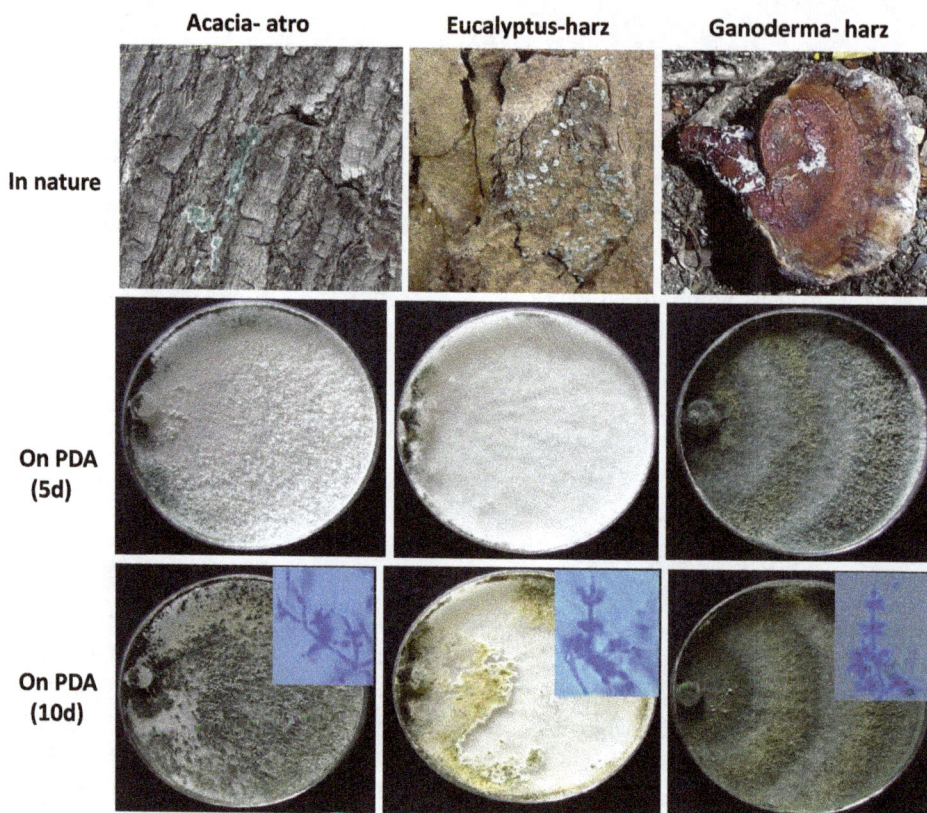

and tree bark and evaluation for biocontrol against a newly reported pathogen (*S. delphinii*, MTCC 11568) of cotton.

Of the three new isolates, *T. harzianum* CICR-G (MTCC 11511) was isolated from a parasitized basidiocarp of a *Ganoderma* sp. that was growing as a parasite on roots of an Acacia tree, *T. harzianum* CICR-E (MTCC 11500) was isolated from the bark of an Eucalyptus tree and *T. atroviride* CICR-A (MTCC 11512) was isolated from the bark of an Acacia tree (Fig. 1). There was large cultural variability among the two *T. harzianum* isolates which were also phylogenetically distantly related (Fig. 2). The Tef1 large (fourth) intron sequence data from all the three isolates have been deposited with GenBank viz. accession nos. KC679853 (*T. harzianum* CICR-G), KC679855 (*T. harzianum* CICR-E) and KC679854 (*T. atroviride* CICR-A).

In confrontation assay, both *T. harzianum* CICR-G and *T. harzianum* CICR-E were able to overgrow the test pathogen, but *T. atroviride* CICR-A failed to overgrow *S. delphinii* colony even after prolonged incubation (Fig. 3). Interestingly, the ability to colonize the sclerotia (resting structures of *S. delphinii*) also differed- *T. harzianum* CICR-E being the most effective while *T. atroviride* CICR-A being unable to colonize the sclerotia (Fig. 4).

The isolates also differed in their ability to suppress *S. delphinii* in sterile soil. *T. harzianum* CICR-G being the

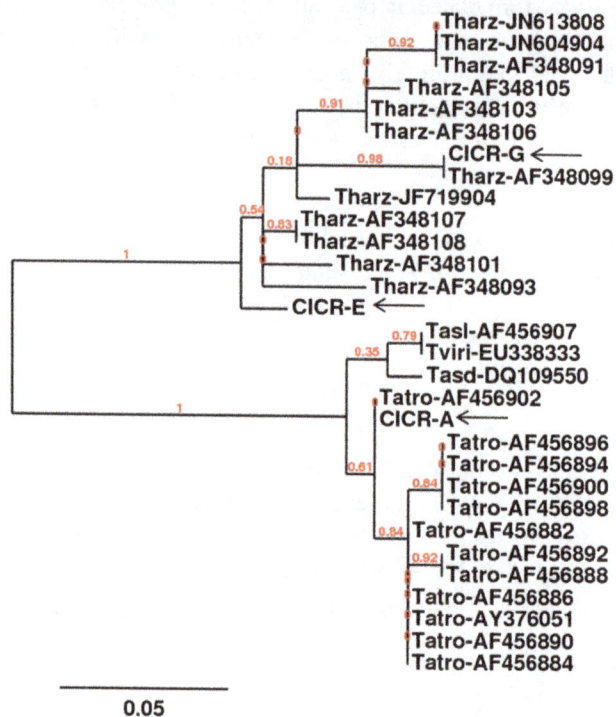

Fig. 2 Phylogenetic analysis of *Trichoderma* isolates based on the sequence of the fourth intron of translation elongation factor 1-alpha gene. The positions of new isolates are indicated with *arrows*

Fig. 3 Confrontation assay for antagonism of *Trichoderma* isolates on *Sclerotium delphinii*

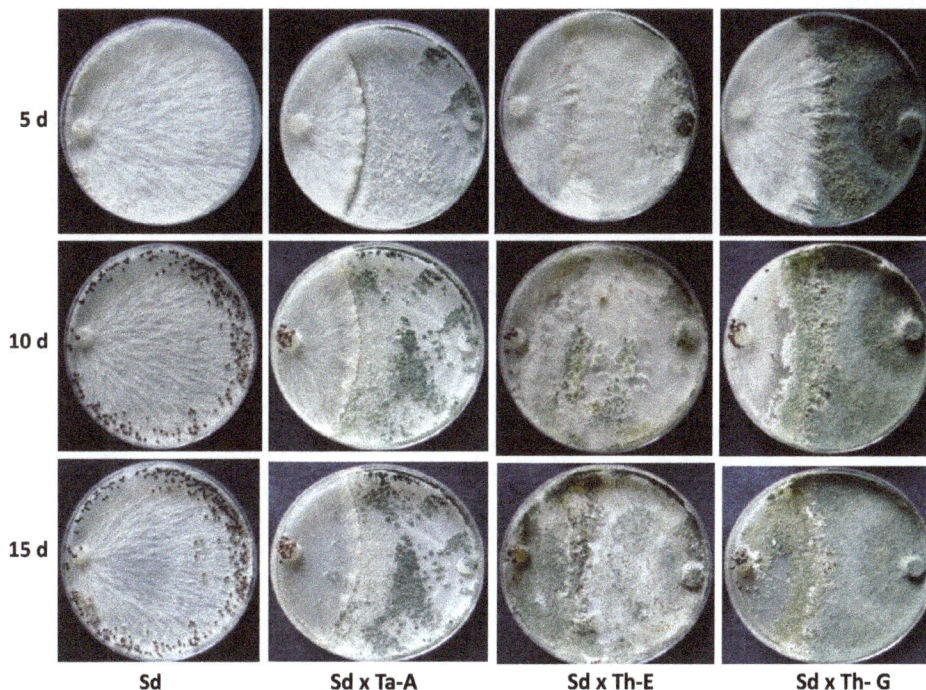

5 d			
10 d			
15 d			
Sd	Sd x Ta-A	Sd x Th-E	Sd x Th- G

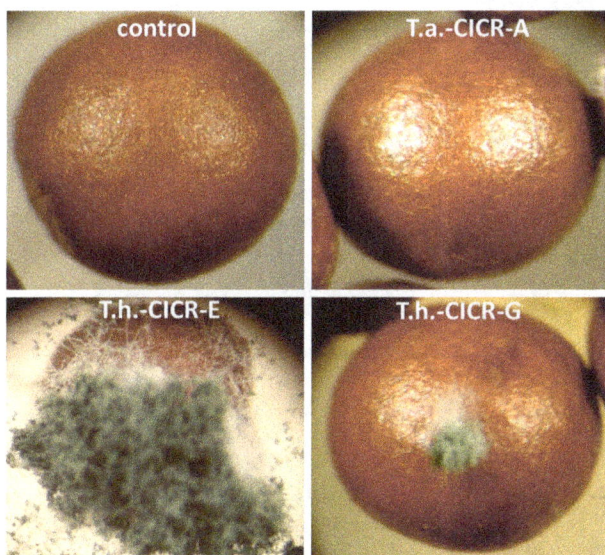

Fig. 4 Colonization of sclerotia of *Sclerotium delphinii* by *Trichoderma* isolates, observed under a stereo binocular microscope. Control: a sclerotium from a pure culture

most effective, while *T. atroviride* CICR-A being the least effective (Fig. 5). Based on this experiment, *T. harzianum* CICR-G was selected for further studies. It may be noted that even though *T. harzianum* CICE-E was more effective in confrontation assay, *T. harzianum* CICR-G was better as a biocontrol agent in pot soil. This is quite common, as the behaviour of an antagonist in pure culture is many a times different from that in soil where the performance of the bioagent is an outcome of interactions of the antagonist with the pathogen under the influence of several biotic and abiotic factors.

For developing a formulation product, we assessed the ability of *T. harzianum* CICR-G to conidiate in PDB of varying strengths. Interestingly, this isolate conidiated profusely within 3 days on PDB of one-fourth strength (Fig. 6). After 7 days, the number of conidia produced was 4.5×10^{10}, 7×10^{10} and 4.7×10^{10}, on $0.25\times$, $0.5\times$ and $1\times$ PDB, respectively (in a flask with 100 ml medium). We mixed the mat from two flasks ($0.5\times$ PDB) per kg talcum powder and after drying and packaging, obtained an initial approximate CFU (colony forming units) count of 10^8/g formulation product, designated as TrichoCASH 1 % WP.

Wet seed treatment (5 g/kg seeds) provided a uniform coating on acid de-linted cotton seeds (Fig. 7) that were used for sowing in non-sterile soil pre-infested with *S. delphinii*. Treating the seeds with TrichoCASH significantly protected seeds and seedlings from *S. delphinii* infection in non-sterile soil (Fig. 8). The seedling stand in non-sterile soil not pre-infested with *S. delphinii* was also significantly higher when seeds were treated with this formulation, as compared to non-treated seeds.

Ability to sporulate under adverse conditions is a desirable trait for biocontrol fungi as this is related to ease of formulation. Of the four different species/strains of *Trichoderma* tested on four different sources of PDA (lab. made, HiMedia, SRL and Titan media), *T. harzianum* CICR-G was least affected by the source of the culture medium (Fig. 9a, b). The laboratory made PDA supported conidiation of all the four species tested, but even on this medium there was wide variation in ability to conidiate;

Fig. 5 Biological suppression
of *Sclerotium delphinii* by
Trichoderma isolates in cotton.
Data are mean of 4
replicates ± SE

Fig. 5 Biological suppression of *Sclerotium delphinii* by *Trichoderma* isolates in cotton. Data are mean of 4 replicates ± SE

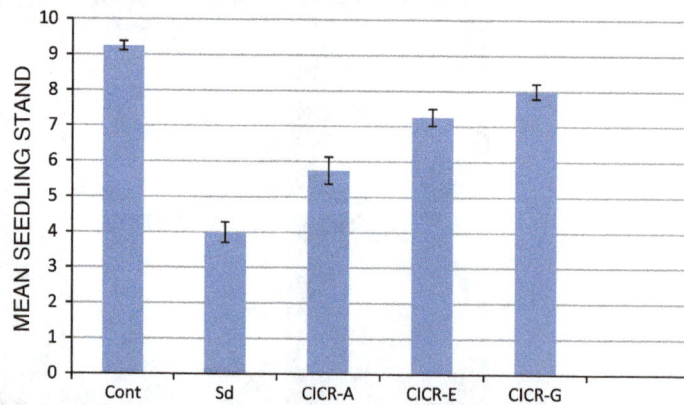

Fig. 6 Conidiation of *Trichoderma harzianum* CICR-G on different strengths of PDB after 3 days of incubation. 1×: full strength, 0.5×: half strength, 0.25×: a quarter strength

T. harzianum CICR-G being the most abundantly sporulating. It is very interesting to note that PDA from different sources have significant effect on conidiation ability of *Trichoderma* spp., and PDA from some commercial sources did not support conidiation of *Trichoderma* (except of *T. harzianum* CICR-G). It needs to be ascertained if the "loss-of-conidiation" of some *Trichoderma* species often observed in laboratories is related to switch to a different source/batch of PDA procured from commercial sources.

S. delphinii, like its close relative *S. rolfsii*, is a soil-borne pathogen that over winters through the production of highly melanised sclerotial bodies (Punja 1985; Xu et al. 2008). These attributes make it difficult to control using conventional practices. This pathogen, mostly associated with ornamental crops and certain field crops like groundnut, was hitherto not known to occur in cotton (Edmunds et al. 2003; Farr et al. 2006; Xu et al. 2010). We

Fig. 7 Seed treatment with TrichoCASH 1 % WP. *Top*: untreated cotton seeds; *bottom*: treated seeds

have recently reported this pathogen to be infecting both seedlings and mature cotton plants in field (Mukherjee et al. 2013a). The soil-borne nature of this pathogen would mean that the pathogen might multiply in soil on crop

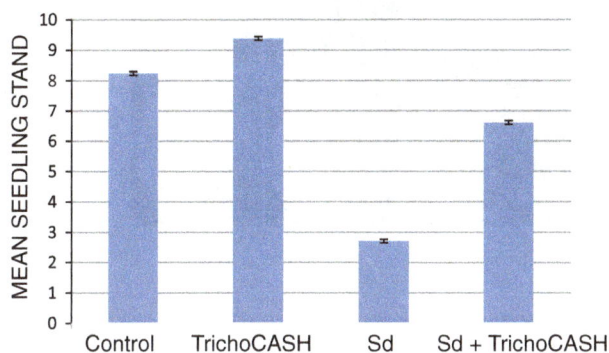

Fig. 8 Biological suppression of *Sclerotium delphinii* in cotton in non-sterile soil. Data are mean of 21 replicates ± SE

residues and thus, the incidence would increase with time. In the present study, we have isolated a novel *T. harzianum* strain (naturally occurring as a mycoparasite) that is effective against this emerging pathogen, and also developed a formulation for field applications. The formulation is being tested in multiple locations across India for biocontrol efficacy against seed rot and seedling diseases in cotton.

Experimental procedure

Fungal strains and growth conditions

All the three *Trichoderma* strains studied here were isolated from Nagpur, Maharashtra, India, either from infected tree-pathogenic *Ganoderma* sp. or from bark of *Eucalyptus* sp. and *Acacia* sp. The fungi were collected with a sterile cotton swab and the conidia suspended in sterile distilled water. The suspension was plated on potato dextrose agar plates after serial dilution and the isolated colonies were further purified by serial dilution and plating (three times). The pathogen *Sclerotium delphinii* (MTCC 11568) was obtained from our previous studies (Mukherjee et al.

2013a). Routinely, the fungi were grown at ambient temperatures (25–30 °C) on potato dextrose medium prepared in laboratory (200 g potatoes, 20 g dextrose, and 20 g agar–agar, when required, per litre), unless otherwise stated.

Identification of fungi

The fungal strains were identified based on the sequences of the large sub-unit of translation elongation factor 1-alpha (Tef1) as per standard methods (http://www.isth.info/tools/blast/markers.php). In brief, the large (4th) intron of *tef1* gene was amplified using the primer pair EF1-728F and EF1-986R as recommended, and the product sequenced using an automated DNA sequencer. The species were identified by BLASTN on the NCBI site and the identity confirmed by comparing the sequences with authentic sequences from the GenBank, and a phylogenetic tree constructed on http://www.phylogeny.fr.

Confrontation assays

Ability of the *Trichoderma* isolates to antagonize the test pathogen *S. delphinii* was assessed using confrontation assay on PDA plates by simultaneous inoculation of both *Trichoderma* and the pathogen near the edge of the plate, placed opposite each other. Ability of *Trichoderma* to overgrow the pathogen colony and also to colonize the sclerotia was recorded.

Comparative evaluation for biocontrol in green house

S. delphinii was grown on autoclaved sorghum grains for 7 days and was inoculated to sterile soil at 2 g per pot containing 2 kg autoclaved black cotton soil. The pots were covered with poly bags for 2 days to facilitate the establishment of the pathogen and after 2 days, 10 seeds of cotton (*Gossypium hirsutum*, variety PKV 081) treated with *Trichoderma* spore suspension (10^7/ml in 0.5 % aqueous carboxy-methyl cellulose) were sown in each pot. Nontreated seeds sown in pathogen-infested soil served as control. Observation on healthy plant stand was recorded after 10 days.

Development of a talc-based formulation product

Based on the performance in green house, the best isolate (*T. harzianum* CICR-G) was selected for formulation development and subsequent evaluation. A mycelia disc (6 mm diameter) was inoculated in 100 ml PDB at three concentrations of the medium ($1\times$, $0.5\times$ and $0.25\times$). Conidiation was counted after 7 days and the mycelial mat along with conidia from $0.5\times$ PDB were mixed thoroughly with

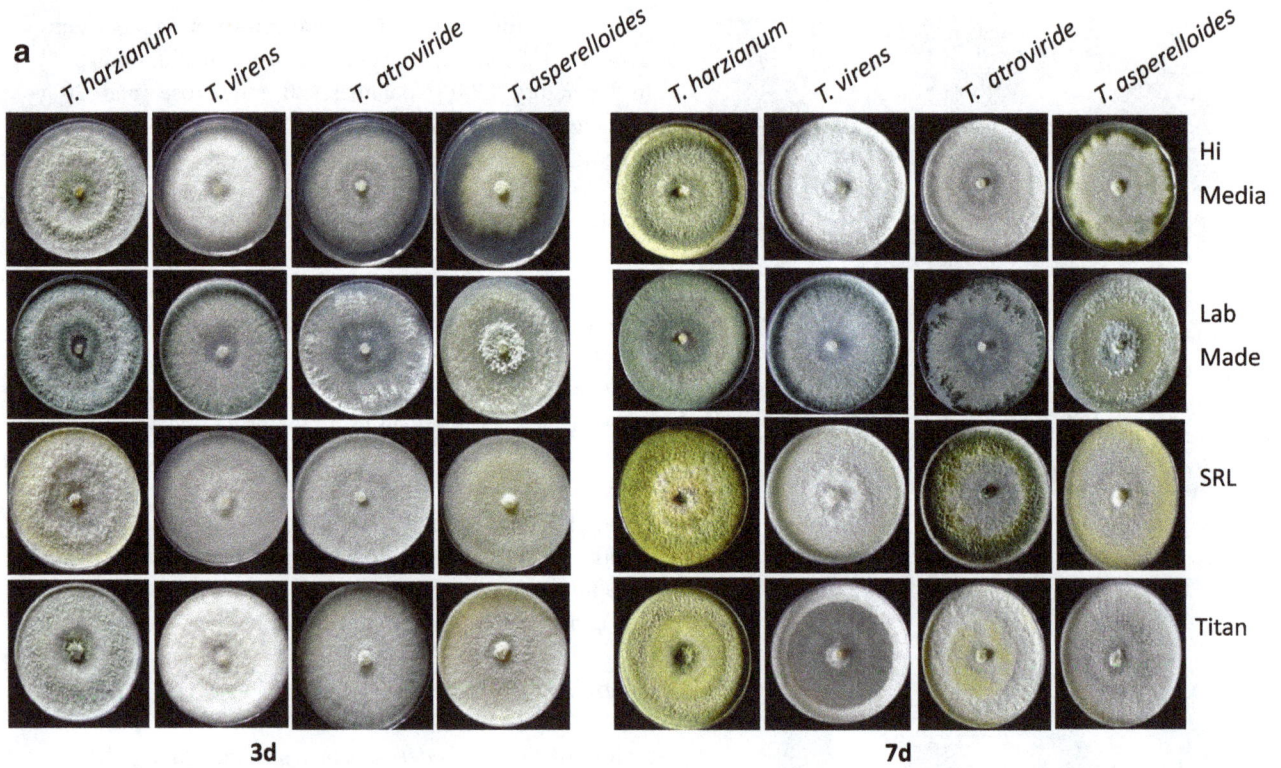

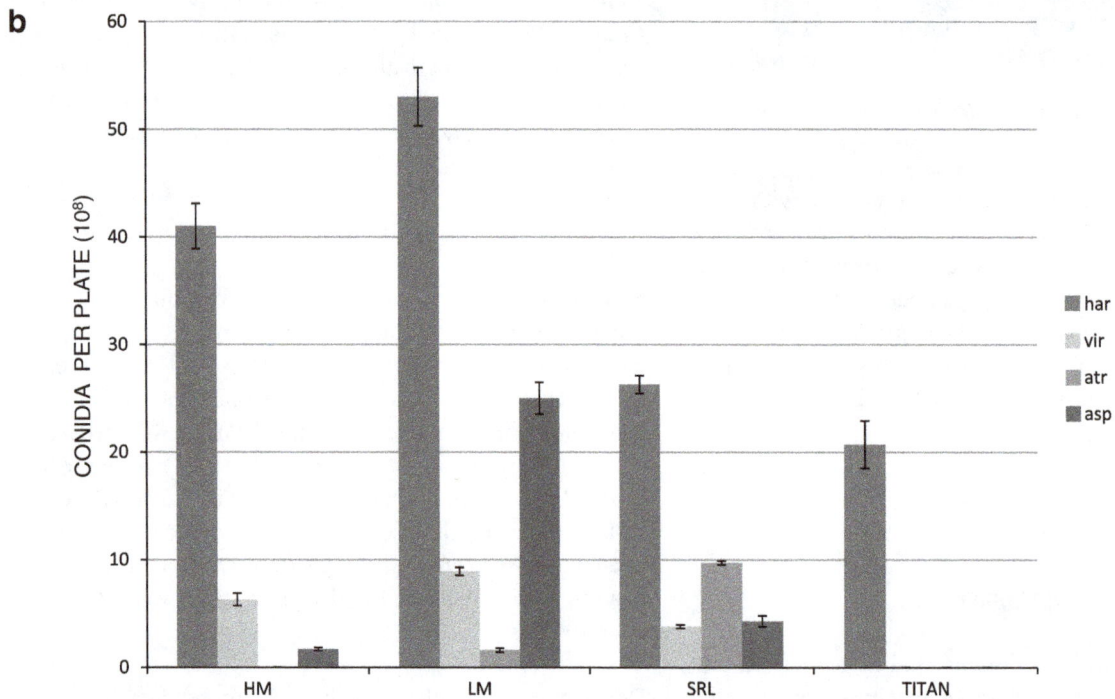

Fig. 9 a Growth and **b** conidiation of four *Trichoderma* species on PDA from various sources. Data are mean of 3 replicates ± SE

autoclaved talcum powder pre-treated with 0.5 % CMC (5 g CMC dissolved in 100 ml water mixed with 1 kg talcum powder). The mix was air-dried in a laminar flow hood and the colony forming units were counted on PDA amended with 100 mg/L rose Bengal, after serial dilution. The formulation was named as TrichoCASH 1 % WP.

Evaluation of TrichoCASH in green house

Five g of TrichoCASH was taken in a poly bag, and 25 ml water was added to make a slurry. Cotton seeds were treated with this slurry (@5 g/kg seeds) and seeds dried in shade before sowing (10 seeds per pot) in non-autoclaved

Table 1 Media/components used in this study

Media/component	Manufacturer	Cat. no.	Batch no.	Date of Mfg	Expiry date
PDA (infusion from 200 g boiled potatoes), 20 g dextrose (Hi Media), 20 g agar–agar (Hi Media) in 1 litre RO water pH 6.5	In-house	NA	NA	NA	NA
PDA 39 g/L RO water pH 6.5	Hi Media Laboratories, Mumbai	MU096	0000138972	March 2012	March 2015
PDA 39 g/L RO water pH 6.5	SRL Laboratories, Mumbai	PM015	10062523	June 2012	March 2015
PDA 39 g/L RO water pH 6.5	Titan Biotech, Bhiwandi, Rajasthan	TMV344	V3411112	Not mentioned	November 2014
Dextrose	Hi Media Laboratories, Mumbai	RM077	0000044659	March 2009	Not mentioned
Agar–Agar type I	Hi Media Laboratories, Mumbai	RM666	0000053188	February 2009	February 2014

pot soil (5 kg capacity pots) pre-infested with *S. delphinii* as described above. Observations on healthy seedlings were taken after 15 days.

Effect of source of medium on conidiation

We have earlier observed that certain strains of *Trichoderma* did not conidiate in PDA from various commercial sources, thus posing a limitation in commercial formulations (Mukherjee PK, unpublished). Hence, we evaluated ability of this strain to conidiate on PDA from different commercial sources, vis-à-vis some other commonly used *Trichoderma* species (*T. atroviride*- this study, *T. asperelloides*, *T. virens*- both kindly gifted by Dr. Ashis Das, NRCC, Nagpur). Mycelial discs were inoculated in the centre of culture plates containing PDA from different commercial sources (Table 1) and observations on conidia production was recorded after 7 days incubation at ambient temperatures. Spores were counted using a hemocytometer, after appropriate dilution.

Acknowledgments The authors thank Dr. K. R. Kranthi, Director, Central Institute for Cotton Research, Nagpur, for his encouragement and support..

Conflict of interest The authors declare no conflict of interest.

References

Druzhinina IS, Seidl-Seiboth V, Herrera-Estrella A, Horwitz BA, Kenerley CM, Monte E, Mukherjee PK, Zeilinger S, Grigoriev IV, Kubicek CP (2011) *Trichoderma*: the genomics of opportunistic success. Nat Rev Microbiol 9:749–759

Edmunds BA, Gleason ML, Wegulo SN (2003) Resistance of hosta cultivars to petiole rot caused by *Sclerotium rolfsii* var. *delphinii*. Hort Technol 13:302–305

Farr DF, Rossman AY, Palm ME, McCray, EB (2006) Fungal Databases. Systematic Botany & Mycology Laboratory, ARS, USDA. http://nt.ars-grin.gov/fungaldatabases/

Harman GE (2006) Overview of mechanisms and uses of *Trichoderma* spp. Phytopathology 96:190–194

Harman GE, Howell CR, Viterbo A, Chet I, Lorito M (2004) *Trichoderma* species: opportunistic, avirulent plant symbionts. Nat Rev Microbiol 2:43–56

Howell CR (2006) Understanding the mechanisms employed by *Trichoderma virens* to effect biological control of cotton diseases. Phytopathology 96:178–180

Lorito M, Woo SL, Harman GE, Monte E (2010) Translational research on *Trichoderma*: from omics to the field. Annu Rev Phytopathol 48:395–417

Mukherjee AK, Mukherjee PK, Kranthi S (2013a) *Sclerotium delphinii* infecting cultivated cotton in India- a first record. New Disease Reports (Submitted)

Mukherjee PK, Mukherjee AK, Kranthi S (2013b) Reclassification of *Trichoderma viride* (TNAU), the most widely used commercial biofungicide in India, as *Trichoderma asperelloides*. Open Biotechnol J 7:7–9.

Punja ZK (1985) The biology, ecology, and control of *Sclerotium rolfsii*. Annu Rev Phytopathol 23:97–127

Sankar P, Jeyarajan R (1996) Seed treatment formulation of *Trichoderma* and *Gliocladium* for biological control of *Macrophomina phaseolina* in sesamum. Indian Phytopath 49:148–151

Schuster A, Schmoll M (2010) Biology and biotechnology of *Trichoderma*. Appl Microbiol Biotechnol 87:787–799

Shoresh M, Harman GE, Mastouri F (2010) Induced systemic resistance and plant responses to fungal biocontrol agents. Annu Rev Phytopathol 48:21–43

Singh HB, Singh BN, Singh SP and Sarma BK (2012) Exploring different avenues of *Trichoderma* as a potent bio-fungicidal and plant growth promoting candidate- an overview. Rev Plant Pathol 5:315–426, Scientific Publishers (India), Jodhpur

Verma M, Brar S, Tyagi R, Surampalli R, Valero J (2007) Antagonistic fungi, *Trichoderma* spp.: panoply of biological control. Biochem Eng J 37:1–20

Xu Z, Gleason ML, Mueller DS, Esker PD, Bradley CA, Buck JW, Benson DM, Dixon PM, Monteiro JEBA (2008) Overwintering of *Sclerotium rolfsii* and *S. rolfsii* var. *delphinii* in different latitudes of the United States. Plant Dis 92:719–724

Xu Z, Harrington TC, Gleason ML, Batzer JC (2010) Phylogenetic placement of plant pathogenic *Sclerotium* species among teleomorph genera. Mycologia 102:337–346

9

Isolation and characterization of a *Bacillus subtilis* strain that degrades endosulfan and endosulfan sulfate

Ajit Kumar · Narain Bhoot ·
I. Soni · P. J. John

Abstract Endosulfan has emerged as a major environmental menace worldwide due to extensive usage and environmental persistence, seeking its remedial by a cheaper and efficient means. Therefore, natural resource (soil) was explored to search a potential candidate for biodegradation of endosulfan. A soil bacterium was enriched and isolated by applying a strong nutritional selection pressure, using a non-sulfur medium supplemented with endosulfan as sole source sulfur. The microbial strain was found to degrade endosulfan as well as its equally toxic metabolite endosulfan sulfate to non-toxic metabolites (endodiol and endosulfan lactone) very efficiently (up to 94.2 %) within 7 days, estimated qualitatively by thin layer chromatography and quantitatively by gas chromatography-electron capture detection methods. The isolate was characterized for its morphological, physiological, biochemical and 16S rRNA sequencing and identified as a new strain of *Bacillus subtilis* with strain designation AKPJ04, which was deposited with accession number Microbial Type Culture Collection and Gene Bank (MTCC) 8561, at MTCC, Institute of Microbial Technology, Chandigarh, India. The partial 16S rRNA sequence was submitted to Genbank, Maryland, USA, with the accession number EU 258611. The primary investigation for endosulfan degrading gene(s) localization suggested its location on chromosomal DNA.

Keywords Biodegradation · Endosulfan · *Bacillus subtilis* · Characterization

Genbank submission: The 16S rDNA sequence of *Bacillus subtilis* AKPJ04 was deposited in GenBank database under accession number EU 258611.

A. Kumar · N. Bhoot · I. Soni · P. J. John
Environment Toxicology Unit, Centre for Advanced Studies in Zoology, University of Rajasthan, Jaipur 302004, India

Present Address:
A. Kumar (✉)
Centre for Bioinformatics, M.D. University,
Rohtak 124001, India
e-mail: akumar.cbt.mdu@gmail.com

Introduction

With the advent of Green Revolution, use of synthetic fertilizers and pesticides has increased at an uncontrolled pace to meet the demands of ever-growing human population. To the darker side of the same, these compounds, particularly, pesticides have posed a serious ecological threat and therefore needs an early scientific attention. Presently, endosulfan is one of the extensively used organochlorine pesticides after the worldwide ban on DDT and BHC. Since then, use of endosulfan has increased dramatically in last three decades. It is used extensively worldwide as a contact and stomach insecticide for Colorado potato beetle, flea beetle, cabbageworm, peach tree borer, tarnished plant bug and as an acaricide on field crop like cotton, paddy, sorghum, oilseeds and coffee (Lee et al. 1995; Kullman and Matsumura 1996; USEPA 2002), apart from its use in vector-control (tsetse fly), as a wood preservative and for the control of home garden pests (C.N.R.C 1975). It is a highly toxic substance and is classified as a Category 1b (highly hazardous) pesticide by USEPA (2002).

There have been several reports of acute poisoning and chronic toxicity of endosulfan. Acute toxicity includes stimulation of Central Nervous System as the major characteristic and is indistinguishable from symptoms of other cyclodienes (USDHHS 1990). Endosulfan has been reported to be highly toxic to aquatic fauna like fish and invertebrates (Sunderam et al. 1992). In addition, there are reported implications in mammalian gonadal toxicity (Sinha et al. 1997), genotoxicity (Chaudhari et al. 1999), teratogenic effects (Yadav 2003) and mutagenic effects (USDHHS 1990). These acute and chronic toxicity and environmental concerns have attracted scientists for an effective and economically viable option search for endosulfan degradation.

Bioremediation has evolved as a very economical and viable process for detoxification of xenobiotics in general and pesticides in particular, as an alternative to existing methods such as incineration and landfill (Gavrilescu and Chisti 2005). Therefore, present investigations were carried out in our laboratory to enrich and isolate endosulfan degrading microorganism from the natural resource (soil), having the past history of endosulfan usage. The isolated strains were studied for their comparative pesticide degradation ability to select the best biodegrader. The best degrading isolate N2 was found to degrade both the isomers of endosulfan (α- and β-isomers) along with the equally toxic metabolite endosulfan sulfate up to 94 % within 7 days (Kumar et al. 2012).

The present report describes the biochemical and molecular characterization and identification of the best endosulfan degrading isolate N2 and comparative analysis of its degradation profile with the standard microorganisms (MOs) reported earlier for endosulfan degradation, *Phanerochaete chrysosporium* (Kullman and Matsumura 1996) and *Mucor thermohyalospora* (Shetty et al. 2000).

Materials and methods

Technical grade endosulfan (99 % pure), an organochlorine insecticide with the chemical name 6,7,8,9,10,10-hexachloro-1,5,5a,6,9,9a-hexahydro-6,9-methano-2,4,3benzo-dioxathiepin-3-oxide for the study, was gifted by Excel India Pvt. Ltd., Ahmedabad, India (Fig. 1). The endosulfan isomers (α- and β-isomers), endosulfan sulfate and other endosulfan metabolite standards for chromatographic analyses were purchased from Hewlett Packard Company, Wilmington, Delaware, USA. Chloroform, *n*-hexane and acetone of chromatographic grade were used for chromatographic studies. All other chemicals used for the study were of analytical grade. The earlier reported endosulfan degraders, *P. chrysosporium* MTCC 4955 (PC) and *M. thermohyalospora* MTCC 1384 (MT), were selected as standards for the comparative study and were purchased from Microbial Type Culture Collection and Gene Bank

(MTCC), Institute of Microbial Technology (IMTECH), Chandigarh.

Culturing of isolate N2 and standard MOs

The isolate N2 for the study of endosulfan degradation profile was cultured in non-sulfur medium (NSM) supplemented with 50-ppm technical endosulfan as sole source of sulfur (Table 1). The other culture conditions included pH 6.5 with rotational agitation of 130 rpm, and incubation temperature and time of 30 °C and 15 days, respectively. The culture conditions were as per optimized parameters reported earlier, with little modification (Kumar et al. 2012). The other two MOs, PC and MT, were cultured as per their corresponding reports (Kullman and Matsumura 1996; Shetty et al. 2000) for comparative study of endosulfan degradation.

Analysis of endosulfan degradation and characterization of metabolites

For endosulfan and its metabolites extraction and analysis, 5 ml of cultured broth of the isolate N2, PC and MT were subjected with equal volumes of ethyl acetate and the organic phase was passed through a 6-cm $MgSO_4$ column in a Pasteur pipette to remove any residual water. The columns were prewashed with ethyl acetate. The extracted elutes containing pesticide and metabolites were gently evaporated at 50 °C in oven and were dissolved in acetone (chromatographic grade) and stored in glass vial at 4 °C for further analysis. A fortification test for recovery of endosulfan isomers was carried out for further chromatographic analysis.

The sample extracts of the N2-culture medium were analyzed for the residues of endosulfan by Gas Chromatography-Electron Capture Detection (GC-ECD) method. The analysis was carried out on a Shimadzu Model 2010 Gas chromatograph (GC) equipped with 63Ni Electron capture detector (ECD), and a capillary column HP ultra 2 (US 4293415) 0.52 × 25 × 0.32. The instrument was supported by Lab Solutions software for the analysis of endosulfan (α- and β-isomers) and endosulfan sulfate. The stock standards (200-ppm) of endosulfan isomers and endosulfan sulfate were obtained from Hewlett Packard Company, USA. Stock standards of 100-ppm were prepared by diluting standard mixture in 1:1 solvent mixture of HPLC grade iso-octane and toluene. These stocks were stored under freezing conditions. Working standard of the mixture was prepared from 100-ppm stock solution. 0.5–1.0 ppm of this mixture of endosulfan isomers and endosulfan sulfate was used for calibrating the GC for analyzing residues of endosulfan in the sample analyzed.

For thin layer chromatography (TLC), the dried endosulfan isomers and metabolites, after extraction from the culture, were dissolved in chromatographic grade acetone and applied to

Fig. 1 Endosulfan and its
isomers (technical grade
composition—α: β::7:3)

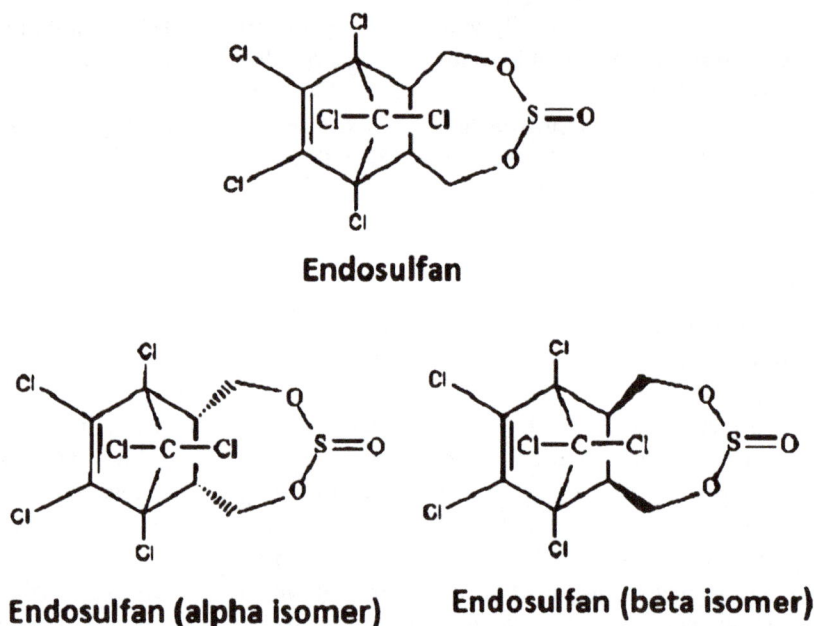

Endosulfan

Endosulfan (alpha isomer)   **Endosulfan (beta isomer)**

Table 1 Composition of (a) non-sulfur medium (NSM) (pH = 6.5),
(b) Trace elements solution

S. no.	Chemical	Amount (g/liter)
a		
1	K_2HPO_4	0.225
2	KH_2PO_4	0.225
3	NH_4Cl	0.225
4	$MgCl_2 \cdot 6H_2O$	0.845
5	$CaCO_3$	0.005
6	$FeCl_2 \cdot 4H_2O$	0.005
7	D-Glucose	1.000
8	Trace element solution	1 mg/lt
b		
1	$MnCl_2 \cdot 4H_2O$	198
2	$ZnCl_2$	136
3	$CuCl_2 \cdot 2H_2O$	171
4	$CoCl_2 \cdot 6H_2O$	24
5	$NiCl_2 \cdot 6H_2O$	24

neutral Silica gel TLC plates (pre-activated at 80 °C for
30 min). The plates were developed in hexane: chloroform:
acetone (9:3:1). The chlorine-containing constituents were
visualized by spraying plates with $AgNO_3$-saturated methanol
and then exposing them to UV-light (Kovacs 1965).

Characterization and identification of isolate N2

The isolate N2 was subjected to different morphological
characterizations (colony morphology, negative staining,
gram staining and endospore staining) along with different
biochemical characterizations (Cappuccino and Sherman
2005). A growth curve was plotted using NSM with

endosulfan as sole source of sulfur for isolate N2 under
optimized culture conditions. Its growth was monitored for
a temperature range of 15–50 °C, a pH range of 4.0–9.0
and salinity range of 2.0 % NaCl–10.0 % NaCl concentrations for further physiological characterization.

The partial 16S rRNA gene sequencing of isolate N2
was carried out by the custom services of IMTECH,
Chandigarh, India, and was subjected to sequence homology search using BLAST (Altshcul et al. 1990) to identify
the isolate. The phylogenetic analysis of the isolate N2 was
also carried out using CLUSTAL-W (Larkin et al. 2007)
for multiple sequence alignment, and a phylogenetic tree
was constructed using Neighbour-Joining method of Phylip
version 3.69 (Felsenstein 2005).

Primary localization of endosulfan degrading gene
in isolate N2

The isolate N2 was also looked for plasmid content by
attempting plasmid isolation by alkaline lysis method (Sambrook et al. 1989) and boiling lysozyme preparation method
(Ausubel et al. 2005). *E. coli* DH5α (plasmid strain) was run as
positive control in both the plasmid isolation protocol and
1.0 % agarose gel was run for visualization of DNA bands.

Results and discussion

Fortification test for endosulfan recovery
from microbial cultures

The fortification tests for recovery of endosulfan isomers
from microbial cultures during degradation analysis were

conducted for five concentrations (0.5, 5.0, 10, 50 and 100 ppm) of endosulfan in the culture broth. The recovery was observed to range from 92.5 to 102.8 % for α-endosulfan and 93.2 to 104.1 % for β-endosulfan with the coefficient of variation ranging from 0.8 to 2.8 % and 0.84 to 4.2 % for α-endosulfan and β-endosulfan, respectively. Based on the fortification analysis of recovery experiments for both endosulfan isomers, the extractions were considered appropriate of residue analysis.

GC-ECD & TLC analysis of endosulfan degradation profile

After GC-separation and ECD analysis, it was found that 8.65 ppm of α-endosulfan, 5.85 ppm of β-endosulfan and 2.98 ppm of endosulfan sulfate remained in the N2-culture system, accounting for about 71.0 % degradation after 3 days of incubation, as compared to 24.82 ppm of α-endosulfan, 15.81 ppm of β-endosulfan and 6.42 ppm of endosulfan sulfate detected in the control sample accounting for 18.74 % abiological degradation. After 7 days of incubation, the degradation of endosulfan in N2-culture system was found to be 94.0 % with 1.79 ppm of α-endosulfan, 1.21 ppm of β-endosulfan and 0.32 ppm of endosulfan sulfate detected by GC-ECD as compared to 23.93 ppm of α-endosulfan, 14.78 ppm of β-endosulfan and 6.98 ppm of endosulfan sulfate detected in the control sample that accounts for 22.58 % abiological degradation (Table 2).

Isolate N2 was also investigated for its comparative degradation with positive control cultures of reported standard microbes, namely, *Phanerochaete chrysosporium* (Kullman and Matsumura 1996) and *Mucor thermohyalospora* MTCC 1384 (Shetty et al. 2000). This comparison

was made on the basis of analyzing endosulfan degradation and its metabolites' profile recovered from the respective culture media after 15 days of incubation, under respective standard culture conditions. The TLC-plate analysis revealed six spots, identified as α-endosulfan, β-endosulfan, endosulfan sulfate, endosulfan lactone, endosulfan hydroxyether and endosulfan diol, on the basis of their respective retention factor (R_f) of 0.69, 0.46, 0.36, 0.29, 0.21 and 0.10. Endosulfan sulfate was the only abiotic degradation metabolite observed in the control, while cultures of N2, PC and MT showed endosulfan diol and endosulfan sulfate as the two common metabolites. Endosulfan diol was produced maximum in the cultures of N2, while PC and MT produced the same as minor metabolite. A spot of endosulfan lactone was observed in the cultures of N2, while no such spot was observed for PC and MT. A spot of endosulfan hydroxyether was observed only for PC-culture while no such metabolite was detected in N2 and MT (Table 3).

The major metabolite detected in the isolate N2 was endosulfan diol with little amount of endosulfan lactone. Endosulfan hydroxyether was not detected in the isolate similar to *M. thermohyalospora* MTCC 1384 and in contrast to *P. chrysosporium*. In *P. chrysosporium*, endosulfan diol and endosulfan hydroxyether were detected, while no trace of endosulfan lactone was observed. *M. thermohyalospora* MTCC 1384 had endosulfan diol as the only metabolite detected apart from spots of α-endosulfan, β-endosulfan and endosulfan sulfate, observed commonly in all the samples. These observations were in accordance with their respective reported findings (Kullman and Matsumura 1996; Shetty et al. 2000). Shetty et al. (2000) reported about degradation of endosulfan by *M. thermohyalospora* to form endosulfan diol as the major

Table 2 Gas chromatography–ECD data of endosulfan isomers and endosulfan sulfate after degradation by isolate N2

Sample	Peak#	Retention time (min)	Compound	Concentration (ppm)	Endosulfan degradation (%)
Control (Day 3)	24	13.818	α-Endosulfan	24.82	18.74
	32	15.751	β-Endosulfan	15.81	
	37	17.207	Endosulfan sulfate	6.42	
Control (Day 7)	25	13.800	α-Endosulfan	23.93	22.58
	32	15.729	β-Endosulfan	14.78	
	37	17.205	Endosulfan sulfate	6.98	
N2 (Day 3)	26	13.757	α-Endosulfan	8.65	71.0
	32	15.708	β-Endosulfan	5.85	
	38	17.207	Endosulfan sulfate	2.98	
N2 (Day 7)	27	13.709	α-Endosulfan	1.79	94.0
	32	15.697	β-Endosulfan	1.21	
	33	17.207	Endosulfan sulfate	0.32	

Culture condition: Media, NSM with 50-ppm technical endosulfan as sole source of sulfur; pH, 6.5; Temperature, 30 °C; Agitation, 130 rpm

Table 3 TLC profile of endosulfan degradation by isolate N2 and standard cultures of *P. chrysoporium* (PC) and *M. thermohyalospora* (MT)

Metabolites	Rf Value§	Control*	N2*	PC	MT
α-Endosulfan	0.69	+++	+	+	+
β-Endosulfan	0.46	+	+	+	+
Endosulfan sulfate	0.36	+	+	+	+
Endosulfan lactone	0.29	−	+	−	−
Endosulfan hydroxyether	0.21	−	−	+	−
Endosulfan diol	0.10	−	+++	+	+

Culture condition: Media, NSM with 50-ppm technical endosulfan as sole source of sulfur; pH, 6.5; Temperature, 30 °C; Agitation, 130 rpm;+++, Formation of major metabolite; +, formation of minor metabolites, −, no metabolite detected

§ TLC solvent system used: hexane:petroleum ether:acetone (9:3:1)

metabolite. Kullman and Matsumura (1996) found that *P. chrysosporium* degraded endosulfan in endosulfan diol and endosulfan hydroxyether by utilizing both oxidative and hydrolytic pathways for metabolism of this pesticide.

Morphological characterization of isolate N2

The isolate N2 was found to be Gram +ve, long, rod-shaped bacilli after gram staining and negative staining. It was also found to be a spore former after performing endospore staining with subterminally/centrally positioned, ellipsoidal and no-swollen spores. Isolate N2 was found to form elevated, large, round and convex colonies with shiny mucoid surface as observed from its colony morphology. The size of the bacterium was found to be 2–4 μ in length and <1.0 μ in width. The bacteria were observed to be arranged in singles and pairs.

Physiological characterization

The isolate N2 was observed to grow in the temperature range of 20–50 °C, while no growth was observed at a temperature of 15 °C (Table 4). Growth of the bacterium was found to occur in the pH range of 6.0–9.0 and a salinity range of 2.0–10.0 % NaCl (Table 4).

The growth curve of N2 was found to be a normal sigmoidal with a delayed lag phase of about 30 h. The log phase was found to extend up to 84 h after which the culture showed stationary phase up to 192 h (8 days). The decline phase initiated after 192 h as evident from growth curve (Fig. 2).

Biochemical characterization

The isolated bacterium tested positive for citrate utilization, TSI test (acid form glucose), Casein hydrolysis, Esculin hydrolysis, Gelatin hydrolysis, Starch hydrolysis, Nitrate hydrolysis, Catalase test, Lysine decarboxylase test and *ortho*-nitrophenyl glucuronide (ONPG) test. The

Table 4 Physiological characterizations of isolate N2

Tests*	Results
Growth at temperatures#	
15 °C	−
20 °C	+
30 °C	+
37 °C	+
42 °C	+
50 °C	+
(# pH 6.5, NaCl 2 %)	
Growth at pH	
4.0	−
6.0	+
7.0	+
8.0	+
9.0	+
Growth on NaCl (%)	
2.0	+
5.0	+
7.0	+
10.0	+

−, No growth; +, Growth

* All tests were conducted in nutrient medium

results were negative for growth on MacConkey agar, Indole test, Methyl red test, Voges–Proskauer tests, H₂S production, gas from glucose, TSI test (acid from lactose), Urea hydrolysis, Arginine dihydrolase, Ornithine decarboxylase and Phosphatase test.

Molecular characterization

The partial nucleotide base sequencing (1,396 base pairs) of 16S rRNA gene of isolate N2 was done to identify the bacterium. After performing BLAST search for sequence homology at GenBank, the bacteria showed 100 % identity with *Brevibacterium halotolerans* strain DSM 8802,

Bacillus subtilis subsp. *subtilis* strain DSM 10, *Bacillus vallismortis* strain DSM11031, *Bacillus mojavensis* strain IFO15718, *Bacillus amyloliquefaciens* strain NBRC 15535 and 99 % identity with *B. subtilis* subsp. *Spizizenii* strain NRRL B-23049 with 100 % query coverage of 16S ribosomal RNA gene. Multiple sequence alignment using CLUSTAL-W was performed for the top scoring sequences

of BLAST results showing 95 % and above maximum indent with the query coverage of 97 %. The phylogenetic tree was obtained using .phy output file of CLUSTAL-W with the help of NJ-method of Phylip (Fig. 3).

On the basis of above characterizations, the isolate was identified as a new strain of *B. subtilis* and was deposited at MTCC, IMTECH, Chandigarh, as strain designation

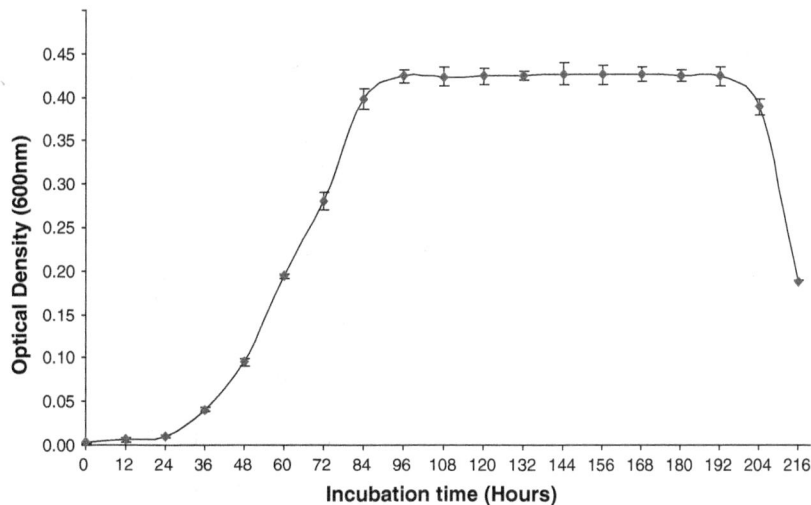

Fig. 2 Growth curve of isolate N2. The values are the mean of triplicate samples, and the *bar* indicates the standard error. The culture conditions include the NSM media with 50-ppm technical endosulfan as sole source of sulfur, pH 6.5, incubation temperature of 30 °C and rotator agitation of 130 rpm

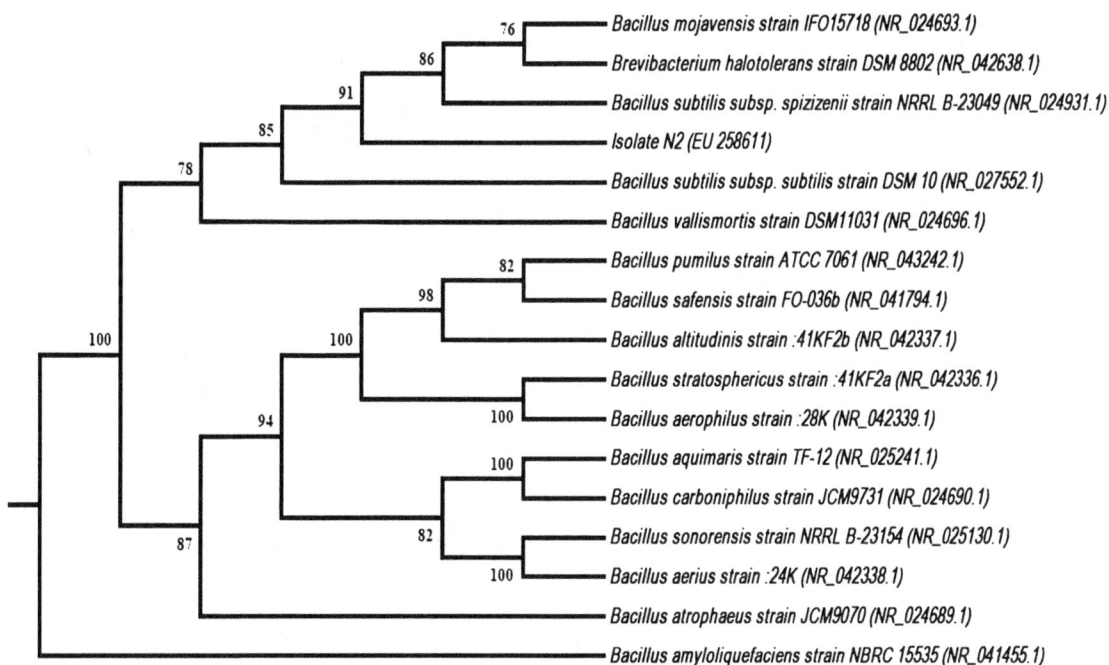

Fig. 3 Phylogenetic analysis of isolate N2. The tree is based on 16S rRNA gene sequences from top scoring homologous strains of *Bacillus* and closely related genera. GenBank accession numbers are indicated in the *parenthesis*. *B. Amyloliquefaciens* strains NBRC 15535 was used as the outgroup. Bootstrap values >70 % are given. The scale used for the distance-based tree was 0.01 substitutions per nucleotide position

AKPJ04 and accession number MTCC 8561, while the 16S rRNA gene sequence of the same was deposited and later published at Genbank with the accession number EU 258611.

The bacterial genus *Bacillus* has also been reported earlier to have endosulfan degradation ability. A bacterial co-culture consisting of two *Bacillus* sp. (MTCC 4444 and MTCC 4445) has been reported to degrade and reduce the toxicity of endosulfan, utilizing the pesticide as carbon source (Awasthi et al. 1997, 2003). The present isolate, *B. subtilis* MTCC 8561, is probably the first *Bacillus* sp. to be reported till date which can utilize endosulfan as sulfur source (Kataoka and Takagi 2013) and degrade it very efficiently up to 94 %. Other bacterial system reported to degrade endosulfan using it as sulfur source are *Alcaligenes faecalis* strain JBW4 (Kong et al. 2013), *Pseudomonas fluorescens* (Giri and Rai 2012), *Achromobacter xylosoxidans* strain C8B (Singh and Singh 2011), *Pseudomonas* and *Burkholderia* (Hussain et al. 2007), *Arthrobacter* (Weir et al. 2006), *Pandoraea* (Siddique et al. 2003), *Mycobacterium* (Sutherland et al. 2000) and *Micrococcus* (Guha et al. 1999). Degradation of endosulfan by bacterial consortia isolated from contaminated soil has also been recently reported (Bhattacharjee et al. 2013).

Primary localization of endosulfan degrading gene(s) of isolate N2

When plasmid isolation from the isolate N2 was attempted by alkaline lysis method, a smear of RNA was obtained when observed under UV-light after agarose gel electrophoresis, while the positive control (*E. coli* DH5α) showed two discrete plasmid bands of closed circular plasmid and nicked plasmids (Fig. 4a). While attempting plasmid isolation, using lysozyme–heat shock treatment method, a smear of RNA was obtained (N2$ lane of Figs. 4b). No band was observed under UV-light when the isolated nucleic acid was treated with RNase and electrophoretic run (Lane N2# of Fig. 4b). It may therefore be inferred that isolate N2 does not harbor any plasmid, as there were negative results of plasmid isolation against the positive control. The isolate N2 (*B. subtilis*) was found to be devoid of any plasmid. The plasmid extraction was carried out by alkaline lysis prep method and confirmed by lysozyme boiling—miniprep method. This observation suggests that the endosulfan degrading gene(s) is located on chromosomal DNA. This finding is in accordance with the reports about chromosome-located *esd*-gene of *Mycobacterium* (Sutherland et al. 2002) and *ese*-gene of *Arthrobacter* (Weir et al. 2006), which have been shown responsible for endosulfan degradation.

Conclusion

The findings of present investigations suggest that the isolate is probably the first *Bacillus* sp. reported till date, to use endosulfan as sulfur source and the extent of biodegradation of the pesticide is also very high as compared to other reported microbes. The major metabolite detected after degradation by the said isolate

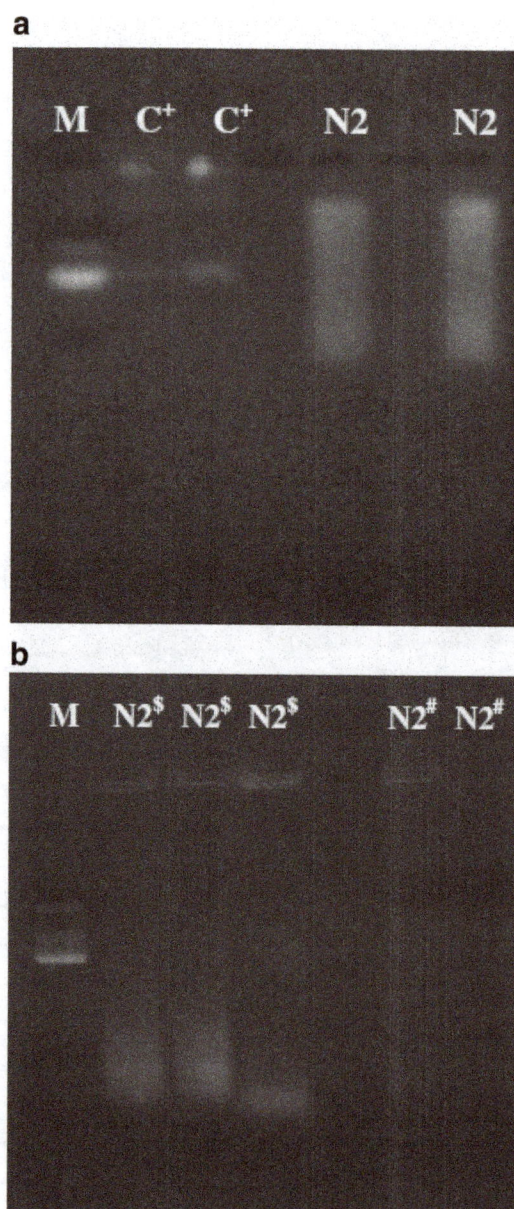

Fig. 4 **a** Plasmid isolation from isolate N2. Agarose gel showing plasmid isolation results of isolate N2 by alkaline lysis method. [*M* marker lane, *C+* positive control lane (*E. coli* DH5α), *N2* isolate N2 lane]. **b** Plasmid isolation from isolate N2. Agarose gel showing plasmid isolation results of isolate N2 by lysozyme treatment method. (*M* Marker lane, *N2$* isolate N2 lane without RNase treatment, *N2#* isolate N2 lane after RNase treatment)

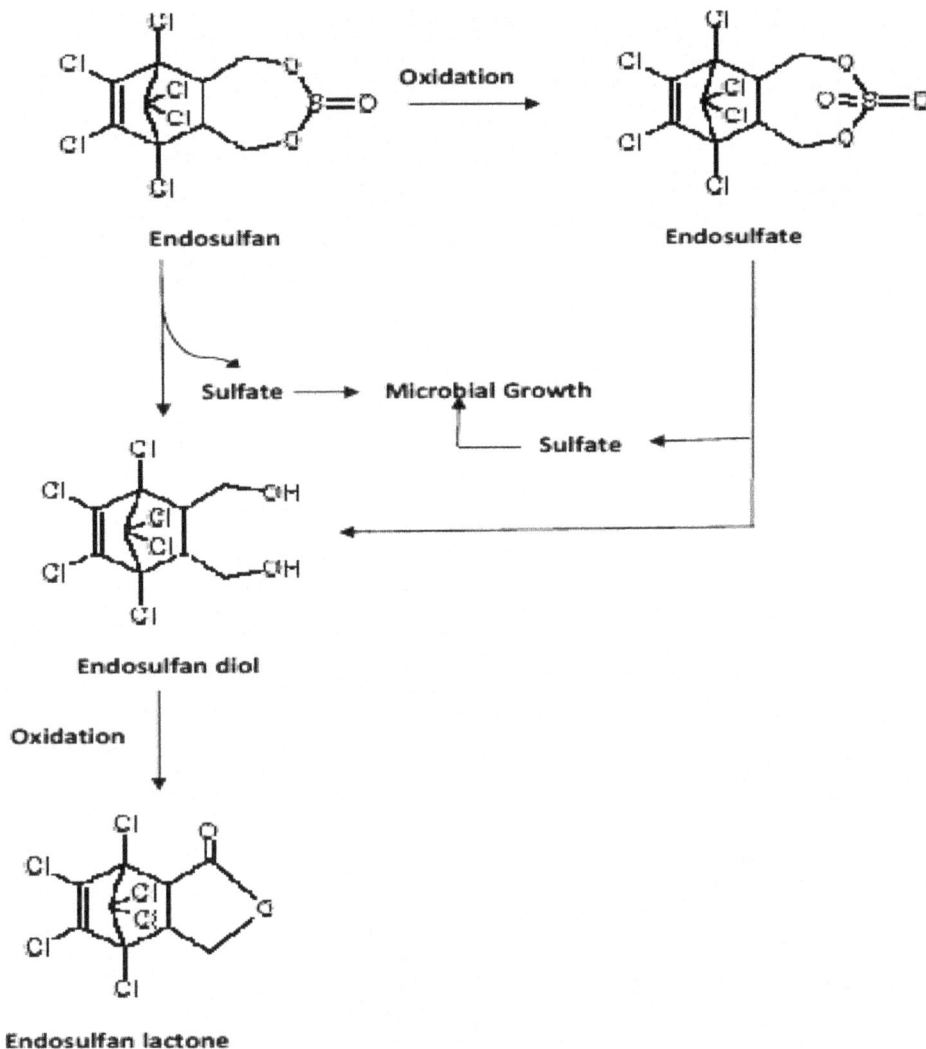

Fig. 5 Proposed pathway for metabolism of endosulfan by *Bacillus subtilis* MTCC 8561

(*B. subtilis*) is endosulfan diol along with endosulfan lactone and endosulfan sulfate detected as minor products. From the metabolites detected after degradation, it has been inferred that the bacterium metabolizes endosulfan and its stable but equally toxic, oxidized product, endosulfan sulfate by directly hydrolyzing it to endosulfan diol followed by oxidation to form endosulfan lactone. The high amount of endosulfan diol, detected after bacterial degradation of the pesticide, suggests that the bacterium is hydrolyzing the compounds to release sulfite group from endosulfan and sulfate group from endosulfan sulfate for using them as sulfur source in metabolism and growth. The proposed degradation pathway of endosulfan by *B. subtilis* AKPJ04 is as represented in Fig. 5. Both the major metabolites produced are non-toxic in nature and thus, the isolate holds promise to be a very good candidate for bioremediation of endosulfan. The present study paves a good platform for scaling and validating the results from shake flask level to soil study. The present work also beckons for the identification and characterization of the gene(s) and enzymes, responsible for endosulfan degradation, that are presently under our investigation.

Acknowledgments The authors wish to thank FASC, MITS, Lakshmangarh, Sikar, Rajasthan, India, for providing laboratory facilities for carrying out the present investigation and IMTECH, Chandigarh, India, for providing facility for 16S rRNA sequencing.

Conflict of interest The authors declare no financial or commercial conflict of interest.

References

Altshcul SF, Gish W, Miller W, Myers EW, Lipman DJ (1990) Basic local alignment search tool. J Mol Biol 215:403–410. (http://www.ncbi.nlm.nih.gov/BLAST/)

Ausubel FM, Brent R, Kingston RE, Moore DD, Seidman JG, Smith JA, Struhl K (eds) (2005) Short protocols in molecular biology: a compendium of methods from current protocols in molecular biology. John Wiley & Sons, New York

Awasthi N, Manickam N, Kumar A (1997) Biodegradation of endosulfan by a bacterial co-culture. Bull Eviron Contam Toxicol 59:928–934

Awasthi N, Singh AK, Jain RK, Khangarot BS, Kumar A (2003) Degradation and detoxification of endosulfan isomers by a defined co-culture of two *Bacillus* strains. Appl Microbiol Biotechnol 62:279–283

Bhattacharjee K, Banerjee S, Bawitlung L, Krishnappa D, Joshi SR (2013) A study on parameters optimization for degradation of endosulfan by bacterial consortia isolated from contaminated soil. Proc Natl Acad Sci India Sect B Biol Sci.

Cappuccino JG, Sherman N (2005) Microbiology A laboratory manual. In: 6th edn. Pearson Education Pvt. Ltd., Delhi

Chaudhari K, Selvaraj S, Pal AK (1999) Studies on the genotoxicity of endosulfan in bacterial system. Mutat Res 439:63–67

C.N.R.C. (1975) Endosulfan: its effects on environmental quality. In: NRC Associate committee on scientific criteria for environmental quality, Canada National Research (CNRC), Report No. 11, Subcommittee of pesticides on related compounds, Subcommittee Report No. 3, Publication No. NRCC 14098 of the Environmental Secretariat

Felsenstein J (2005) PHYLIP: (Phylogeny inference package) version 3.6. (http://evolution.genetics.washington.edu/phylip.html)

Gavrilescu M, Chisti Y (2005) Biotechnology- a sustainable alternative for chemical industry. Biotechnol Adv 23:3339–3348

Giri K, Rai JPN (2012) Biodegradation of endosulfan isomers in broth culture and soil microcosm by *Pseudomonas fluorescens* isolated from soil. Int J Environ Stud 69(5):729–742

Guha A, Kumari B, Bora TC, Roy MK (1999) Degradation of endosulfan by *Micrococcus* sp. and partial characterization of metabolites. Asian J Microbiol Biotech Environ Sci 1:29–32

Hussain S, Arshad M, Saleem M, Zahir ZA (2007) Biodegradation of α- and β-endosulfan by soil bacteria. Biodegradation 18(6):731–740

Kataoka R, Takagi K (2013) Biodegradability and biodegradation pathways of endosulfan and endosulfan sulphate. Appl Microbiol Biotechnol 97:3285–3292

Kong L, Zhu S, Zhu L, Xie H, Wei K, Yan T, Wang J, Wang J, Wang F, Sun F (2013) Colonization of *Alcaligenes faecalis* strain JBW4 in natural soils and its detoxification of endosulfan. Appl Microbiol Biotechnol

Kovacs MF (1965) Thin layer chromatography for pesticide residue analysis. J Assoc Off Agric Chem 97:982–988

Kullman SW, Matsumura F (1996) Metabolic pathways utilized by *Phanerochaete chrysosporium* for degradation of the cyclodiene pesticide endosulfan. Appl Environ Microbiol 62:593–600

Kumar A, John PJ, Soni I (2012) Enrichment and isolation of endosulfan degrading microorganism from natural resource. J Bio Environ Sci 2(6):41–53

Larkin MA, Blackshields G, Brown NP, Chenna R, McGettigan PA, McWilliam H, Valentin F, Wallace IM, Wilm A, Lopez R, Thompson JD, Gibson TJ, Higgins DG (2007) ClustalW and ClustalX version 2. Bioinformatics 23(21):2947–2948. (http://www.ebi.ac.uk/clustalW/)

Lee N, Sherrit JH, Mc Adam DP (1995) Hapten synthesis and development of ELISAs for the detection of endosulfan in water and soil. J Agric Food Chem 43:1730–1739

Sambrook J, Fritsch EF, Maniatis T (1989) Molecular cloning: a laboratory manual, 2nd edn. Cold Spring Harbor Laboratory Press, New York

Shetty PK, Mitra J, Murthy NBK, Namitha KK, Savitha KN, Raghu K (2000) Biodegradation of cyclodiene insecticide endosulfan by *Mucor thermohyalospora* MTCC 1384. Curr Sci 79:1381–1383

Siddique T, Okeke BC, Arshad M, Frankenberger WT Jr (2003) Enrichment and isolation of endosulfan-degrading microorganisms. J Environ Qual 32:47–54

Singh NS, Singh DK (2011) Biodegradation of endosulfan and endosulfan sulphate by *Achromobacter xylosoxidans* strain C8B in broth medium. Biodegradation 22:845–857

Sinha N, Narayan R, Saxena DK (1997) Effect of endosulfan on testis of growing rats. Bull Environ Contam Toxicol 58:79–86

Sunderam RIM, Cheng DMH, Thompson GB (1992) Toxicity of endosulfan to native and introduced fish in Australia. Environ Toxicol Chem 11:1469–1476

Sutherland TD, Horne I, Harcourt RL, Russell RJ, Oakeshott JG (2000) Enrichment of an endosulfan-degrading mixed bacterial culture. Appl Environ Microbiol 66:2822–2828

Sutherland TD, Weir KM, Lacey MJ, Horne I, Russell RJ, Oakeshott JG (2002) Enrichment of a microbial culture capable of degrading endosulfate, the toxic metabolite of endosulfan. J Appl Microbiol 92:541–548

USDHHS-United States Department of Health and Human Services (1990) Toxicological profile for endosulfan. Agency for Toxic Substances and Disease Registry, Atlanta

USEPA (2002) Re-registration eligibility decision (RED) fact sheet. United States Environmental Protection Agency (USEPA), EPA-738-F-02-012 for Endosulfan (Case 0014)

Weir KM, Sutherland TD, Horne I, Russell RJ, Oakeshott JG (2006) A single monooxygenase, *Ese*, is involved in the metabolism of organochlorides endosulfan and endosulfate in an *Arthrobacter* sp. Appl Environ Microbiol 72:3524–3530

Genbank: http://www.ncbi.nlm.nih.gov/Web/Genbank/

Yadav KPS (2003) Weight lent to endosulfan study discarded by officials. Down Earth 12:7

10

Phorbol ester degradation in Jatropha seedcake using white rot fungi

Anjali Bose · Haresh Keharia

Abstract White rot fungi are well known for their ability to degrade a wide range of xenobiotics due to their enzymatic systems. Therefore, the present investigation was aimed at screening ten different white rot fungi for degradation of phorbol esters from Jatropha seedcake (JSC). The JSC was fermented with pure cultures of white rot fungi for 20 days under solid state condition. All the white rot fungi tested exhibited degradation of phorbol ester during fermentation of JSC without adversely influencing the nutritional properties of the seedcake. *Ganoderma lucidum* and *Trametes zonata* were found to degrade phorbol ester in JSC to undetectable levels. This study demonstrates the potential of white rot fungi for degradation of phorbol esters, a major anti-nutritional factor, in JSC preventing its utilization as cattle feed.

Keywords Jatropha seedcake · White rot fungi · Phorbol ester

The bio-diesel production from *Jatropha curcas* L. gained momentum due to its inedible oil content that can be converted to fuel without competing with the food market. The extraction of oil from Jatropha seeds is associated with generation of substantial amount of seedcake waste at an average rate of 500 g cake per kg of seeds used (Zanzi et al. 2008). Inspite of its high protein content along with presence of all essential amino acids, except lysine (Makkar and Becker 1997b), it cannot be used in feed formulation due to the presence of potential anti-nutritional components like phorbol esters (PE), lectins and trypsin inhibitors (Makkar et al. 1997). The PEs, have been identified as main toxicants in JSC, which could not be destroyed even by heating at 160 °C for 30 min (Makkar et al. 1998) and, therefore, its removal is currently an important issue to be addressed. Several physico-chemical methods have been developed for PE removal, but none has proved to be economically feasible (Aregheore et al. 2003; Martínez-Herrera et al. 2006; Rakshit et al. 2008).

In this context, use of JSC as substrates for microbial fermentation would not only add to its utility, but during the process of fermentation there exists a possibility of degradation of anti-nutritional factors present in it, thereby solving its subsequent disposal issues. The feasibility of this approach has been demonstrated for reducing gossypol in cotton seed meal using *Geotrichum candidum* (Sun et al. 2008) and ricin in castor seedcake by *Paecilomyces variotii* (Madeira et al. 2011).

As white rot fungi are well known for their ability to degrade a wide range of xenobiotics, such as polyphenolic compounds and synthetic dyes, due to the secretion of extracellular enzymes (Asgher et al. 2008; Alberts et al. 2009), the present study was undertaken to investigate the ability of ten different white rot fungi, namely, *Ganoderma lucidum* (GL), *Pleurotus florida* (PF), *Pleurotus sapidus* (PS), *Pleurotus sajor-caju* (PSC), *Pleurotus ostreatus* (PO), *Phanerochaete chrysosporium* (PC), *Trametes hirsute* (TH), *Trametes zonata* (TZ), *Trametes gibbosa* (TG) and *Trametes versicolor* (TV) for degradation of PE in deoiled JSC. They were grown on 2 % (w/v) malt extract agar plates at 27 °C, preserved at 4 °C on malt extract agar slopes and maintained by subculturing once in 2 months.

A. Bose · H. Keharia (✉)
BRD School of Biosciences, Sardar Patel Maidan, Sardar Patel
University, Satellite Campus, Vadtal Road, P.O. Box 39,
Vallabh Vidyanagar 388 120, Gujarat, India
e-mail: haresh970@gmail.com

Twenty-five grams of JSC (obtained from Food Processing and Bioenergy division, Anand Agriculture University, Gujarat) was taken in 250-mL Erlenmeyer flask, moistened with 30 mL of distilled water and autoclaved. The flasks were inoculated with two blocks (1 cm × 1 cm) of actively growing individual fungal culture, followed by incubation at 30 °C and 70 % relative humidity. An uninoculated flask served as experimental control. After 20 days of fermentation, the content of the flasks were extracted and analyzed for PE and nutrients. All the experiments were done in triplicates.

Phorbol esters were extracted by following the method described by Joshi et al. (2011) and quantified using C-18 reverse-phase HPLC employing a Luna 18 column (250 × 4.6 mm, octadecyl group, particle size 5 μm) procured from Phenomenex (USA). The separation was carried out with the solvent system: water and acetonitrile (40 % acetonitrile for 15 min followed by gradient of 40–75 % acetonitrile for 20 min and then to 100 % acetonitrile for 5 min and finally returned to 40 % for the next 5 min) with a flow rate 1.3 mL/min at 25 °C. The detector (Photo Diode Array) wavelength was set on 280 nm and 5 μL of sample was injected for analysis.

Normally, *Jatropha* sp. is reported to have four to six PEs or its derivatives (Haas et al. 2002; Barros et al. 2011), out of which, phorbol-12-myristate 13-acetate (PMA) is the major PE in *J. curcas* (Makkar and Becker 1997a). However, we observed three major peaks at 25.1, 25.26 and 25.6 min (Fig. 1a), which appeared at a lower retention time as compared to external standard PMA (Sigma Chemical Co., USA) dissolved in absolute methanol that appeared at 33.09 min. The type and quantity of individual PEs in Jatropha seed depends on the genotype of plant and the prevailing soil/climatic conditions (Martínez-Herrera et al. 2006). Makkar et al. (1997) reported varying concentration of PE from 0.8 to 3.3 mg/g in *J. curcas* kernel meal from different geographical sites. Ahmed and Salimon (2009) analyzed three different varieties of tropical *J. curcas* from Malaysia, Indonesia and India and observed two, five and four PE peaks, respectively.

The fermentation of JSC with edible white rot fungi lowered PEs content from 1.072 mg/g in unfermented control up to undetectable level in fermented seedcake (Fig. 1 and supplementary figure). The degradation of PE was observed by all the white rot fungi tested but the extent of degradation varied with the fungal culture (Fig. 1b). *G. lucidum* and *T. zonata* were found to completely degrade PEs in JSC. Trace amounts of PEs could be detected in JSC upon fermentation by *T. versicolor* and *T. gibbosa*, whereas *Pleurotus* sp. degraded ~70 % of initial PE present. However, *Ph. chrysosporium* could only reduce PE content up to ~45 %. The extent of reduction in PE content by *G. lucidum*, *T. zonata*, *T. versicolor* and *T. gibbosa* was

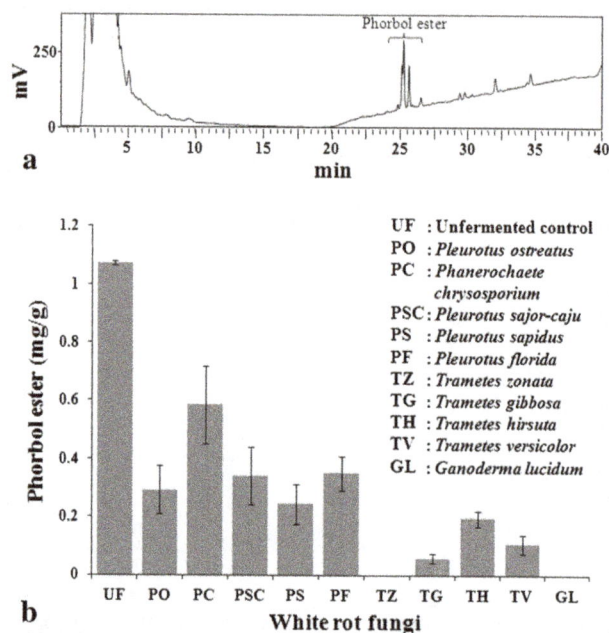

Fig. 1 a HPLC chromatogram of phorbol esters from unfermented Jatropha seedcake and **b** Effect of fungal treatment on phorbol ester concentrations

comparable to the reports by Barros et al. (2011) and Belewu and Sam (2010).

Further the nutritive value was evaluated using the protocols of Indrayan et al. (2005) with few modifications. The analysis involved the determination of moisture content by drying samples at 105 °C to constant weight. Ash content and the total organic matter were estimated by determining the loss in weight after igniting the samples in a muffle furnace at 550 °C for 12 h. Dry fermented seedcake was digested using concentrated H_2SO_4 followed by distillation with NaOH in Kel Plus Nitrogen estimation system (Classic DX, Pelican Equipments) prior to determination of total nitrogen. Total nitrogen was estimated through titration of the distillate collected from the Kel Plus distillation unit against 0.1 (N) H_2SO_4. The total protein content was determined by multiplying total Kjeldahl nitrogen with 6.25. Total fat was determined by extracting 2 g sample with hexane in a Soxhlet extractor for 6 h. The hexane extract was then evaporated and residue left was weighed to determine total fat. The total carbohydrate and nutritive value was then calculated using following equations:

Carbohydrate (%) was given by: 100
− (Ash % + Fat % + Protein %).   (1)

Nutritive value was determined by: 4 × Protein % + 9 × Fat % + 4 × Carbohydrate %.

(2)

To determine total *P* and *K*, the samples were first digested with a mixture of nitric acid and perchloric acid

Table 1 Nutritive composition of unfermented (control) and fermented Jatropha seedcake

Treatments	Protein (%)	Fat (%)	Ash (%)	P (ppm)	K (ppm)	Moisture (%)	Carbohydrate (%)	Nutritive value
UF	23.33 ± 0.48	0.65	5.83 ± 0.17	56	3,250	58.3 ± 0.28	11.99	146.73
PO	25.92 ± 0.06	0.7	7.55 ± 0.16	189	4,830	55.7 ± 0.41	10.13	150.5
PC	25.26 ± 0.25	0.65	7.89 ± 0.31	175	6,350	51.57 ± 0.22	14.63	165.41
PSC	26.44 ± 0.3	0.7	7.64 ± 0.18	168	6,650	53.95 ± 0.48	11.27	157.14
PS	30.15 ± 0.39	0.65	7.47 ± 0.01	189	6,750	54.73 ± 0.75	7	154.45
PF	25.33 ± 0.25	0.7	7.63 ± 0.58	189	4,730	53.45 ± 0.15	12.89	159.18
TZ	27.33 ± 0.12	0.6	6.98 ± 0.3	182	6,300	56.65 ± 0.19	8.44	148.48
TG	28.05 ± 0.15	0.5	7.43 ± 0	161	5,230	55.48 ± 0.31	8.54	150.86
TH	29.46 ± 0.63	0.55	7.28 ± 0.16	175	2,550	53.64 ± 0.43	9.07	159.07
TV	29.39 ± 0.07	0.55	7.74 ± 0.2	175	5,830	52.44 ± 0.72	9.88	162.03
GL	28.07 ± 0.12	0.5	8.24 ± 0.18	175	6,680	54.86 ± 0.05	8.33	150.1

UF unfermented Jatropha seedcake as experimental control, *PO Pleurotus ostreatus, PC Phanerochaete chrysosporium, PSC Pleurotus sajor-caju, PS Pleurotus sapidus, PF Pleurotus florida, TZ Trametes zonata, TG Trametes gibbosa, TH Trametes hirsuta, TV Trametes versicolor* and *GL Ganoderma lucidum* fermented Jatropha seedcake

(9:4). Phosphorus in the digestate was estimated through formation of vanado-molybdo-phosphoric–heteropoly complex followed by absorbance measurement at 420 nm (Gupta 2004). The potassium was measured in the digestate using flame photometer (Singh et al. 2007).

The results of proximate composition of fungal fermented and unfermented JSC (Table 1) exhibited slight increase in protein content (2.03–6.92 %). The increment in the protein content could be due to the addition of microbial protein during the process of fermentation. Ash content was found to increase significantly in JSC upon fermentation by all the white rot fungi tested. The increase in ash content of fermented seedcake may be considered as an indicator of mineralization (Table 1). Similar observation regarding increase in ash content has been reported for cotton waste upon fermentation by *Volvariella volvacea* (Akinyele and Akinyosoye 2005). Further, the fat content remained almost unchanged in most of the treatments, however, 4–9 % decrease in carbohydrate content was observed. Such decrease in carbohydrate content may be attributed to its utilization as nutrient source during fungal colonization. In present study, the overall nutritive value of the fermented seedcake remained almost unchanged. However, increase in nutritive value of agrowastes upon fermentation by fungi has been reported by Akinyele and Akinyosoye (2005).

Thus, the present investigation clearly demonstrated that solid state fermentation of JSC by white rot fungi could totally remove PE content and could be applied for large scale detoxification. Apart from this, the fermented seedcake would then retain high protein content and other nutritional values applicable to the animal feed industry.

Acknowledgments A. Bose acknowledges the University Grants Commission, New Delhi for financial support in form of meritorious fellowship. Authors thank Central Salts and Mineral Chemicals Research Institute, Bhavnagar for providing the analytical facility toward HPLC analysis.

Conflict of interest The authors declare that they have no conflict of interest.

References

Ahmed WA, Salimon J (2009) Phorbol esters as constituents of tropical *Jatropha curcas* seed oil. Eur J Sci Res 31:429–436

Akinyele BJ, Akinyosoye FA (2005) Effect of *Volvariella volvacea* cultivation on chemical composition of agrowastes. Afr J Biotechnol 4:979–983

Alberts JF, Gelderblom WCA, Botha A, van Zyl WH (2009) Degradation of aflatoxin B1 by fungal laccase enzymes. Int J Food Microbiol 135:47–52

Aregheore EM, Becker K, Makkar HPS (2003) Detoxification of a toxic variety of *Jatropha curcas* using heat and chemical treatments, and preliminary nutritional evaluation with rats. S Pac J Nat Sci 21:50–56

Asgher M, Bhatti HN, Ashraf M, Legge RL (2008) Recent developments in biodegradation of industrial pollutants by white rot fungi and their enzyme system. Biodegradation 19:771–783

Barros CRM, Ferreira LMM, Nunes FM, Bezerra RFM, Dias AA, Guedes CV, Cone JW, Marques GSM, Rodrigues MAM (2011) The potential of white-rot fungi to degrade phorbol esters of *Jatropha curcas* L. seed cake. Eng Life Sci 11:107–110

Belewu MA, Sam R (2010) Solid state fermentation of *Jatropha curcas* kernel cake: proximate composition and antinutritional components. J Yeast Fungal Res 1:44–46

Gupta PK (2004) Soil, plant, water and fertilizer analysis. Agrobios, India

Haas W, Sterk H, Mittelbach M (2002) Novel 12-deoxy-16-hydroxyphorbol diesters isolated from the seed oil of *Jatropha curcas*. J Nat Prod 65:1334–1440

Indrayan AK, Sharma S, Durgapal D, Kumar N, Kumar M (2005) Determination of nutritive value and analysis of mineral elements for some medicinally valued plants from Uttaranchal. Current Sci 89:1252–1255

Joshi C, Mathur P, Khare SK (2011) Degradation of phorbol esters by *Pseudomonas aeruginosa* PseA during solid-state fermentation of deoiled *Jatropha curcas* seed cake. Bioresour Technol 102:4815–4819

Madeira JV Jr, Macedo JA, Macedo GA (2011) Detoxification of castor bean residues and the simultaneous production of tannase and phytase by solid-state fermentation using *Paecilomyces variotii*. Bioresour Technol 102:7343–7348

Makkar HPS, Becker K (1997a) *Jatropha curcas* toxicity: identification of toxic principles. In: Proceedings of the 5th international symposium on poisonous plants. San Angelo, 19–23 May

Makkar HPS, Becker K (1997b) Potential of Jatropha seedcake as a protein supplement in livestock feed and constraints to its utilization. In: Proceedings of the Jatropha 97: international symposium on biofuel and industrial products from *Jatropha curcas* and other tropical oil seed plants. Managua, Nicaragua, 23–27 February

Makkar HPS, Becker K, Sporer F, Wink M (1997) Studies on nutritive potential and toxic constituents of different provenances of *Jatropha curcas*. J Agric Food Chem 45:3152–3157

Makkar HPS, Aderibigbe AO, Becker K (1998) Comparative evaluation of nontoxic and toxic varieties of *Jatropha curcas* for chemical composition, digestibility, protein degradability and toxic factors. Food Chem 62:207–215

Martínez-Herrera J, Siddhuraju P, Francis G, Dávila-Ortíz G, Becker K (2006) Chemical composition, toxic/antimetabolic constituents, and effects of different treatments on their levels, in four provenances of *Jatropha curcas* L. from Mexico. Food Chem 96:80–89

Rakshit KD, Darukeshwara J, Raj KR, Narasimhamurthy K, Saibaba P, Bhagya S (2008) Toxicity studies of detoxified Jatropha meal (*Jatropha curcas*) in rats. Food Chem Toxicol 46:3621–3625

Singh D, Chhonkar PK, Dwivedi BS (2007) Manual on soil, plant and water analysis. Westville Publishing House, New Delhi

Sun ZT, Liu C, Du JH (2008) Optimisation of fermentation medium for the detoxification of free gossypol in cottonseed powder by *Geotrichum candidum* G07 in solid-state fermentation with response surface methodology. Ann Microbiol 58:683–690

Zanzi R, Perez JAS, Soler PB (2008) Production of biodiesel from *Jatropha curcas* in the region of Guantanamo in Cuba. In: Proceedings of the 3rd international congress university-industry cooperation, Ubatuba, Brazil

11

In silico study on Penicillin derivatives and Cephalosporins for upper respiratory tract bacterial pathogens

K. M. Kumar · P. Anitha · V. Sivasakthi ·
Susmita Bag · P. Lavanya · Anand Anbarasu ·
Sudha Ramaiah

Abstract Upper respiratory tract infection (URTI) is an acute infection which involves the upper respiratory tract: nose, sinuses, tonsils and pharynx. URT infections are caused mainly by pathogenic bacteria like *Streptococcus pneumoniae, Haemophilus influenzae* and *Staphylococcus aureus*. Conventionally, β-lactam antibiotics are used to treat URT infections. Penicillin binding proteins (PBPs) catalyze the cell wall synthesis in bacteria. β-Lactam antibiotics like Penicillin, Cephalosporins, Carbapenems and Monobactams inhibit bacterial cell wall synthesis by binding with PBPs. Pathogenic bacteria have efficiently evolved to resist these β-lactam antibiotics. New generation antibiotics are capable of inhibiting the action of PBP due to its new and peculiar structure. New generation antibiotics and Penicillin derivatives are selected in this study and virtually compared on the basis of interaction studies. 3-Dimensional (3D) interaction studies between Lactivicin, Cefuroxime, Cefadroxil, Ceftaroline, Ceftobiprole and Penicillin derivatives with PBPs of the above-mentioned bacteria are carried out. The aim of this study was to suggest a potent new generation molecule for further modification to increase the efficacy of the drug for the URTI.

Keywords Upper respiratory tract infections ·
Penicillin binding proteins · β-Lactam antibiotics ·
Docking

K. M. Kumar · P. Anitha · V. Sivasakthi · S. Bag ·
P. Lavanya · A. Anbarasu · S. Ramaiah (✉)
School of Biosciences and Technology, VIT University,
Vellore 632014, Tamil Nadu, India
e-mail: sudhaanand@vit.ac.in

Introduction

The respiratory tract is a frequent site of infection because it comes in direct contact with the physical environment and is exposed to airborne microorganisms. Worldwide, approximately 4 million children under 5 years of age die each year from respiratory tract infections (RTIs) (Garenne et al. 1992). It is estimated that throughout the world 1.9 million children <5 years old died from acute respiratory infection in 2001, 70 % of them in Africa and South East Asia (Williams et al. 2001). Nasopharyngitis, pharyngitis, tonsillitis and otitis media are common upper respiratory tract (URT) infections which constitute 87.5 % of the total episodes of respiratory infections. URT infections can be caused by a variety of bacteria like *Chlamydia pneumoniae, Mycoplasma pneumoniae, Streptococcus pyogenes, Streptococcus pneumoniae, Bordetella pertussis, Staphylococcus aureus, Escherichia coli* and *Haemophilus influenzae* (Peter et al. 1985). The majority of URT infections are caused by only three species *S. pneumoniae, S. aureus* (Gram-positive bacteria) and *H. influenzae* (Gram-negative bacteria). The treatments of these three bacterial infections have been more complicated by the emergence and spread of multi-drug resistant strains (Doern et al. 1988, 1997). Two mechanisms have been reported to be responsible for antibiotic resistance: structural modification in Penicillin binding protein (PBP) targets and production of β-lactamase, first identified in 1972 (Williams and Moosdeen 1986; Reid et al. 1987; Jorgensen 1992). PBPs are the membrane bound enzymes which catalyze the steps involved in bacterial cell wall biosynthesis and are the target enzymes of β-lactam antibiotics (Ghuysen 1991; Goffin and Ghuysen 1998; Macheboeuf et al. 2006; Sauvage et al. 2008). Peptidoglycan is the major component of bacterial cell wall synthesized by PBPs. Every bacterial species has more than two PBPs.

S. pneumoniae, the major human pathogen causing URT infections is responsible for over 1.6 million deaths every year (Lynch and Zhanel 2005). It has six PBPs, PBP1a, PBP1b, PBP2a, PBP2b, PBP2x and PBP3, which are highly conserved. Penicillin resistance in *S. pneumoniae* has been reported in many countries. The mechanism of Penicillin resistance is due to the modification of active site motif in PBPs of *S. pneumoniae*. Penicillins and extended spectrum Cephalosporins have high level of resistance to PBP1a, PBP2x and PBP2b of *S. pneumoniae* (Sheldon and Mason 1998). *S. aureus* is a potent pathogen that can cause respiratory tract infections (Ragle et al. 2010). It has PBP1b, PBP2a and PBP3. The resistance of *S. aureus* to Penicillin was identified in 1940 and 1965, but recently it has become a major threat to public health concern (Metan et al. 2005), alteration in PBP2a encoded gene decreases the affinity of most β-lactam antibiotics. *H. influenzae* is a common and exclusively human commensal of the nasopharynx. *H. influenzae* colonizes in the nasal cavity of approximately 80 % of the human population. *H. influenzae* has PBP4 and PBP5 which are low molecular weight proteins. The treatment of *H. influenzae* infections has been more complicated by the emergence and spread of multi-drug resistant strains (Doern et al. 1988, 1997). Several computational investigations have been done on β-lactam antibiotics and PBPs. Yoshida et al. reported the crystal structures of PBP3 in methicillin-resistant *S. aureus* (MRSA) and nature of its interactions with Cefotaxime. The study explains in detail about the hydrophobic and hydrogen bond interaction of Cefotaxime with the active sites of the PBP3 and PBP2 of *S. aureus*. Experimentally they proved it with nanoelectrospray mass spectrometry and ultracentrifugation to measure its sensitivity to different types of Penicillin derivatives (Yoshida et al. 2012). Samo Turk et al. study mainly focused to discover non-covalent inhibitor for PBP2x and PBP2a experimentally and computationally. The study reported the minimum inhibitory concentration of non-covalent inhibitor against several Gram-positive bacterial strains, including MRSA and analyzed the binding affinity of inhibitor with PBP2a and PBP2x (Turk et al. 2011). Another computational study investigated the interaction of Carbenicillin, Ceftazidime and Cefotaxime with binding site of PBP1b and PBP3 (Sainsbury et al. 2011). Sainsbury et al. reported the crystal structures of apo-PBP and complexes with Ceftazidime and Carbenicillin and investigated the similarities and differences between these structures. Fumihiro Kawai et al. determined the high-resolution apo crystal structures of two-low molecular weight PBPs, PBP4 and PBP5 from *H. influenzae*. They demonstrated the binding affinity of designed β-lactam antibiotics and Amoxicillin with PBP4 and PBP5 (Kawai et al. 2010). Though Penicillin derivatives and Cephalosporins have been used for bacterial infections over a period of time, many bacterial pathogens have become

resistant to these antibiotics. One major mode of resistance is by the alternation of PBPs resulting in low affinity to β-lactam antibiotics. Researchers have explored the mechanism of resistance to β-lactam antibiotics using only a few Penicillin derivatives or Cephalosporins (Sainsbury et al. 2011; Turk et al. 2011; Yoshida et al. 2012). This prompted us to investigate in detail using a wide spectrum of β-lactam antibiotics (both Penicillin derivatives and Cephalosporins). Our results indicate that of 19 β-lactam antibiotics, Ceftobiprole and Ceftaroline might have better affinity to PBPs and hence it may be effective in the treatment of URT bacterial infections. Our results are also comparable to previous experimental findings (Hebeisen et al. 2001; Sader et al. 2005; Jones et al. 2005; Kosowska et al. 2005; Davies et al. 2006; Citron and Goldstein 2008; Estrada et al. 2008; Henry et al. 2010; Kosowska et al. 2010; Mosian et al. 2010; Dauner et al. 2010) and the findings of our research might provide clues as to how Ceftobiprole and Ceftaroline exert their inhibitory action on bacterial pathogens.

Methods

Preparation of macromolecular and small molecular models

PBP was thought to be essential for the synthesis of bacterial cell wall. All types of the PBPs (PBP1a, PBP1b, PBP2a, PBP2b, PBP2x, PBP3, PBP4, PBP5 and PBP6) were selected for this study. 3-Dimensional (3D) structures of the PBPs were obtained from Protein Data Bank (PDB) (Berman et al. 2000). 3D structures of PBPs were visualized through PyMOL viewer (Lill and Danielson 2010). Co-crystallized ligands were identified and removed from the target proteins then water molecules removed and H atoms were added to the structure and minimizations were performed using Swiss pdb viewer (Guex and Peitsch 1997). The 3D coordinates of the Penicillin derivatives and Cephalosporins were obtained from NCBI PubChem Compound database (Li et al. 2010) and constructed using chemsketch (Li et al. 2004). Hydrogen atoms were added to all the structures and gasteiger atomic partial charges were computed. A geometry optimization of all the compounds was performed using chimera (Pettersen et al. 2004) for flexible conformations of the compounds during the docking.

PDB ID of every PBP was depicted in Table 1 and two-dimensional structures of Penicillin derivatives and Cephalosporins are shown in Fig. 1.

Active site identification

The catalytic binding site was believed to be a small region, a cleft or pocket, where lead molecules can bind to

Table 1 Active site residues of PBPs

PBPs	PDB ID	Name of the organism	Active site residues
PBP1a	2C6W	*Streptococcus pneumoniae*	Ala270, Tyr271, Asp273, Asn274, Trp311, Asn315, Leu345, Gly346, Ala347, Arg348, His349, Hln350, Ser351
PBP1b	2Y2Q	*Staphylococcus aureus*	Asp337, Phe341, Thr342, Ala345, Glu346, Glu349, Tyr443, Gln447, Asn448, Asn449, Phe452, Asp453, Glu540
PBP2a	1VQQ	*Staphylococcus aureus*	Ser403, Lys406, Arg445, Tyr446, Glu447, Ile459, Glu460, Ser403, Ser462, Asp463, Asn464
PBP2b	2WAE	*Streptococcus pneumoniae*	Thr55, Thr56, Ser57, Ser81, Gln180, Ala183, Val184, Gly185, Ala188, Thr189, Gly190, Thr191, Ser218, Ser258, Leu259, Asn260, Asp261, Arg 262, Arg280
PBP2x	1PYY	*Streptococcus pneumoniae*	Lys420, Val423, Pro424, Thr425, Arg426, Arg463, Glu476, Glu497, Ile498, Val499, Gly500, Ala650, Arg654, Pro660, Ile661, Val662, Gly664
PBP3	3OC2	*Streptococcus pneumoniae*	Ala162, His163, Gly166, Phe167, Arg175, Glu176, Gly177, Leu180, Tyr268, Pro278, Met281, Arg282, Asn283, Met286, Ile287, Phe383, Pro384, Gly385, Glu386, Arg387
PBP4	1TVF	*Staphylococcus aureus*	Gln133, Val136, Ser137, Asn138, Ser139, Phe225, Phe225, Thr226, Lys227, Gln228, Tyr239, Thr240, Phe241, Asn242, Leu245, Leu258, Lys259, Thr260
PBP5	3A3J	*Haemophilus influenzae*	Val75, Val77, Leu79, Lys80, Asn86, Asn121, Asp193, Leu194, Leu194, Pro195, Glu196, Glu197, Ile200
PBP6	3ITB	*Escherichia coli*	Ser40, Ile103, Ile104, Gln105, Ser106, Pro192, Asn193, Arg194, Asn195, Met208, Lys209, Thr210, Gly211, Thr212

stimulate the target protein and produce the desirable effect. Thus, recognizing the catalytic binding site residues in the protein structure was of high importance in computer-aided drug designing. Identification of accurate catalytic binding site was difficult because the target proteins were capable of undergoing conformational changes (Liao and Andrews 2007). Qsite finder (Laurie and Jackson 2005) recognizes the possible ligand binding sites using the van der Waal's probes and interaction energy. In the present study, Qsite finder was employed for locating the active sites in PBP1a, PBP1b, PBP2a, PBP2b, PBP2x, PBP3, PBP4, PBP5 and PBP6 proteins.

Virtual screening of β-lactam antibiotics

iGEMDOCK (A Generic Evolutionary Method for molecular DOCKing) automated docking program (Yang and Chen 2004). iGEMDOCK integrated the structure-based virtual screening, molecular docking, post screening analysis and visualization steps. We selected all types of PBPs (PBP1a, PBP1b, PBP2a, PBP2b, PBP2x, PBP3, PBP4, PBP5 and PBP6) to carry out the structure-based virtual screening study of penicillin derivatives and Cephalosporins. The 3D coordinates of each therapeutic target protein and ligand molecules were implemented through the GEMDOCK graphical environment interface. Before docking, the output path was set. GEMDOCK default parameters included the population size ($n = 200$), generation ($g = 70$) and number of solutions ($s = 10$) to compute the probable ligand binding mechanism for each target protein. Then the docking run was

started using GEMDOCK scoring function. After docking, the individual binding pose of each ligand was observed and their binding affinity with the target proteins was analyzed. In the post docking screening the best binding pose and total energy of each ligand was analyzed. The details of best binding pose and total energy values were saved in output folder. Protein–ligand binding site was analyzed and visualized using PyMOL (Lill and Danielson 2010).

Docking

The automated docking studies were carried out using Auto-Dock version 4.0 (Morris et al. 2009). 3D structure of each PBPs were implemented through the graphical user interface AUTODOCKTOOLS (ADT 1.4.6). The graphical user interface AUTODOCKTOOLS was performed to set up the enzymes: all hydrogens were added, Kollman United Atoms charges loaded and non-polar hydrogens were merged to carbon atoms. The initial parameters and van der Waals well depth of 0.100 kcal/mol for macromolecules, generated PDBQT files were saved. The 3D structures of ligand molecules were constructed, optimized, and converted into Mol2 file format with the help of the chimera. The charges of the non-polar hydrogen atoms are assigned to the atom to which the hydrogen is attached. The resulting files were saved as PDBQT files. The drug binding site for the ligands on PBP1a, PBP1b, PBP2a, PBP2b, PBP2x, PBP3, PBP4, PBP5 and PBP6 were identified using Qsite finder online server. The grid point was set at the ligand binding site in each one of the obtained

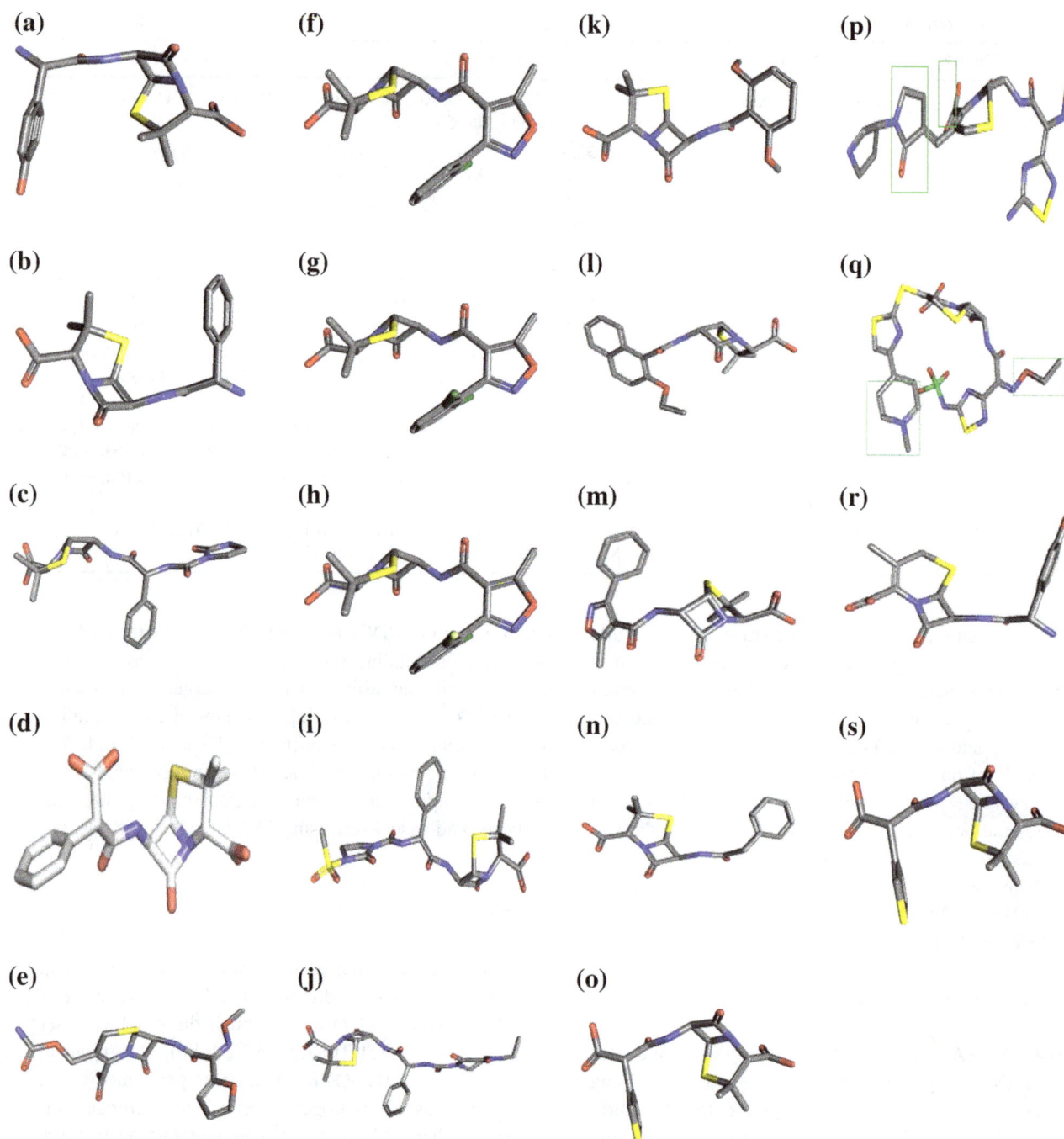

Fig. 1 3-Dimensional structures of Penicillin derivatives and Cephalosporins: **a** Amoxicillin, **b** Ampicillin, **c** Azlocillin, **d** Carbenicillin, **e** Cefuroxime, **f** Cloxacillin, **g** Dicloxacillin, **h** Flucloxacillin, **i** Mezlocillin, **j** Piperacillin, **k** Methicillin, **l** Nafcillin, **m** Oxacillin, **n** Penicillin G, **o** Ticarcillin, **p** Ceftobiprole, **q** Ceftaroline, **r** Cefadroxil and **s** Lactivicin (The *highlighted boxes* indicate the non-essential components in **p** Ceftobiprole and **q** Ceftaroline respectively)

PDB structures. AUTODOCK 4.0 was performed for all docking calculations. The AUTODOCKTOOLS was used to generate the grid parameter files and docking parameter files. The docking parameters were also used to calculate docking scores for β-lactam antibiotics and Penicillin derivatives. Protein–ligand docking calculations were carried out on PBPs. Lamarckian genetic algorithm (Morris et al. 1998) was used to generate possible protein–ligand binding conformations.

ADME screening

The molinspiration (Jarrahpour et al. 2011) server was used to predict the ADME properties of the antibiotics. It predicted both physiochemical and pharmacological properties. Smiles (Simplified Molecule Input Line Entry Specification) of the antibiotics was submitted. It predicted the properties of the drug such as molecular volume, number of hydrogen bond donors and acceptors, LogP and rotatable bonds. It provided high-speed molecular properties calculated and drug likeness for a given compound. The acceptability of the analogs is evaluated based on Lipinski's rule of 5 (Lipinski et al. 2006), which is essential for structure-based drug design.

Results and discussion

The 3D structures of PBP1a, PBP1b, PBP2a, PBP2b, PBP2x, PBP3, PBP4, PBP5 and PBP6 are analyzed and 19 β-lactam antibiotics are optimized to have minimal potential energy using chimera and then the virtual screening study is carried out for ligand molecules. From the virtual screening analysis, we list binding mode of Penicillin derivatives and Cephalosporins based on total energy (Table 2). The best binding poses for each ligand molecule into each target protein are determined and the one having lowest binding energy among the different poses generated. The lower energy scores represent better protein–ligand binding affinity compared to higher energy values. Among the 19 ligands, Cephalosporins are found to have lower binding energy value than the Penicillin

derivatives. Especially the fifth generation Cephalosporins, Ceftaroline and Ceftobiprole has least binding energy value. Ceftobiprole shows best binding pose with PBP1b, PBP2a, PBP2b and PBP2x (total energy value for PBP1b = −110.7 kcal/mol, PBP2a = −108.2 kcal/mol, PBP2b = −110.4 kcal/mol, PBP2x = −116 kcal/mol). The Ceftaroline shows best binding conformation with PBP3, PBP4, PBP5 and PBP6 (total energy for PBP3 = −114 kcal/mol, PBP4 = −104.8 kcal/mol, PBP5 = −131.2 kcal/mol and PBP6 = −118.0 kcal/mol). On comparing the binding mode of Penicillin derivative, Azlocillin shows higher binding affinity with PBP1a (total energy value = −122.1 kcal/mol). These compounds have more stable ligand–receptor complex amongst other compounds. We further analyzed the docked conformation for finding the binding mode of fifth generations Cephalosporins, Ceftaroline and Ceftobiprole into selected target proteins to validate the position obtained likely to represent reasonable binding modes or conformations.

Docking of Ceftobiprole into PBPs

Docking simulation of Ceftobiprole is performed for PBP1a, PBP1b, PBP2a, PBP2b, PBP2x, PBP3, PBP4, PBP5 and PBP6. From the docking result, we identified that Ceftobiprole has best binding affinity with the PBP2x of *S. aureus*. Docking of Ceftobiprole results in the formation of more than five hydrogen bonds with PBP1b,

Table 2 Virtual screening results of β-lactam antibiotics by iGEMDOCK

S. no	#Ligand	PBP-1A	PBP-1B	PBP-2A	PBP-2B	PBP-2X	PBP-3	PBP-4	PBP-5	PBP-6
1	Amoxicillin	−121.1	−103.7	−89.8	−88.2	−84.1	−94.5	−67.6	−98.7	−79.9
2	Ampicillin	−89.3	−69.3	−87.3	−76.1	−82.1	−86.0	−64.8	−94.4	−84.4
3	Azlocillin	**−122.1**	−83.3	−99.8	−93.9	−84.4	−100.4	−72.9	−91.4	−85.0
4	Carbenicillin	−86.6	−74.9	−91.8	−84.7	−97.6	−98.2	−75.5	−107.6	−90.4
5	Cefadroxil	−107.5	−87.4	−90.1	−100.3	−101.7	−112.9	−72.3	−88.6	−83.4
6	Ceftobiprole	−104.6	**−110.7**	**−108.2**	**−110.4**	**−116.0**	−113.0	−83.2	−113.0	−102.0
7	Ceftaroline	−104.3	−97.1	−79.4	−97.0	−104.4	**−114.1**	**−104.8**	**−131.2**	**−118.0**
8	Cefuroxime	−103.4	−104.0	−94.6	−110.3	−94.3	−86.1	−78.2	−91.0	−80.0
9	Cloxacillin	−91.4	−90.9	−83.3	−89.4	−82.0	−89.4	−69.0	−94.3	−82.0
10	Dicloxacillin	−95.9	−83.9	−85.7	−82.6	−99.1	−97.1	−62.7	−85.2	−84.0
11	Flucloxacillin	−89.6	−75.4	−96.6	−88.9	−74.9	−89.8	−62.1	−90.8	−82.1
12	Lactivicin	−91.1	−95.9	−87.1	−90.9	−95.8	−95.3	−65.6	−98.1	−89.0
13	Methicillin	−95.5	−102.1	−102.0	−97.0	−92.3	−109.6	−74.9	−93.5	−94.0
14	Mezlocillin	−88.6	−101.9	−102.1	−92.9	−97.4	−105.0	−72.4	−112.1	−97.1
15	Nafcillin	−89.3	−100.8	−77.3	−82.0	−111.4	−101.5	−67.1	−97.8	−84.3
16	Oxacillin	−101.6	−88.5	−94.2	−86.9	−78.1	−90.3	−67.5	−97.4	−87.1
17	Penicillin G	−84.3	−74.4	−77.3	−79.3	−83.1	−89.5	−66.2	−81.6	−72.2
18	Piperacillin	−89.2	−86.4	−81.3	−97.5	−97.1	−103.7	−73.4	−99.8	−88.0
19	Ticarcillin	−99.8	−79.1	−81.6	−85	−80.9	−86.3	−64.7	−95.2	−92.0

The values in bold font indicate best binding energies

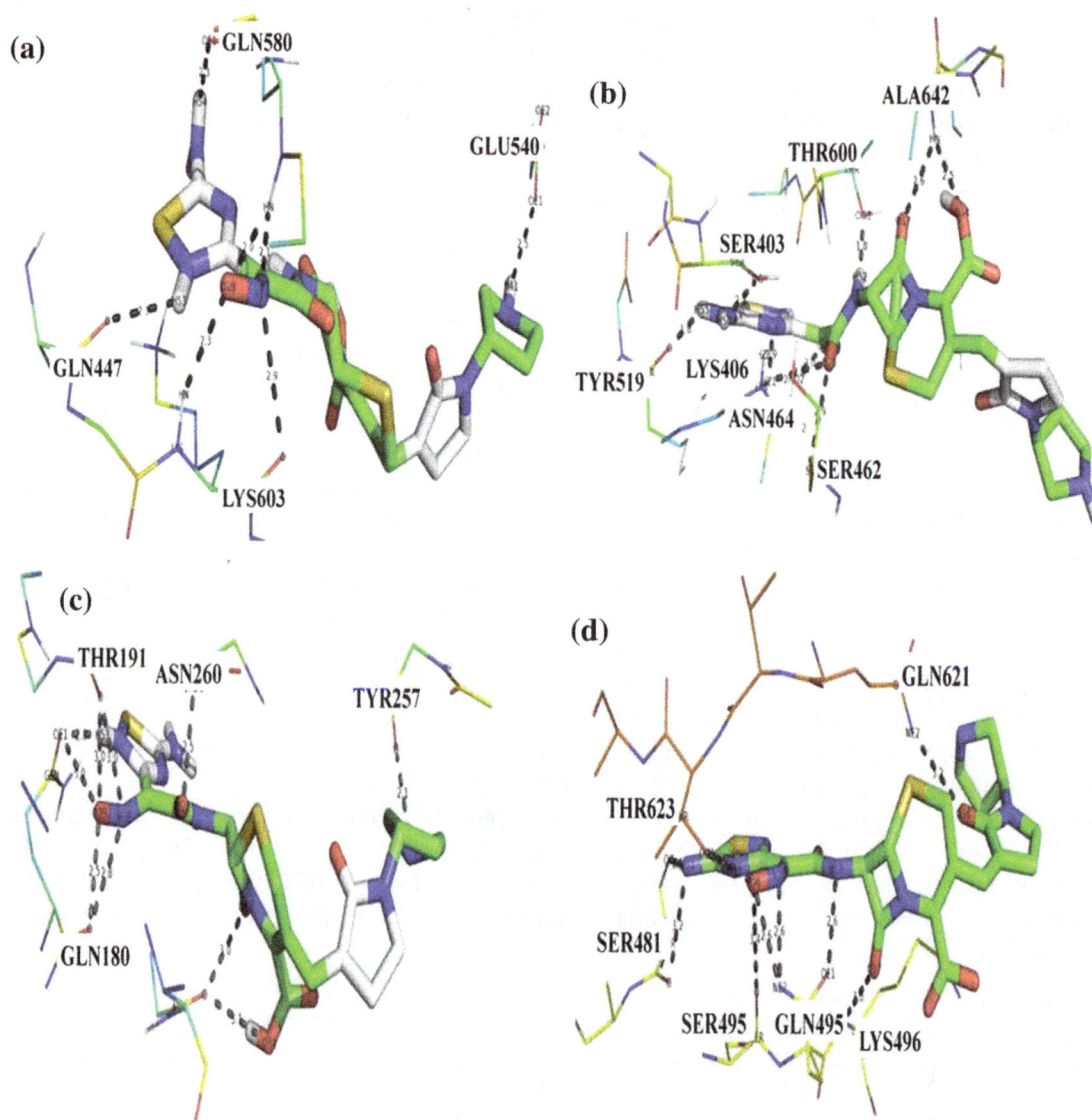

Fig. 2 Docking results of Ceftobiprole against PBP1b, PBP2a, PBP2b and PBP2x. **a** Binding mode of Ceftobiprole in PBP1b. **b** A close-up view of the binding site of Ceftobiprole in PBP2a. **c** Ceftobiprole interaction with PBP2b. **d** Binding mode of Ceftobiprole with PBP2x. Ligand atoms are *colored* by its type. The interacted amino acids residues, hydrogen bond networks in the binding pocket and the distance (in Å units) of bonds are all shown

PBP2a, PBP2b and PBP2x (Fig. 2). Amino acid residues Gln582, Glu540, Lys603 and Gln601 are involved in interaction with PBP1b; in PBP2a, the interacting amino acids are Ala642, Thr600, Tyr519, Ser403, Ser462, Asn464 and Lys406. In PBP2b, Asn260, Tyr257, Thr191 and Gln180 are involved in the interaction with Ceftobiprole. In close assessment of this binding mode, binding docking energies are calculated for PBP1b, PBP2a, PBP2b, and PBP2x (Table 3). In PBP2x, the amino acid residues Gln621, Lys496, Gln495, Ser481 and Thr623 interact with Ceftobiprole (Table 4). Davies et al. (2006) report that Ceftobiprole itself inhibits PBP1a, PBP2b and PBP2x, which are responsible for Penicillin resistance in *S.*

pneumoniae. Our results are similar to the findings of Davies et al. Ceftobiprole, a fifth generation Cephalosporin in phase 3 clinical trials, exhibits a broad spectrum of activities against many clinically important Gram-positive and Gram-negative pathogens, such as *S. pneumoniae, H. influenzae*, and *S. aureus* (Hebeisen et al. 2001; Jones et al. 2002; Kosowska et al. 2005; Zbinden et al. 2002). Docking analysis of Ceftobiprole shows best results against *S. pneumoniae* and *S. aureus*. Our results are similar to previous studies (Hebeisen et al. 2001; Jones et al. 2002; Kosowska et al. 2005). Lovering et al. (2012) report that the affinity of Ceftobiprole to PBP2a of MRSA is high. Henry et al. (2010)

report that PBP5 has less sensitivity to Ceftobiprole than PBP2a. Another study reveals that Ceftobiprole is a novel broad-spectrum antibiotic that inhibits PBP2a and

Table 3 AutoDock estimated docked energies of Ceftobiprole and Ceftaroline

S. no	Target	Ceftobiprole (kcal/mol)	Ceftaroline (kcal/mol)
1	PBP1a	−5.1	−5.2
2	PBP1b	−6.76	−4.12
3	PBP2a	−6.12	−3.43
4	PBP2b	−7.04	−5.1
5	PBP2x	−7.32	−5.3
6	PBP3	−6.1	−7.42
7	PBP4	−4.34	−5.65
8	PBP5	−6.21	−9.2
9	PBP6	−5.3	−8.3

Table 4 H-bond interactions and bond length obtained for Ceftobiprole with PBP1b, PBP2a, PBP2b and PBP2x

Protein–ligand complex	H-bond interactions	Bond length (Å)
Ceftobiprole-PBP1b	(Gln 582)O-H54	2.3
	(Gln 582)NH-O29	2.0
	(Gln 582)NH-N28	2.1
	(Glu 540)O-H41	2.5
	(Lys 603)O-N28	2.9
	(Lys603)NH-O29	2.3
Ceftobiprole-PBP2a	(Ala642)NH-O34	2.5
	(Ala642)NH-O17	2.6
	(Thr600)O-H52	1.8
	(Tyr 519)O-H55	1.8
	(Ser403)O-H54	2.1
	(Ser462)O-O20	2.9
	(Asn 464)H-O29	2.0
	(Lys 406)H-O20	2.8
Ceftobiprole-PBP2b	(Asn260)H-O20	2.5
	(Tyr257)O-H41	2.1
	(Thr191)O-O29	3.0
	(Thr191)O-N28	3.0
	(Gln180)O-O29	3.0
	(Gln180)O-N28	2.8
Ceftabiprole-PBP2x	(Gln621)N-O36	3.2
	(Lys496)N-O17	3.0
	(Gln495)O-N18	2.6
	(Gln495)N-N28	2.8
	(Gln180)N-O29	2.6
	(Ser 495)O-O29	3.4
	(Thr 623)O-O29	3.2
	(Ser 481)O-N27	3.2

PBP2x, which are responsible for the resistance in *S. pneumoniae* and *S. aureus*, respectively (Dauner et al. 2010). Though many reports on the inhibitory activity of Ceftobiprole for specific PBPs are available in literature, none of the studies have focused on the binding pattern of Ceftobiprole to all type of PBPs. Our study reveals the binding pattern of Ceftobiprole with all type of PBPs. The possible binding mode of Ceftobiprole in the PBP1b, PBP2a, PBP2b, PBP2x binding site and corresponding 2D interaction models along with hydrogen bonds and bond distance are shown in Fig. 2.

Docking of Ceftaroline into PBPs

Ceftaroline is a antibiotic of the Cephalosporin type among the majority of currently available β-lactam antibiotics. Cephalosporins are used for effective treatment of bacterial respiratory tract infections. In our results on the binding conformation modes of Penicillin derivatives and Cephalosporins with PBPs, Ceftaroline shows higher affinity with the PBP3, PBP4, PBP5 and PBP6 than the other PBPs. In examining the interaction and position of the Ceftaroline in PBP3, PBP4, PBP5 and PBP6 active site predicted by our docking procedure, it is observed that multiple hydrogen

Table 5 H-bond interactions and bond length obtained for Ceftaroline with PBP3, PBP4, PBP5 and PBP6

Protein–ligand complex	H-bond interactions	Bond length (Å)
Ceftaroline-PBP3	(Arg54)NH-O15	2.7
	(Gln121)N-O13	3.2
	(Tyr124)OH-O43	1.8
	(Tyr124)O-O23	3.2
Ceftaroline-PBP4	(Asn260)O-N18	3.1
Ceftaroline-PBP5	(Ala311)N-O42	3.1
	(Gln366)N-N30	3.1
	(Phe312)N-O43	2.7
	(Arg192)N-O15	2.8
	(Asn47)N-O13	2.6
	(Asn47)O-N11	3.1
Ceftaroline-PBP6	(Thr270)N-O14	3.0
	(Thr270)O-O14	2.7
	(Arg194)NH-O43	3.2
	(Arg194)NH-O23	3.2
	(Asn193)N-N10	3.4
	(Asn193)O-N11	3.1
	(Asn193)O-O13	2.6
	(Ile104)O-N11	2.7
	(Met208)O-O13	3.5
	(Lys209)N-O15	3.0
	(Ser106)O-O5	3.1

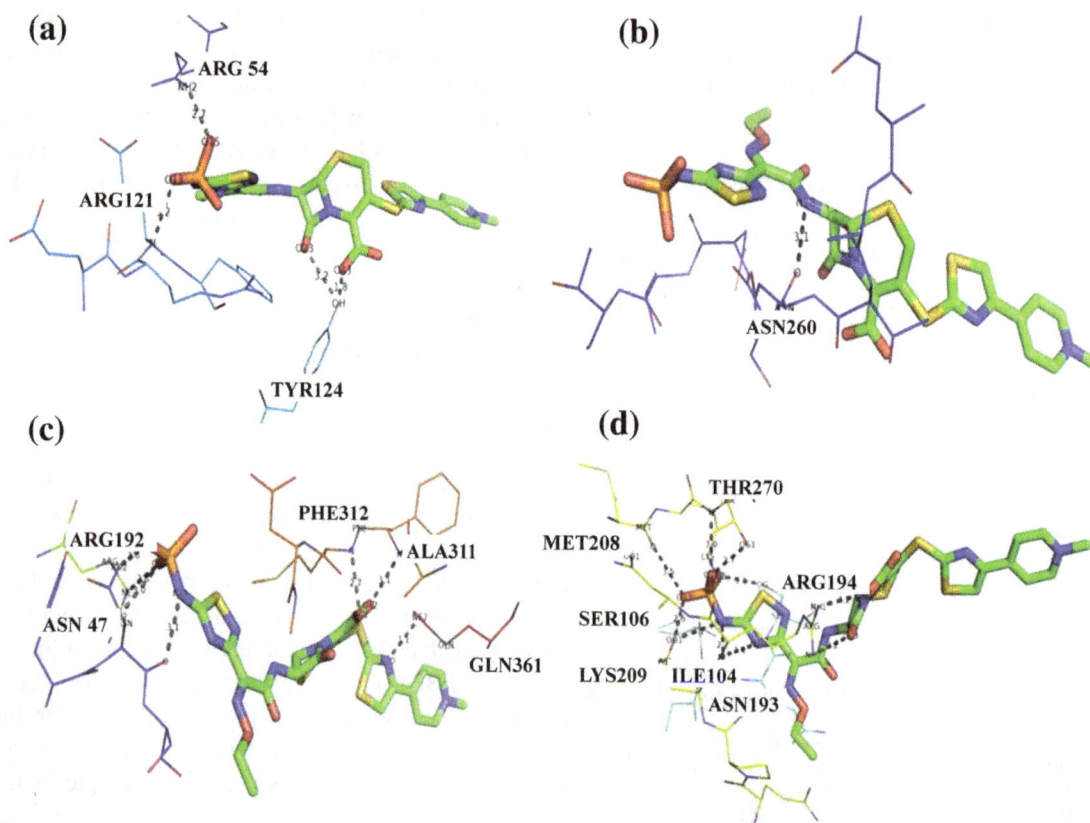

Fig. 3 Docked complex of Ceftaroline–PBP3, PBP4, PBP5 and PBP6. **a** A close-up view of the predicted binding site for Ceftaroline in PBP3. **b** Binding mode of Ceftaroline with PBP4. **c** Ceftaroline binding site in PBP5. (3D) Interaction of Ceftaroline with PBP6. Ligand atoms are *colored* by its type. The interacted amino acids residues, hydrogen bond networks in the binding pocket and the distance (in Å units) of bonds are all shown

bonds are formed (Table 5). In addition, the amino acid residues Arg54, Glu121 and Tyr124 of PBP3 are involved in van der Waals' interactions. In PBP4, only one amino acid residue Asn260 is involved in interaction with Ceftaroline. Binding of Ceftaroline to PBP5 and PBP6 involves more than six hydrogen bonds. The binding affinity of Ceftaroline for MRSA PBP2a, methicillin-susceptible *S. aureus* (MSSA) PBPs 1 to 3, and *S. pneumoniae* PBP2x/2a/2b correlates well with its low MICs and bactericidal activity against these resistant organisms (Kosowska et al. 2010; Moisan et al. 2010). Citron et al. report the effects of Ceftaroline activity against Gram-positive and Gram-negative pathogens, including MSSA, MRSA, *E. faecalis*, *S. pyogenes*, *S. pneumoniae*, *H. influenzae*, *M. catarrhalis*, *K. pneumonia*, *E. coli*, *P. aeruginosa*, and *A. baumannii* (Citron and Goldstein 2008; Jones et al. 2005). Other studies reveal that Ceftaroline has potent activity against MRSA and *S. pneumoniae*. The Gram-negative spectrum of Ceftaroline is similar to that of other broad-spectrum Cephalosporins (Estrada et al. 2008; Moisan et al. 2010; Kosowska et al. 2010). Morrissey et al. report that the Ceftaroline has excellent activity against MRSA and

Penicillin-resistant *S. pneumoniae*. Furthermore, Ceftaroline maintains good activity against *H. inlfuenzae* (Sader et al. 2005; Mushtaq et al. 2007; Morrissey et al. 2009). Our results are consistent with the previously studied ones (Kosowska et al. 2010; Moisan et al. 2010; Citron and Goldstein 2008; Jones et al. 2005; Estrada et al. 2008; Kosowska et al. 2010; Sader et al. 2005; Mushtaq et al. 2007; Morrissey et al. 2009). Although many studies have been reported the inhibitory action of Ceftaroline to specific PBPs, no studies have been done for the binding pattern of Ceftaroline with all type of PBPs. Our results clearly explain the binding pattern of Ceftaroline with all type of PBPs. The binding energy calculated by AutoDock for Ceftaroline–PBP complexes is shown in Table 3. The best possible binding mode of Ceftaroline in PBP4, PBP5 and PBP6 and their corresponding 2D interaction models are displayed in Fig. 3.

ADME screening

For each of the Penicillin derivatives and Cephalosporins, we analyzed for a number of physiochemical properties and pharmaceutically relevant properties, such as molecular weight, H-bond donors, H-bond acceptors, logP (octanol/water), and their position according to Lipinski's rule of 5 (Table 6). Lipinski's rule of 5 is a rule of thumb to predict drug likeness, or determine if a compound with a certain biological or pharmacological activity has properties that would make it a likely orally active drug in humans. The rule describes physiochemical properties important for a drug's pharmacokinetics in the human body, including its ADME. The drug molecule shows poor absorption and permeation when they have more than 5 hydrogen bond donors, molecular weight over 500, logP is over 5 and more than 10 hydrogen bond acceptors. In this study, of the 19 ligands, 16 structures showed possible values for the properties analyzed and exhibited drug-like characteristics based on Lipinski's rule of 5. Methicillin has more than 7 rotatable bonds. Rotatable bond more than 10 and molecular weight more than 500 can lead to decreased permeability and oral bioavailability. But Ceftobiprole and Ceftaroline show molecular weight more than 500. Hence to improve the action of these two drugs, we have highlighted the non-essential regions (Fig. 1) that may possibly be spliced to reduce the molecular mass. However, the effectiveness of these low molecular mass compounds has to be tested in both in vivo and in vitro.

Table 6 Molecular properties of Penicillin derivatives and Cephalosporins obtained from Molinspiration

S. no	Antibiotics	LogP (<5)	Molecular weight (<500 dalton)	HBA count (<10)	HBD count (<5)	Rotatable bond count (<7)
1	Amoxicillin	2.31	365.40	6	4	4
2	Ampicillin	−2.00	349.40	5	3	4
3	Azlocillin	0.20	461.49	6	4	5
4	Carbenicillin	1.13	378.39	6	3	5
5	Cloxacillin	2.61	435.88	5	2	4
6	Dicloxacillin	2.90	470.32	5	2	3
7	Flucloxacillin	2.69	453.87	5	2	3
8	Methicillin	0.85	380.41	10	3	11
9	Mezlocillin	0.21	539.58	8	3	5
10	Nafcillin	3.21	414.47	5	2	5
11	Oxacillin	2.05	401.43	5	2	4
12	Penicillin G	1.5	334.39	4	2	4
13	Piperacillin	1.2	517.55	7	2	6
14	Ticarcillin	0.99	384.42	6	3	5
15	Ceftobiprole	−1.68	564.16	11	7	4
16	Ceftaroline	2.43	699.03	16	5	2
17	Cefadroxil	−1.22	377.10	7	5	3
18	Lactivicin	−0.60	296.14	5	1	2
19	Cefuroxime	−0.2	424.39	10	3	7

Conclusion

In the present study, molecular docking studies were performed to explore possible binding modes of Penicillin derivatives and Cephalosporins into all types of PBPs, PBP1a, PBP2b, PBP2x and PBP3 of *S. pneumoniae,* PBP1b, PBP2a and PBP4 of *S. aureus*, PBP5 of *H. influenzae,* as these organisms are most frequently found pathogens in the URT. The molecular docking study revealed that the Cephalosporins show higher affinity with PBPs than the Penicillin derivatives. Especially the fifth generation Cephalosporins, Ceftobiprole and Ceftaroline show best results to all types of PBPs. The binding affinity was evaluated by the binding free energies (DGb, Kcal/mol) and hydrogen bonding. The compounds which revealed the highest binding affinity are the ones with lowest binding free energy. On comparing the binding energy and the binding site residues, we found that all compounds differ in their binding modes or binding site residues for hydrogen bond formation. The conclusion drawn from this virtual screening and docking result was that the Ceftobiprole has highest binding affinity with the PBP2x of *S. pneumoniae*. The Ceftaroline has maximum number of interaction with PBP5 of *H. influenzae*. The above results suggest that the Ceftobiprole and Ceftaroline can be potent inhibitors for all types of PBPs. From ADME screening of all the 19 compounds, 16 compounds satisfied Lipinski's rule of 5. Ceftobiprole and Ceftaroline show molecular weight more than 500 which decreases their permeability and bioavailability. These drugs can further be modified to satisfy Lipinski's rule of 5. Though, there are a few reports on the in vitro analysis of Ceftobiprole and Ceftaroline, there are no in silico studies that predict the binding and active regions in these molecules. Our study is probably the first such attempt and we infer that our results will throw light for the future development of more potent next generation antibiotics for the treatment of upper respiratory infections and counter the emergence of antibiotic resistant strains.

Acknowledgments Dr. Anand Anbarasu gratefully acknowledges the Indian council of Medical Research (ICMR), Government of India Agency for the research grant [IRIS ID: 2011-03260]. P. Lavanya thanks ICMR for the Research fellowship through the ICMR grant [IRIS ID: 2011-03260]. The authors would also like to thank the management of VIT University for providing the necessary facilities to carry out this research project.

Conflict of interest The authors declare that there is no conflict of interest.

References

Berman HM, Westbrook J, Feng Z, Gilliland G, Bhat TN, Weissig H, Shindyalov IN, Bourne PE (2000) The protein data bank. Nucleic Acids Res 28:235–242

Citron DM, Goldstein EJC (2008) Effects of in vitro test method variables on ceftaroline activity against aerobic Gram-positive and Gram-negative pathogens, poster D-2232. In: Interscience Conference on Antimicrobial Agents and Chemotherapy. Infectious Disease Society of America. American Society for Microbiology, Washington, DC

Dauner DG, Nelson RE, Taketa DC (2010) Ceftobiprole: a novel, broad-spectrum cephalosporin with activity against methicillin-resistant Staphylococcus aureus. Am J Health Sys Pharm 67(12):983–993

Davies TA, Shang W, Bush K (2006) Activities of Ceftobiprole and other β-lactams against Streptococcus pneumoniae clinical isolates from the United States with defined substitutions in penicillin-binding proteins PBP 1a, PBP 2b, and PBP 2x. Antimicrob Agents Chemother 50:2530–2532

Doern GV, Jorgensen JH, Thornsberry C (1988) National collaborative study of the prevalence of antimicrobial resistance among clinical isolates of Haemophilus influenzae. Antimicrob Agent Chemother 32:180–185

Doern GV, Brueggmann AB, Pierce G, Holley HP, Rauch A (1997) Antibiotic resistance among clinical isolates of in the United States in 1994 and 1995 and detection of β-lactamase-positive strains resistant to amoxicillin-clavulanate: results of a national multicenter surveillance. Antimicrob Agents Chemother 41:292–297

Estrada VA, Lee M, Hesek D, Vakulenko SB, Mobashery S (2008) Co-opting the cell wall in fighting methicillin-resistant Staphylococcus aureus: potent inhibition of PBP 2a by two anti-MRSA b-lactam antibiotics. J Am Chem Soc 130:9212–9213

Garenne M, Ronsmans C, Campbell H (1992) The magnitude of mortality from acute respiratory infections in children under 5 years in developing countries. World Health Stat Q 45:180–191

Ghuysen JM (1991) Serine β-lactamases and Penicillin-binding proteins. Annu Rev Microbiol 45:37–67

Goffin C, Ghuysen JM (1998) Multimodular Penicillin-binding proteins: an enigmatic family of orthologs and paralogs. Microbiol Mol Biol Rev 62:1079–1093

Guex N, Peitsch MC (1997) SWISS-MODEL and the Swiss-PdbViewer: an environment for comparative protein modeling. Electrophoresis 18:2714–2723

Hebeisen PI, Krauss H, Angehrn P, Hohl P, Page MGP, Then RL (2001) In vitro and in vivo properties of Ro 63–9141, a novel broad-spectrum cephalosporin with activity against methicillin-resistant staphylococci. Antimicrob Agents Chemother 45:825–836

Henry X, Amoroso A, Coyette J, Joris B (2010) Interaction of Ceftobiprole with the low-affinity PBP 5 of Enterococcus faecium Antimicrob. Agents Chemother 54:953955

Jarrahpour A, Fathi J, Mimouni M, Benhadda T, Sheikh J, Chohan ZH, Petra AP (2011) Osiris and molinspiration (POM) together as a successful support in drug design: antibacterial activity and biopharmaceutical characterization of some azo schiff bases. Med Chem Res 19(7):1–7

Jones RN, Deshpande LM, Mutnick AH, Biedenbach BJ (2002) In vitro evaluation of BAL9141, a novel parenteral cephalosporin active against oxacillin-resistant staphylococci. J Antimicrob Chemother 50:915–932

Jones RN, Fritsche TR, Ge Y, Kaniga K, Sader HS (2005) Evaluation of PPI-0903 M (T91825), a novel cephalosporin: bactericidal activity, effects of modifying in vitro testing parameters and optimization of disc diffusion tests. J Antimicrob Chemother 56:1047–1052

Jorgensen JH (1992) Update on mechanisms and prevalence of antimicrobial resistance in Haemophilus influenzae. Clin Infect Dis 14:1119–1123

Kawai F, Clarke TB, Roper D, Han GJ, Hwang KY, Unzai S, Obayashi E, Park SY, Tame RH (2010) Crystal Structures of Penicillin-Binding Proteins 4 and 5 from Haemophilus Influenzae. J Mol Biol 396:634–645

Kosowska SK, Hoellman DB, Lin G, Clark C, Credito K, McGhee P, Dewasse B, Bozdogan B, Shapiro S, Appelbaum PC (2005) Anti pneumococcal activity of ceftobiprole, a novel broad-spectrum cephalosporin. Antimicrob Agents Chemother 49:1932–1942

Kosowska SK, McGhee PL, Appelbaum PC (2010) Affinity of ceftaroline and other β-lactams for penicillin-binding proteins from Staphylococcus aureus and Streptococcus pneumoniae. Antimicrob Agents Chemother 54:1670–1677

Laurie A, Jackson R (2005) Q-SiteFinder: an energy-based method for the prediction of protein-ligand binding sites. Bioinformatics 21:1908–1916

Li Z, Wan H, Shi Y, Ouyang P (2004) Personal experience with four kinds of chemical structure drawing software: review on ChemDraw, ChemWindow, ISIS/Draw, and ChemSketch. J Chem Inf Comput Sci 44:1886–1890

Li Q, Cheng T, Wang Y, Bryant SH (2010) PubChem as a public resource for drug discovery. Drug Discov Today 15(23–24):1052–1057.

Liao JJ, Andrews RC (2007) Targeting protein multiple conformations: a Structure-based strategy for kinase drug design. Curr Top Med Chem 7:1394–1407

Lill MA, Danielson ML (2010) Computer-aided drug design platform using PyMOL. J Comput Aided Mol Des 25:13–19

Lipinski CA, Lombardo F, Dominy PW, Feeney PJ (2006) Experimental and computational approaches to estimate solubility and permeability in drug discovery and development settings. Adv Drug Deliv Rev 23:3–25

Lovering AL, Gretes MC, Safadi SS, Danel F, Castro LD, Page MGP, Strynadka NCG (2012) Structural insights into the anti-methicillin-resistant Staphylococcus aureus (MRSA) activity of ceftobiprole. J Biol Chem 287:32096–32102

Lynch JP, Zhanel GG (2005) Escalation of antimicrobial resistance among Streptococcus pneumoniae: implications for therapy. Semin Respir Crit Care Med. 26:575–616

Macheboeuf P, Martel CC, Job V, Dideberg O, Dessen A (2006) Penicillin binding proteins: key players in bacterial cell cycle and drug resistance processes. FEMS Microbiol Rev 30:673–691

Metan G, Zarakolu P, Unal S (2005) Rapid detection of antibacterial resistance in emerging Gram-positive cocci. J Hosp Infect 61:93–99

Moisan H, Pruneau M, Malouin F (2010) Binding of Ceftaroline to penicillin binding proteins of Staphylococcus aureus and Streptococcus pneumoniae. J Antimicrob Chemother 65:713–716

Morris GM, Goodsell DS, Halliday RS, Huey R, Hart WE, Belew RK, Olson AJ (1998) Automated docking using a Lamarckian genetic algorithm and an empirical binding free energy function. J Comput Chem 19:1639–1662

Morris GM, Huey R, Lindstrom W, Sanner MF, Belew RK, Goodsell DS, Olson AJ (2009) AutoDock4 and AutoDockTools4: automated docking with selective receptor flexibility. J Comput Chem 30:2785–2791

Morrissey I, Ge Y, Janes R (2009) Activity of the new cephalosporin Ceftaroline against bacteraemia isolates from patients with community-acquired pneumonia. Int J Antimicrob Agents 33(6):515–519

Mushtaq S, Warner M, Ge Y, Kaniga K, Livermore DM (2007) In vitro activity of Ceftaroline (PPI-0903 M, T-91825) against bacteria with defined resistance mechanisms and phenotypes. J Antimicrob Chemother 60:300–311

Peter E, Derek RF, Brown J (1985) Penicillin-binding proteins of β-lactam-resistant strains of *Staphylococcus aureus*. FEBS Lett 192:28–32.

Pettersen EF, Goddard TD, Huang CC, Couch GS, Greenblatt DM (2004) UCSF Chimera -a visualization system for exploratory research and analysis. J Comput Chem 25:1605–1612

Ragle BE, Karginov VA, Wardenburg JB (2010) Prevention and treatment of *Staphylococcus aureus* with a β-cyclodextrin derivative. Antimicrob Agents Chemother 54(1):298–304

Reid AJ, Simpson IN, Harper PB, Amyes SG (1987) Ampicillin resistance in *Haemophilus influenzae*: identification of resistance mechanisms. J Antimicrob Chemother 20:645–656

Sader HS, Fritsche TR, Kaniga K, Jones RN (2005) Antimicrobial activity and spectrum of PPI-0903 M (T-91825), a novel cephalosporin, tested against a worldwide collection of clinical strains. Antimicrob Agents Chemother 49:3501–3512

Sainsbury S, Bird L, Rao V, Sharon M, Shepherd, Stuart DI, Hunter WN, Owens RJ, Ren J (2011) Crystal structures of Penicillin-binding protein 3 from *Pseudomonas aeruginosa*: comparison of native and antibiotic-bound forms. J Mol Biol 405:173–184

Sauvage E, Kerff F, Terrak M, Ayala JA, Charlier P (2008) The Penicillin-binding proteins: structure and role in peptidoglycan biosynthesis. FEMS Microbiol Rev 32:234–258

Sheldon LK, Mason OE (1998) Management of infections due to antibiotic-resistant *Streptococcus pneumoniae*. Clin Microbiol Rev 11:628–644

Turk S, Verlaine O, Gerards T, Zivec M, Humljan J, Sosic I, Amoroso A, Zervosen A, Luxen A, Joris B, Gobec S (2011) New noncovalent Inhibitors of Penicillin-Binding Proteins from Penicillin-Resistant Bacteria. Plos one 6:5-e19418

Williams JD, Moosdeen F (1986) Antibiotic resistance in: epidemiology, mechanisms and therapeutic possibilities. Rev Infect Dis 8(Suppl 5):S555–S561

Williams BG, Gouws E, Boschi-Pinto C, Bryce J, Dye C (2001) Estimates of world-wide distribution of child deaths from acute respiratory infections. Lancet Infect Dis 2(1):25–32

Yang M, Chen CC (2004) GEMDOCK: a generic evolutionary method for molecular docking. Protein Structure Funct Bioinfo 55:288–304

Yoshida H, Kawai F, Obayashi E, Akashi S, Roper DI, Tame JR, Park SY (2012) Crystal structures of penicillin-binding protein 3 (PBP3) from methicillin-resistant staphylococcus aureus in the apo and cefotaxime bound forms. J Mol Biol 423:351–364

Zbinden R, Punter V, Graevenitz AV (2002) In vitro activities of BAL9141, a novel broad-spectrum pyrrolidinone cephalosporin, against Gram-negative non fermenters. Antimicrob Agents Chemother 46:871–874

12

Purification and characterization of detergent-compatible protease from *Aspergillus terreus* gr.

Francois N. Niyonzima · Sunil S. More

Abstract The possibility of using *Aspergillus terreus* protease in detergent formulations was investigated. Sodium dodecyl sulfate (SDS) and native polyacrylamide gel electrophoresis indicated that the purified alkaline protease (148.9 U/mg) is a monomeric enzyme with a molecular mass of 16 ± 1 kDa. This was confirmed by liquid chromatography–mass spectrometry. The active enzyme degraded the copolymerized gelatin. The protease demonstrated excellent stability at pH range 8.0–12.0 with optimum at pH 11.0. It was almost 100 % stable at 50 °C for 24 h, enhanced by Ca^{2+} and Mg^{2+}, but inhibited by Hg^{2+}, and strongly inhibited by phenylmethyl sulfonyl fluoride. It showed maximum activity against casein followed by gelatin; its V_{max} was 12.8 U/ml with its corresponding K_M of 5.4 mg/ml. The proteolytic activity was activated by Tween-80, Triton-100 and SDS, and remained unaltered in the presence of H_2O_2 and NaClO. The enzyme exhibited higher storage stability at 4, 28 and -20 °C. It was stable and compatible to the desired level in the local detergents. The addition of the protease to the Super wheel improved its blood stain removal. The isolated protease can thus be a choice option in detergent industry.

Keywords Alkaline protease · *Aspergillus terreus* gr. · Purification · Detergent

Introduction

Proteases (EC 3.4.21–24 and 99) are hydrolytic enzymes that cleave peptide bonds of proteins. The extracellular proteases of microbial origin are important enzymes and account for approximately 60 % of the total worldwide enzyme sale (Rao et al. 1998). Alkaline proteases are primarily used in the detergent industry and need not to be in pure form. However, proteases that are used in other areas such as pharmaceutical and medical applications require high purity (Kumar and Takagi 1999). Choudhary and Jain (2012) reported that enzyme purification is tedious, time consuming and very expensive. The precipitation is the most common method used to partially purify and concentrate the protein from fermentation crude extract (Bell et al. 1983). It is performed by the addition of inorganic salt like ammonium sulfate or organic solvent such as acetone or ethanol (Kumar and Takagi 1999). Alkaline proteases generally do not bind to anion exchangers as they are generally positively charged. The cation exchangers can thus be used and the elution of the bound molecules can be done by increasing the salt or pH gradient (Fujiwara et al. 1993). The lectin-agarose affinity chromatography is also used to separate glycoproteins from non-glycosylated proteins (Kobayashi et al. 1996).

The study of alkaline protease properties is important from the point of view of its practical applicability. Alkaline proteases used in detergent preparations must have higher activity at alkaline pH, broad temperature range, broad substrate specificity, stability in the presence of surfactants, oxidizing agents, and compatibility with detergents (Kumar and Takagi 1999; Adinarayana et al. 2003; Choudhary and Jain 2012). The cations like Ca^{2+}, Mg^{2+} and Mn^{2+} usually increased the activity and stabilized the enzyme (Sharma et al. 2006; Anandan et al. 2007; Kalpana devi et al. 2008; Dubey et al. 2010). The nature of the enzyme and its active site as well as its cofactor requirements can be deduced from inhibition studies (Sigma and Mooser 1975). Alkaline proteases are differently affected by various inhibitors.

F. N. Niyonzima · S. S. More (✉)
Department of Biochemistry, Center for Post Graduate Studies, Jain University, Bangalore 560011, India
e-mail: sunilacr@yahoo.co.in

However, most of the fungal alkaline proteases are generally serine proteases (Coral et al. 2003). The alkaline proteases have in general the broad substrate specificity and breakdown a variety of natural substrates and synthetic substrates (Kumar and Takagi 1999).

Enzymes are used in a very small amount in detergent preparations to increase the cleaning ability of detergents (Bajpai and Tyagi 2007). If a detergent does not contain an enzyme, it may not completely remove the stains resulting in permanent residues (Hasan et al. 2010). The performance of an enzyme in a detergent is based on the detergent composition, type of stains to be removed, water hardness, washing temperature and procedure (Hasan et al. 2010). Kirk et al. (2002) emphasized that the detergent protease must work at room temperature to save energy. The cleansing process is the reverse process of coagulation and adhesion and requires energy from external sources such as human hands and washing machine. In the presence of a detergent, the energy gets reduced (Bajpai and Tyagi 2007).

The industrial demand of proteases with novel and better properties continues to stimulate the researchers in this area. For the production of proteases for industrial use, isolation, purification and characterization of new promising strain are continuous processes (Kumar et al. 2002). Although different alkaline proteases have been isolated from several bacteria and fungi, few have better properties that can be commercially exploited. *Aspergillus terreus* gr. was recently identified as a potent producer of an extracellular stable alkaline protease (Niyonzima and More 2013). The aim of the present study was, therefore, to purify and characterize an alkaline protease of *A. terreus* as well as to ascertain its suitability as detergent additive.

Materials and methods

Fungal strain and fermentation conditions

The alkaline protease-producing fungus used was isolated from potato grown soil fields of Bangalore and identified as *A. terreus* gr. based on morphological and microscopic features and it has been deposited into the National Fungal Culture Collection of India (Agarkar Research Institute, Pune). The composition of the fermentation medium found after optimization consisted of (w/v) 1.5 % casein, 0.1 % KH_2PO_4, 0.1 % K_2HPO_4, 0.02 % $MgSO_4$ and 2 % soybean meal, pH 10.0. The optimized fermentation medium was inoculated with 2 % (v/v) inoculum (2×10^6 spores/ml) and incubated at 37 °C for 5 days. After incubation, the culture broth was centrifuged (REMI C-30 BL Cooling centrifuge, India) at 10,000 rpm for 10 min at 4 °C. The supernatant was used as a crude enzyme (Niyonzima and More 2013).

Alkaline protease activity and protein estimation

The proteolytic activity was determined as per Niyonzima and More (2013) using casein as the substrate. Protein concentration was estimated as per Lowry et al. (1951).

Purification of alkaline protease

The method of Kim et al. (1996) with slight modifications was followed to partially purify the alkaline protease of *A. terreus* gr. To one volume of supernatant, 3 volumes of cold acetone (−20 °C) was slowly added. The acetone precipitate was then collected by centrifugation at 13,000 rpm for 10 min after an incubation period of 3 h at −20 °C. A minimum volume of 0.1 M Tris–HCl buffer (pH 9.0) was used to dissolve the precipitate (0.5 ml for 100 mg). The resulted enzyme solution was subjected to lyophilization (Freeze dryer, Model LY3TTE, Snijders Scientific, Tilburg Holland) and the resulted powder served as the partially purified alkaline protease. Affinity chromatography method using agarose-bound lectins (Spivak et al. 1977) was used with minor modifications to completely purify the protease. The lectin concanavalin A (Con A) was chosen as the ligand. A 0.5 × 9 cm Con A-agarose column (2-ml bed volume) was prepared (10 mg/ml) and equilibrated with 0.1 M Tris–HCl buffer (pH 9.0). The lyophilized sample collected in the above step was dissolved in the same buffer and loaded into the column. Fractions were collected for the bound and unbound proteins and subjected to the lyophilization. The bound protein was eluted with buffer containing sucrose of 1 M strength. The protein content and enzyme activity were determined for both samples as described earlier. The adsorbed protein showing higher activity was used as the purified alkaline protease.

SDS-PAGE

Sodium dodecyl sulfate- and native polyacrylamide gel electrophoresis (SDS-PAGE) was performed under non-reducing conditions (Laemmli 1970). Electrophoresis was carried out at 50 V in the stacking gel (6 %) and at 100 V in the separating gel (15 %). The gel was stained (0.25 % coomassie brilliant blue R 250, 15 % methanol, 7.5 % glacial acetic acid) for 2 h and destained overnight with stain solution excluding the dye. The protease was sized with the help of the protein molecular weight marker.

Native PAGE

The purity and the nature of the protease were studied as per Laemmli (1970) using 15 % native PAGE. The procedure used was the same as described for SDS-PAGE except that the SDS was not included in the gel and electrophoretic solutions, and the samples were not heated.

Gelatin zymography

The zymography was carried out in 15 % polyacrylamide gel containing 0.1 % (w/v) gelatin as a co-polymerized substrate (Heussen and Dowdle 1980). The native PAGE band was extracted and mixed with the sample buffer without heat denaturation prior to electrophoresis. Electrophoresis was carried out as described for SDS-PAGE. After electrophoresis, the gel was placed in a tray on a gel rocker and washed twice with a wash buffer (2.5 % Triton X-100 in distilled H_2O, 0.02 % NaN_3) for 20 min each to remove the SDS. It was then rinsed for 10 min in a 50 ml incubation buffer (50 mM Tris–HCl, pH 8.0, 5 mM $CaCl_2$, 0.02 % NaN_3). The gel was placed in 150 ml fresh incubation buffer and incubated at 37 °C for 24 h to allow the gelatin degradation. The staining and destaining were the same as described for SDS-PAGE. The activity band was indicated by a clear, colorless zone against the blue background.

LC–MS

LC–MS of the purified protein was carried using an Agilent 1290 Infinity LC system coupled to an Agilent 6530 Q TOF with an Agilent Poroshell C18, 75 × 2.1 mm, 5 μm analytical column. 0.1 % formic acid in water (A) and 90 % acetonitrile in water with 0.1 % formic acid (B) served as solvents for LC. Before MS analysis, 3 % acetonitrile/water with 0.1 % formic acid solution was utilized to dilute the protein samples. The LC gradient started with 3 % B and in 15 min it went up to 95 % B and then returned back to 3 % B by 16 min. Spectra were noted in positive ion and in profile mode. Data were acquired on standard (3,200 m/z), 2 GHz, MS only mode, range 500–3,200 m/z. Agilent Mass Hunter Qualitative Analysis software was used to analyze the data. The protein masses were obtained by deconvoluting mass spectrum using the maximum entropy algorithm in Bioconfirm module. The molecular mass was determined by a minimum of five peaks well above the baseline.

Effect of pH on activity and stability of the alkaline protease

The effect of pH on alkaline protease activity was studied by pre-incubating 0.05 ml of enzyme in 0.95 ml of the concerned buffer at room temperature (28 ± 2 °C) for 30 min. After pre-incubation, the assay was continued as reported earlier. pH 10.0, 11.0 and 12.0 were used to study the stability of the enzyme. For each pH, 5.7 ml of a given buffer was mixed with 0.3 ml of the enzyme and mixed well. The enzyme reaction was followed at room temperature and the residual activity was noted after 0, 30 min, 1, 2, 4 and 24 h by considering maximum activity as 100 %.

Effect of temperature on activity and stability of the alkaline protease

The mixture of 0.05 ml of alkaline protease and 0.95 ml of 0.1 M glycine–NaOH buffer (pH 11.0) was treated at different temperatures ranging from 0 to 90 °C for 30 min. The enzyme activity was determined as above. Thermal stability was examined by incubating 0.3 ml of the enzyme with 5.7 ml of 0.1 M glycine–NaOH buffer (pH 11.0) at 50, 60, 70 and 80 °C. The residual activity was recorded after 0, 30 min, 1, 2, 4 and 24 h and was expressed as percentage of initial activity taken as 100 %.

Effect of various metal ions on alkaline protease activity

0.05 ml of alkaline protease was mixed with 0.95 ml of 10 mM of the chloride under study (chloride of Ca, Mg, Co, Cu, Fe, Hg, K, Mn, Zn, Co and Na) and incubated for 30 min at 50 °C. The residual alkaline protease activity was assessed by the standard assay procedure and was expressed as percentage of activity without metal ion, considered as 100 %.

Effect of different group-specific reagents on the activity of alkaline protease

The specific reagents evaluated were EDTA, urea, sodium azide (NaN_3), tosyl-L-lysylchloromethyl ketone (TLCK), dithiothreitol (DTT), iodoacetamide (IAA), phenylmethyl sulfonyl fluoride (PMSF), N-ethylmaleimide (NEM), N-Diazoacetylnorleucine methyl ester (DAN) and N-bromosuccinimide (NBS). The purified alkaline protease (0.05 ml) was mixed with 0.95 ml of 5 mM specific inhibitor and pre-incubated for 30 min at 50 °C. The residual activity was assessed using control which was without inhibitor and was taken as 100 %.

Broad substrate specificity

Casein, gelatin, BSA, egg albumin and fibrinogen were used to study the substrate specificity of the alkaline protease. 0.95 ml of 0.1 M glycine–NaOH buffer (pH 11.0) was mixed with 0.05 ml of the purified enzyme. The mixture was incubated at 50 °C in the presence of 2 ml of 0.65 % (w/v) substrate for 10 min. The activity was determined as earlier.

Determination of kinetic parameters

The various casein concentrations (0–20 mg/ml) in 0.1 M glycine–NaOH buffer (pH 11.0) were incubated for 10 min with equal amounts of the alkaline protease (0.1 ml of 1 mg/ml) at 50 °C. The enzyme activity was then recorded as earlier. The K_M and the maximum rate of reaction (V_{max}) were evaluated graphically (Lineweaver and Burk 1934).

Compatibility of alkaline protease with detergent components

The enzyme was tested for its stability in the presence of detergent components. The surfactants used were SDS, Triton X-100 and Tween-80, whereas the oxidizing and bleaching agents were H_2O_2 and NaClO, respectively (Pathak and Deshmukh 2012). 0.95 ml of 1 % (and 5 %) surfactant/oxidizing agent was mixed with 0.05 ml of enzyme in 0.1 M glycine–NaOH buffer (pH 11.0) and pre-incubated for 30 min at 50 °C. The residual activity of protease was assessed as earlier. The control was without any additive.

Storage stability of alkaline protease

The storage stability of the enzyme at −20, 4 and 28 °C was analyzed. The enzyme preparation (30 mg/ml) was stored at the corresponding temperature and the residual activity was recorded by the standard assay procedure as described previously after exactly 10 days for a period of 40 days. The relative activity at day one was considered as 100 % (Beena et al. 2012).

Compatibility of alkaline protease with local detergents

Ariel (Procter and Gamble Home Products Ltd-Mumbai, India), Henko (Henkel Spic, India), More choice (San Soaps and Detergents, Bengaluru, India), Super wheel and Surf excel (Hindustan Unilever Limited-Mumbai, India) were used as local detergents. They were diluted in tap water to a final concentration of 0.7 % (w/v) and heated at 95 °C for 10 min (Adinarayana et al. 2003; Kalpana devi et al. 2008; Dubey et al. 2010). A reaction mixture comprising 0.3 ml of enzyme and 5.7 ml detergent was pre-incubated at room temperature (28 ± 2 °C). The residual activity was recorded at 0, 30 min, 1, 2, 4 and 24 h. The procedure was repeated at 60 and 90 °C. The relative activity was expressed as percentage enzyme activity taking the activity of the control sample without a detergent as 100 % (Ali 2008).

Cleansing potential of the alkaline protease as a detergent additive

The small white cloth pieces (4 × 4 cm) were stained with blood and air dried for a week. The stained clothes were incubated in 50 ml tap water + 1 ml of 7 mg/ml detergent, and in a mixture of 50 ml tap water, 1 ml detergent solution (0.7 %, w/v) and 1 ml of alkaline enzyme (20 mg/ml). The visual examination of air dried cloth pieces after clean tap water rinse was used to assess the stain removal. The unused stained cloth was used as the control (Adinarayana et al. 2003).

Statistical analysis

Three independent experiments for each treatment were performed to determine enzyme activity. The means were compared by ANOVA and means for groups in homogeneous subsets were given by Duncan's multiple range test (DMRT) at the 5 % significance level. The SPSS statistical package (PASW Statistics 18) was used for statistical evaluations.

Results and discussion

Purification of alkaline protease of *Aspergillus terreus* gr.

The fungus was grown under submerged fermentation since it was found to be appropriate and simple method of producing in bulk the extracellular alkaline protease (Niyonzima and More 2013). The enzyme was partially purified by acetone precipitation from crude enzyme and further totally purified by the affinity chromatography with agarose-bound lectins. A 28.6 purification fold with a yield in enzyme activity of 38.5 % and a specific activity of 148.9 U/mg was recorded. The purification results are summarized in Table 1. Anandan et al. (2007) reported 21-purification fold for an alkaline protease of *Aspergillus tamarii* with 905.7 U/mg and 8 % yield. Charles et al. (2008) recovered 58 % alkaline protease of *Aspergillus nidulans* HA-10 with a specific activity of 424 U/mg. A low recovery of 3.2 % alkaline protease of *Aspergillus flavus* with 170 U/mg and 5.8-purification fold was noted by Muthulakshmi et al. (2011). A better yield was also recorded for the alkaline protease when acetone was used as a primary precipitating agent (Kim et al. 1996). Concanavalin A affinity chromatography also gave a significant yield for the protease purification from *Clitocybene bularis* (Sabotiè et al. 2009).

Table 1 Summary of purification of alkaline protease of *A. terreus* gr.

Fraction	Total activity (U)	Total protein (mg)	Specific activity (U/mg)	Purification fold	Yield of activity (%)
Cell-free extract	240	46	5.2	1	100
Acetone precipitate	174	12.8	13.6	2.6	72.5
Purified enzyme by affinity chromatography	92.3	0.62	148.9	28.6	38.5

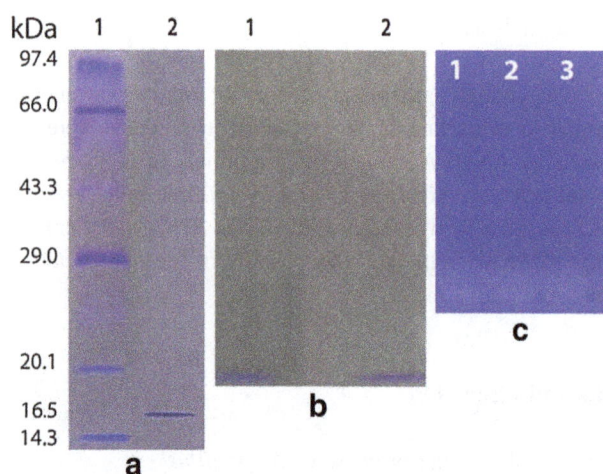

Fig. 1 a SDS-PAGE of the purified protease: *Lane 1* protein molecular weight marker, *Lane 2* purified alkaline protease. **b** Native PAGE of purified alkaline protease: *lane 1* and 2. **c** Activity staining of the native PAGE gel: gelatin degradation is shown by a clear colorless zone against the *blue* background

SDS-PAGE and molecular weight determination

The molecular weight of the protein band was 16 ± 1 kDa. The purified enzyme was homogenous showing a single band (Fig. 1a). SDS-PAGE is the most widely used method to monitor protein purification and to size proteins. However, the determination of the relative molecular weight must be confirmed by another method like mass spectrometry. Figure 2 shows the deconvoluted mass spectrum of the alkaline protease of *A. terreus* as analyzed by the LC–MS. The protein showed a mass of 16.24 kDa. This also correlated with the SDS-PAGE analysis. A low molecular weight of 15 kDa was also reported for bacterial alkaline protease (Adinarayana et al. 2003). From the viewed reports, the molecular weight of *Aspergillus* proteases ranges from 23 to 124 kDa. For examples, 124 kDa was recorded for serine protease from *Aspergillus fumigatus* by Wang et al. (2005), 68 kDa from *Aspergillus niger* Z1 (Coral et al. 2003), 45 kDa from *A. tamarii* (Anandan et al. 2007), 33 kDa from *Aspergillus oryzae* AWT 20

(Sharma et al. 2006), and 23 kDa from *Aspergillus parasiticus* (Tunga et al. 2003). The molecular weights of alkaline proteases of *Aspergillus* species are thus different from one another which may be ascribed to genetic differences.

Native PAGE and gelatin zymography

A single band at 16 ± 1 kDa was also seen (Fig. 1). The appearance of a single band for both native and SDS (Fig. 1a, b) gels suggested the purified alkaline protease to be a pure monomeric enzyme. To be convinced that the native PAGE band was not a contaminant; the gelatin zymography was carried out to show the activity of alkaline protease band. A clear colorless zone (of gelatin hydrolysis) against the blue background was observed (Fig. 1c). This further confirmed the homogeneity observed in both native and SDS-PAGE gels. The colorless zone resulted from the substrate degradation by proteases of *Aspergillus* species was also reported (Anandan et al. 2007; Charles et al. 2008).

Effect of pH on activity and stability of the alkaline protease

The enzyme was inactive at low pH and active over a broad alkaline pH range (8.0–12.0) with maximum activity at pH 11.0 (Table 2). A high protease activity in the alkaline region can be attributed to the better binding of the enzyme to the substrate since the pH strongly decides the binding of the enzyme to the substrate (Takami et al. 1990). The optimum pH in the 7.5–10.0 was reported for most of the alkaline proteases of *Aspergillus* species (Coral et al. 2003; Hossain et al. 2006; Sharma et al. 2006; Anandan et al. 2007; Ali 2008; Charles et al. 2008; Dubey et al. 2010; Choudhary and Jain 2012). The pH 10.5 was also observed for important enzymes used in detergents (subtilisin Carlberg and subtilisin Novo or BPN17) (Adinarayana et al. 2003). The variation in alkaline protease activities at different pH optima may be ascribed to the genetic variability among the different *Aspergillus* species.

The stability of the protease was studied at three different pHs viz. pH 10.0, 11.0 and 12.0. The enzyme remained unaltered at pH 11.0 and 12.0 for 24 h at 28 °C and a slight decrease in relative activity was seen at pH 10.0 after 24 h (data not shown). Good stability was also seen for an alkaline protease of *Aspergillus* species in the range of pH 8.0–9.0 for 20 min at 40 °C (Choudhary and Jain 2012), 7.0–11.0 at 50 °C for 1 h (Kalpana devi et al. 2008) and 6.0–12.0 for 30 min at 30 °C (Dubey et al. 2010). The enzyme recovered from the present study appeared to have higher stability than others purified from *Aspergillus* species since no one retained a significant

Fig. 2 Deconvoluted mass spectrum of alkaline protease of *A. terreus* gr.

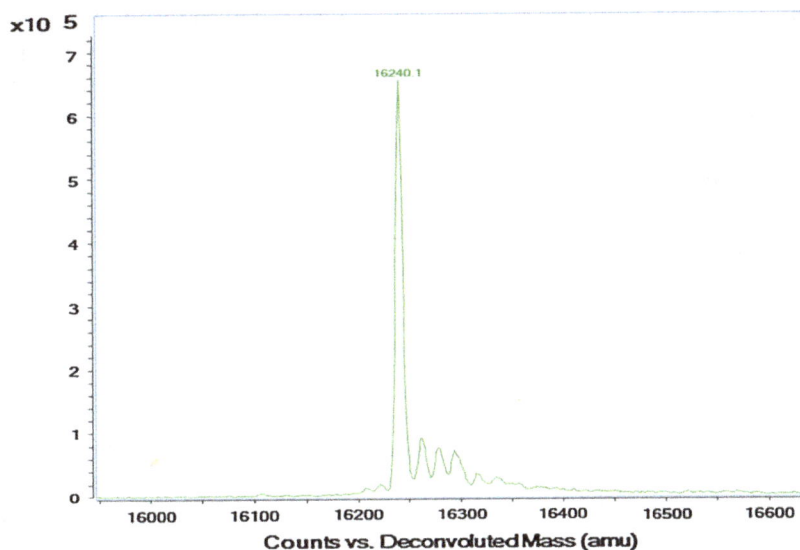

Table 2 Effect of pH, temperature, cations and specific inhibitors on the alkaline protease activity

pH	Enzyme activity (U/ml)	Temp. (°C)	Enzyme activity (U/ml)	Cation (10 mM)	Residual activity (%)	Specific inhibitor (5 mM)	Residual activity (%)
2	2.5 ± 0.5[e]	0	40.1 ± 2.7[b]	Control	100.0 ± 0.0[c,d]	Control	100.0 ± 0.0[a]
3	4.5 ± 0.6[e]	10	40.4 ± 1.3[b]	Mn^{2+}	85.9 ± 14.8[d,e]	PMSF	16.3 ± 3.7[e]
4	6.3 ± 1.5[e]	20	42.6 ± 2.3[b]	Fe^{3+}	81.1 ± 9.7[d,e]	TLCK	93.3 ± 3.9[a,b]
5	13.5 ± 0.6[d]	30	43.2 ± 0.2[b]	Hg^{2+}	74.8 ± 9.6[e]	DTT	90.1 ± 2.4[a,b]
6	28.7 ± 5.0[c]	40	45.2 ± 1.6[b]	Mg^{2+}	152.2 ± 17.6[b]	IAA	94.8 ± 2.1[a,b]
7	33.3 ± 1.1[b]	50	52.8 ± 2.5[a]	Ca^{2+}	182.6 ± 9.9[a]	NBS	49.7 ± 9.7[d]
8	37.3 ± 0.7[a,b]	60	52.3 ± 0.3[a]	Co^{2+}	121.1 ± 9.8[c]	NEM	88.8 ± 9.7[a,b]
9	38.7 ± 3.5[a]	70	52.1 ± 1.8[a]	Zn^{2+}	89.1 ± 12.8[d,e]	NaN_3	78.0 ± 11.0[c]
10	40.0 ± 0.9[a]	80	51.8 ± 0.8[a]	Cu^{2+}	100.67 ± 1.2[c,d]	EDTA	87.3 ± 5.7[b,c]
11	40.9 ± 3.0[a]	90	35.5 ± 2.3[c]	Fe^{2+}	99.7 ± 8.1[c,d]	DAN	91.7 ± 5.3[a,b]
12	40.6 ± 2.1[a]			Ba^{2+}	102.88 ± 11.0[c,d]	Urea	95.5 ± 4.1[a,b]
13	24.9 ± 4.0[c]			K^+	97.4 ± 4.4[d]		
				Na^+	92.54 ± 13.2[d]		

The values bearing the same letters in a column do not differ significantly at $P_{0.05}$

activity after 24 h, suggesting the enzyme to be used in detergent preparations.

Effect of temperature on activity and stability of the alkaline protease

The enzyme showed higher activity for most of the temperatures tested with an optimum at 50 °C although statistically at par with 60, 70 and 80 °C (Table 2). The appreciable activity of the alkaline protease at low, ambient and higher temperatures highlights its capacity to be used at any washing temperature. Beena et al. (2012) also purified the protease that was capable of being used in cold and hot washing conditions. The maximum activity at 50 °C was also recorded, while working with the protease

of *Aspergillus* species (Kalpana devi et al. 2008; Muthulakshmi et al. 2011). Most of the optimum temperatures reported for alkaline proteases of *Aspergillus* species are in the 30–45 °C range (Coral et al. 2003; Hossain et al. 2006; Anandan et al. 2007; Ali 2008; Charles et al. 2008; Dubey et al. 2010; Choudhary and Jain 2012).

When the thermostability profile examined, the retention of full activity was seen for over 4, 2, 1 h and 30 min at 50, 60, 70 and 80 °C, respectively. However, a loss in relative activity in the range of 1.3 % (50 °C) to 19.6 % (at 80 °C) was noted after 24 h (data not shown). The alkaline protease secreted by *A. terreus* retained full activity in the 40–90 °C range, but only for 1 h (Ali 2008). The alkaline protease produced by *A. niger* was 100 % stable at 40 °C for 1 h (Kalpana devi et al. 2008). The enzyme of the present study had a higher stability demonstrating its suitability in detergent formulations. This thermostability can be attributed to its glycoproteinic nature. An increased thermostability of many glycosylated proteins at elevated temperatures was due to the presence of the carbohydrate moiety (Ahmed et al. 2007).

Effect of various metal ions on alkaline protease activity

Ca^{2+} and Mg^{2+} enhanced the alkaline protease activity, while Co^{2+}, Fe^{2+}, Ba^{2+}, K^+, Na^+ and Cu^{2+} did not show any alteration in enzyme activity. A slight reduction in alkaline protease activity was observed with Mn^{2+}, Fe^{3+} and Zn^{2+}; however, a loss in enzyme activity of 25 % was observed with Hg^{2+} (Table 2). The stimulation of alkaline protease of *Aspergillus* species by Ca^{2+} and Mg^{2+} has been previously described (Sharma et al. 2006; Anandan et al. 2007; Kalpana devi et al. 2008; Dubey et al. 2010) suggesting the vital role of these cations in maintaining the active site and thus improving the enzyme thermostability. Similar to the alkaline protease of the present study, Mn^{2+}, Fe^{2+} and Zn^{2+} also inhibited the alkaline protease purified from *Aspergillus* species (Anandan et al. 2007; Kalpana devi et al. 2008). The enzyme was labile in the presence of Hg^{2+} as 25 % of the activity was lost. Hg^{2+} was also inhibited by the alkaline proteases of *Aspergillus* species (Sharma et al. 2006; Anandan et al. 2007).

Effect of different group-specific reagents on the activity of alkaline protease

The alkaline protease was unaltered by TLCK, DTT, IAA, NEM, EDTA, urea and DAN. A slight, moderate and strong inhibition was observed with NaN_3, NBS and PMSF, respectively (Table 2). The strong inhibition of the protease by PMSF suggested the serine residue in the active site. The essential serine residue was reported for the alkaline proteases of *Aspergillus* species (Coral et al. 2003; Hossain et al. 2006; Sharma et al. 2006; Anandan et al. 2007; Charles et al. 2008; Dubey et al. 2010). The moderate inhibition of alkaline protease by NBS may indicate that the tryptophan is near to the active site. Since the alkaline protease retained full activity in presence of TLCK, DTT, IAA, NEM, urea and DAN, the histidine and cysteine residues as well as sulfhydryl groups are not involved in the alkaline protease catalysis. Similarly, the –SH group was not involved in the alkaline protease of *A. tamarii* (Anandan et al. 2007). In this work, the enzyme activity was not lost in the presence of EDTA. Lack of inhibition by EDTA suggested that the enzyme not to be a metalloprotein. A similar result has been reported for alkaline proteases recovered from a fungal crude extract

(Sharma et al. 2006; Anandan et al. 2007). The stability of protease in the presence of chelating agents like EDTA is a requirement for any detergent enzyme since EDTA is used in detergent formulation as a water softener (Beg and Gupta 2003).

Broad substrate specificity

The enzyme was able to hydrolyze all the tested substrates with the greatest activity against casein although statistically at par with gelatin (data not shown). Similarly, the alkaline protease of *Aspergillus tamarii* was active towards casein than hemoglobin (Anandan et al. 2007). The alkaline protease of *A. niger* was also more active against casein than BSA and gelatin (Dubey et al. 2010). In contrast, the alkaline protease of *A. flavus* was active against gelatin than casein, egg albumin, and the BSA (Hossain et al. 2006). Since the alkaline protease of *A. terreus* gr. has the ability to hydrolyze various protein substrates, it can be a candidate of choice in detergent formulations as protein stain removal.

Determination of kinetic parameters

The K_M and V_{max} of the purified alkaline protease were determined from Lineweaver–Burk plot as 5.4 mg/ml and 12.8 U/ml, respectively (data not shown). Similarly, the K_M of 5 mg/ml was calculated for the alkaline protease purified from *A. oryzae* AWT 20 (Sharma et al. 2006). 0.6 mg/ml and 60 U/mg were recorded for the alkaline protease of *A. flavus* (Muthulakshmi et al. 2011), while 0.8 mg/ml and 85 U/mg of protein were seen for the alkaline protease of *A. niger* (Kalpana devi et al. 2008). The lower K_M seen indicated the higher affinity of the protease of *Aspergillus* species towards their substrates.

Compatibility of alkaline protease with detergent components

The stability of the enzyme in the presence of NaClO, H_2O_2, Tween 80, Triton X-100 and SDS was analyzed. The surfactants increased the activity (Fig. 3). The activation of alkaline protease in the presence of SDS indicates its suitability as detergent constituent. The mechanism proposed by Bajpai and Tyagi (2007) is that the negative charges of the SDS react with positive charges of calcium and magnesium present in the hard water/wash water thereby deactivating them. A better performance of protease was also observed with nonionic surfactant (Tween 80) (Beena et al. 2012). The enzyme was 100 % stable in the presence of bleaching and oxidizing agents (Fig. 3). It can thus withstand bleaching and oxidizing reactions taking place during washing. As the concentration of the

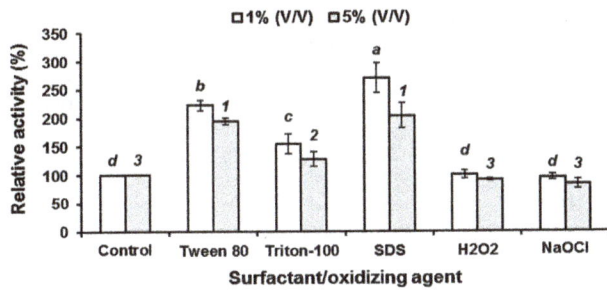

Fig. 3 Stability of the alkaline protease in various detergent components. The values bearing the same letters or numbers do not differ significantly at $P_{0.05}$

concerned compound increased from 1 to 5 %, a slight to moderate decrease in enzyme activity was noted (Fig. 3). Low concentration of surfactants and oxidizing agents is thus necessary for alkaline protease-based detergent formulations.

Storage stability of alkaline protease

The alkaline protease was stable for over the whole storage period studied as only 4, 7 and 9 % of its initial activity was lost after 40 days at 4, 28 and -20 °C, respectively

(data not shown). The protease inhibitors would have contributed to the marginal decline in enzyme activities on prolonged storage conditions. Low value was observed at -20 °C which may be due to ice crystal formation that inactivates enzymes (Beena et al. 2012). The long shelf life of the protease suggested its utilization in detergent industry.

Compatibility of alkaline protease with local detergents

To check if the enzyme can be commercially exploited in detergent industry, its stability and compatibility with 5 local powder detergents were investigated at room temperature (28 ± 2 °C), 60 and 90 °C. The protease was 100 % stable and compatible for 2 h at 60 °C with all the detergents except for Super wheel, and a retention of 83.98, 85.50, 85.57, 87.14 and 89.16 % was recorded after 24 h for Super wheel, More choice, Ariel, Henko and Surf excel, respectively (Fig. 4). The alkaline protease was also active and retained 79.06–83.2 and 55.75–75.14 % with tested detergents at 28 and 90 °C, respectively, after 24 h (data not shown).The detergent compatibility of the alkaline protease of *Aspergillus* species has been reported (Ali 2008; Kalpana devi et al. 2008; Dubey et al. 2010;

Fig. 4 Stability and compatibility of enzyme with detergents. The values bearing the same letters or numbers for each detergent do not differ significantly at $P_{0.05}$

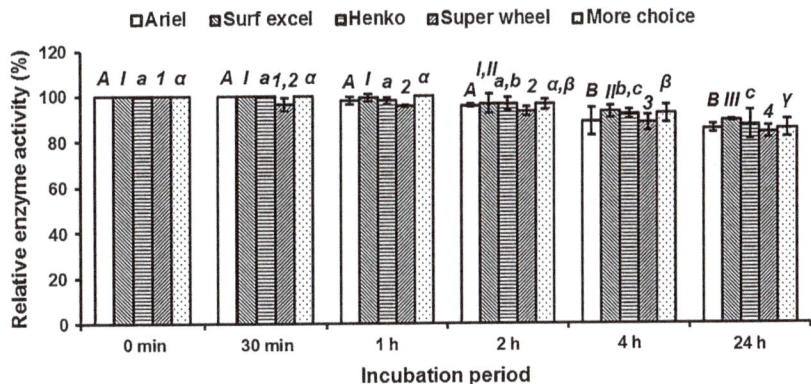

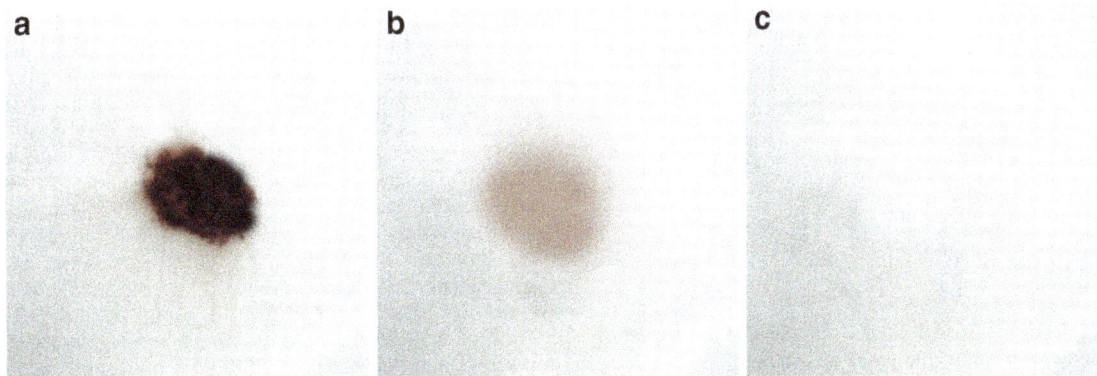

Fig. 5 Blood stain removal analysis of enzyme preparation. Unwashed stained cloth (**a**), blood stained cotton cloth washed with Super wheel and tap water (**b**), blood stained cloth washed with alkaline protease, super wheel and tap water (**c**)

Choudhary and Jain 2012), but no one was able to retain maximum activity after 1 h. The difference in enzyme stability of *Aspergillus* species in the presence of local detergents may be ascribed to the detergent composition. The enzyme was more active at room temperature and at 60 °C than at 90 °C. The high stability seen at room temperature had a good impact in saving electricity and in maintaining the quality of clothes.

Cleansing potential of the alkaline protease
as a detergent additive

The protease-Super wheel (Non enzymatic commercial detergent) preparation was able to completely remove blood stain from white cotton fabric (Fig. 5). Similarly, the blood stain removal was significantly improved by the alkaline protease supplementation to a detergent (Adinarayana et al. 2003; Kalpana devi et al. 2008). As the blood stained was removed at a lesser time when a detergent solution was with an enzyme, it can be inferred that the addition of an enzyme improved and accelerated the performance of the detergent solution.

Conclusion

The properties of the protease of *A. terreus* were investigated. The enzyme was a thermostable alkaline serine protease and showed high affinity for various protein substrates, stability and compatibility when mixed with surfactants, bleaches, oxidizing agents and local powder detergents. In addition, it was able to completely destain blood on white cloth. These properties suggested the present enzyme to be commercially exploited as an ideal candidate for detergent preparations.

Acknowledgments The authors acknowledge the Jain University for financial support.

Conflict of interest The authors declare that they have no conflict of interest in the publication.

References

Adinarayana K, Ellaiah P, Prasad DS (2003) Purification and partial characterization of thermostable serine alkaline protease from a newly isolated *Bacillus subtilis* PE-11. AAPS PharmSciTech 4(4):1–9

Ahmed SA, Saleh SA, Abdel-Fattah AF (2007) Stabilization of *Bacillus licheniformis* ATCC 21415 alkaline protease by immobilization and modification. Aust J Basic Appl Sci 1(3):313–322
Ali UF (2008) Utilization of whey amended with some agro-industrial by-products for the improvement of protease production by *Aspergillus terreus* and its compatibility with commercial detergents. Res J Agric Biol Sci 4(6):886–891
Anandan D, Marmer WN, Dudley RL (2007) Isolation, characterization and optimization of culture parameters for production of an alkaline protease isolated from *Aspergillus tamari*. J Ind Microbiol Biotechnol 34(5):339–347
Bajpai D, Tyagi VK (2007) Laundry detergents: an overview. J Oleo Sci 56(7):327–340
Beena AK, Geevarghese PI, Jayavardanan KK (2012) Detergent potential of a spoilage protease enzyme liberated by a psychrotrophic spore former isolated from sterilized skim milk. Am J Food Technol 7(2):89–95
Beg QK, Gupta R (2003) Purification and characterization of an oxidation-stable, thiol-dependent serine alkaline protease from *Bacillus mojavensis*. Enzyme Microbiol Technol 32(2):294–304
Bell DJ, Hoare M, Dunnill P (1983) The formation of protein precipitates and their centrifugal recovery. Adv Biochem Eng Biotechnol 26:1–72
Charles P, Devanathan V, Anbu P, Ponnuswamy MN, Kalaichelvan PT, Hur B (2008) Purification, characterization and crystallization of an extracellular alkaline protease from *Aspergillus nidulans* HA-10. J Basic Microbiol 48(5):347–352
Choudhary V, Jain PC (2012) Screening of alkaline protease production by fungal isolates from different habitats of Sagar and Jabalpur district (MP). J Acad Indus Res 1(4):215–220
Coral G, Arikan B, Unaldi MN, Guvenmez H (2003) Thermostable alkaline protease produced by an *Aspergillus niger* strain. Ann Microbiol 53(4):491–498
Dubey R, Adhikary S, Kumar J, Sinha N (2010) Isolation, production, purification, assay and characterization of alkaline protease enzyme from *Aspergillus niger* and its compatibility with commercial detergents. Dev Microbiol Mol Biol 1(1):75–94
Fujiwara N, Masui A, Imanaka T (1993) Purification and properties of the highly thermostable alkaline protease from an alkaliphilic and thermophilic *Bacillus* sp. J Biotechnol 30(2):245–256
Hasan F, Shah AA, Javed S, Hameed A (2010) Enzymes used in detergents: lipases. Afr J Biotechnol 9(2):4836–4844
Heussen C, Dowdle EB (1980) Electrophoretic analysis of plasminogen activity in polyacrylamide gels containing sodium dodecyl sulfate and copolymerized substrates. Anal Biochem 102:196–202
Hossain MT, Das F, Marzan LW, Rahman MS, Anwar MN (2006) Some properties of protease of the fungal strain *Aspergillus flavus*. Int J Agric Biol 8(2):162–164
Kalpana devi M, Banu AR, Gnanaprabhal GR, Pradeep BV, Palaniswamy M (2008) Purification, characterization of alkaline protease enzyme from native isolate *Aspergillus niger* and its compatibility with commercial detergents. Indian J Sci Technol 1(7):1–6
Kim W, Choi K, Kim Y, Park H, Choi J, Lee Y, Oh H, Kwon I, Lee S (1996) Purification and characterization of a fibrinolytic enzyme produced from *Bacillus* sp. strain CK 11-4 screened from Chungkook-Jang. Appl Environ Microbiol 62(7):2482–2488
Kirk O, Borchert TV, Fuglsang CC (2002) Industrial enzymes applications. Curr Opin Biotechnol 13(4):345–351
Kobayashi T, Hakamada Y, Hitomi J, Koike K, Ito S (1996) Purification of alkaline proteases from a *Bacillus* strain and their possible interrelationship. Appl Microbiol Biotechnol 45:63–71
Kumar CG, Takagi H (1999) Microbial alkaline protease: from bioindustrial viewpoint. Biotechnol Adv 17(7):561–594

Kumar A, Sachdev A, Balasubramanyam SD, Saxena AK, Lata A (2002) Optimization of conditions for production of neutral and alkaline protease from species of *Bacillus* and *Pseudomonas*. Ind J Microbiol 42:233–236

Laemmli UK (1970) Cleavage of structural proteins during the assembly of the head of bacteriophage T4. Nature 227:680–685

Lineweaver H, Burk D (1934) The determination of enzyme dissociation constants. J Am Chem Soc 56:658–666

Lowry OH, Rosenberg WJ, Farr AL, Randell RJ (1951) Quantitation of protein using Folin Ciocalteu reagent. J Biol Chem 193:265–275

Muthulakshmi CD, Gomathi DG, Kumar G, Ravikumar G, Kalaiselvi M, Uma C (2011) Production, purification and characterization of protease by *Aspergillus flavus* under solid state fermentation. JJBS 4(3):137–148

Niyonzima FN, More SS (2013) Screening and optimization of cultural parameters for an alkaline protease production by *Aspergillus terreus* gr under submerged fermentation. Int J Pharm Bio Sci 4(1):1016–1028

Pathak AP, Deshmukh KB (2012) Alkaline protease production, extraction and characterization from alkaliphilic *Bacillus licheniformis* KBDL4: a Lonar soda lake isolate. Indian J Exp Biol 50(8):569–576

Rao MB, Tanksale AM, Ghatge MS, Deshpande VV (1998) Molecular and biotechnological aspects of microbial protease. Microbiol Mol Bio Rev 62(3):597–635

Sabotiè J, Popoviè T, Brzin J (2009) Aspartic proteases from Basidiomycete *Clitocybe nebularis*. Croat Chem Acta 82(4):739–745

Sharma J, Singh A, Kumar R, Mittal A (2006) Partial purification of an alkaline protease from a new strain of *Aspergillus oryzae* AWT 20 and its enhanced stabilization in entrapped Ca-alginate beads. Internet J Microbiol 2(2):1–14.

Sigma DS, Mooser G (1975) Chemical studies of enzyme active sites. Annu Rev Biochem 44:889–931

Spivak JL, Small D, Hollenberg MD (1977) Erythropoietin: isolation by affinity chromatography with lectin-agarose derivatives. Proc Natl Acad Sci USA 74(10):4633–4635

Takami H, Akiba T, Horikoshai K (1990) Characterization of an alkaline protease from *Bacillus* sp. no. AH-101. Appl Microbiol Biotechnol 33(5):519–523

Tunga R, Shrivastava B, Banerjee R (2003) Purification and characterization of a protease from solid state cultures of *Aspergillus parasiticus*. Process Biochem 38(11):1553–1558

Wang SL, Chen YH, Wang CL, Yen YH, Chern MK (2005) Purification and characterization of a serine protease extracellularly produced by *Aspergillus fumigatus* in a shrimp and crab shell powder medium. Enzyme Microbiol Technol 36:660–665

13

Statistical approach to optimize production of biosurfactant by *Pseudomonas aeruginosa* 2297

**Arthala Praveen Kumar · Avilala Janardhan ·
Seela Radha · Buddolla Viswanath ·
Golla Narasimha**

Abstract The main objective of this paper is to optimize biosurfactant production by *Pseudomonas aeruginosa* 2297 with statistical approaches. Biosurfactant production from *P. aeruginosa* 2297 was carried out with different carbon sources, and maximum yield was achieved with sawdust followed by groundnut husk and glycerol. The produced biosurfactant has showed active emulsification and surface-active properties. From the kinetic growth modeling, the specific growth rate was calculated on sawdust and it was 1.12 day^{-1}. The maximum estimated value of product yield on biomass growth ($Y_{p/x}$) was 1.02 g/g. The important medium components identified by the Plackett–Burman method were sawdust and glycerol along with culture parameter pH. Box–Behnken response surface methodology was applied to optimize biosurfactant production. The obtained experimental result concludes that Box–Behnken designs are very effective statistical tools to improve biosurfactant production. These results may be useful to develop a high efficient production process of biosurfactant. In addition, this type of kinetic modeling approach may constitute a useful tool to design and scaling-up of bioreactors for the production of biosurfactant.

Keywords *Pseudomonas aeruginosa* · Biosurfactant · Kinetic growth modeling · Plackett–Burman · Response surface methodology (RSM)

A. P. Kumar · A. Janardhan · S. Radha · B. Viswanath ·
G. Narasimha (✉)
Applied Microbiology Laboratory, Department of Virology,
Sri Venkateswara University, Tirupati 517 502, India
e-mail: gnsimha123@rediffmail.com

Introduction

Microbial surfactants are structurally different group of surface-active biomolecules produced by a variety of microorganisms and are receiving considerable attention due to their unique properties such as higher biodegradability, lower toxicity, and greater stability (Mukherjee et al. 2006; Mulligan 2005). Biosurfactants are predominantly produced by bacteria, fungi, and yeasts include glycolipids, lipoaminoacids, lipopeptides, lipoproteins, lipopolysaccharides, phospholipids, monoglycerides, and diglycerides. Among these, the glycolipids produced by strains of *Pseudomonas* have received much attention due to their notable tensioactive and emulsifying properties (Maier and Soberon-Chavez 2000; Mulligan 2005). However, biosurfactants have limited applications owing to their high production costs, which can be lowered by process optimization, downstream processing strategies, agro-industrial waste fermentation, and use of hyper-producer strains (e.g., mutant and recombinant strains) (Wei et al. 2004; Perfumo et al. 2010). One of the methods which accomplished the above objective is the selection of suitable media components and optimal culture conditions to enhance biosurfactant productivity. The limitations of classical method of media optimization can be overcome by the application of single factor optimization process by statistical experimental design using Plackett–Burman design and response surface methodology (RSM) (Lotfy et al. 2007; Tanyildizi et al. 2005).

The Plackett–Burman design is a widely used statistical design technique for the screening of the medium components, and the variables screened by Plackett–Burman design were further optimized in a 2^3 factorial Box–Behnken design methodology (Plackett and Burman 1944; Box 1952). Response surface methodology (RSM) is the extensively used statistical technique for media optimization and for designing experiments, evaluating the effects of factor and relative significance and searching the optimum factors related to desired response. It has the intense ability to interpret the interactive effects among input variables are some attractive features of RSM (Al-Araji et al. 2007; Montgomery 1997). In the present study, we have applied response surface methodology (RSM) to enhance the production of biosurfactant by *Pseudomonas aeruginosa* 2297.

Materials and methods

Test organism

Pseudomonas aeruginosa (MTCC 2297) was obtained from the microbial type culture collection (MTCC), Chandigarh, India, and it was maintained on nutrient slants at 4 °C.

Examination of biosurfactant produced by *Pseudomonas aeruginosa*

Oil spreading method and oil collapse method

Oil spreading technique and oil collapsed method were carried out according to the method (Youssef et al. 2004).

Optimization of carbon and renewable sources for enhanced production of biosurfactant

P. aeruginosa was cultivated in 250 ml Erlenmeyer flasks containing 50 ml mineral salt medium (MSM) (g/l) (Camilios et al. 2008). To this medium, different carbon sources (2 %), i.e., glycerol, glucose, coconut oil, and groundnut oil, and different renewable sources (10 %), i.e., rice bran, sawdust, groundnut husk, and wheat bran, were added and sterilized at 121 °C, 15 lbs pressure for 15 min. After sterilization, the flasks were cooled to room temperature, and then, 2 % of overnight culture was inoculated and incubated at 30 °C in an orbital shaker for 6 days. For every 24 h, the broth was collected to analyze growth and biosurfactant production and their emulsification and surface activities.

Experimental design and statistical analysis

Plackett–Burman design

To find out the important medium components, a Plackett–Burman design was applied and it is a design of fractional plan. It allows the investigation of up to $N - 1$

variables with N experiments and assumes that there are no interactions between the different media components. For this study, six components were selected to evaluate their effect on biosurfactant production in 12 experiments, and surface tension was used as a response. Each column represents a different experimental trial, and each row represents different variables. Each variable was tested at two levels, a higher (+) and a lower (–) level (Table 1).

Response surface methodology

The optimized concentrations and interactions between the significant factors were identified by Plackett–Burman design and studied by using response surface methodology. RSM was used for experimental design based on the Box–Behnken design algorithm. The factors settings were tabulated (Table 3). MATLAB version 7.7.0 (R2008b) was used to create a 3-factor Box–Behnken design. The generalized polynomial model of three factors was as follows:

$$Y = \beta_0 + \beta_1 X_1 + \beta_2 X_2 + \beta_3 X_3 + \beta_{12} X_1 X_2 + \beta_{13} X_1 X_3 \\ + \beta_{23} X_2 X_3 + \beta_{11} X_1^2 + \beta_{22} X_2^2 + \beta_{33} X_3^2 \quad (1)$$

where Y—predicted response of fermentation X_1, X_2 and X_3 are the coded settings for three factors β_0—value of fitted response at the center point of the design β_1, β_2, and β_3—linear coefficients β_{12}, β_{13}, and β_{23}—interaction coefficients β_{11}, β_{22}, and β_{33}—quadratic coefficients.

Emulsification index (E_{24})

The emulsifying capacity was evaluated by an emulsification index (E_{24}). The E_{24} of culture samples was determined by adding 2 ml of oil and 2 ml of the cell-free broth to a test tube, vortexed at high speed for 2 min, and allowed to stand for 24 h. The E_{24} index is given as percentage of the height of emulsified layer (cm) divided by the total height of the liquid column (cm). The percentage of emulsification index was calculated by using the following equation (Tabatabaee et al. 2005; Sarubbo et al. 2006).

$$E_{24} = \text{Height of emulsion formed} \\ \times 100/\text{total height of solution}.$$

Extraction of biosurfactant

Initially, culture supernatant was adjusted to pH of 2.0 by adding 5 mol/l H_2SO_4 for precipitation of biosurfactant. The precipitates were extracted with two volumes of ethyl acetate. After vacuum evaporation of the solvents using rotary evaporator, crude biosurfactant was extracted

Table 1 High and low levels of factors with coded settings by PBD

Trial	X_1 Sawdust (gm)	X_2 Groundnut husk (g)	X_3 Glycerol (ml)	X_4 Groundnut oil (ml)	X_5 pH	X_6 Inoculum level (ml)	Surface tension (mN/m)
R1	+1 (10)	+1 (10)	+1 (3)	+1 (3)	+1 (9)	+1 (5)	41.32
R2	−1 (5)	+1 (10)	−1 (1)	+1 (3)	+1 (9)	+1 (5)	39.11
R3	−1 (5)	−1 (5)	+1 (3)	−1 (1)	+1 (9)	+1 (5)	40.02
R4	+1 (10)	−1 (5)	−1 (1)	+1 (3)	−1 (5)	+1 (5)	69.31
R5	−1 (5)	+1 (10)	−1 (1)	−1 (1)	+1 (9)	−1 (1)	49.03
R6	−1 (5)	−1 (5)	+1 (3)	−1 (1)	−1 (5)	+1 (5)	68.23
R7	−1 (5)	−1 (5)	−1 (1)	+1 (3)	−1 (5)	−1 (1)	70.62
R8	+1 (10)	−1 (5)	−1 (1)	−1 (1)	+1 (9)	−1 (1)	43.24
R9	+1 (10)	+1 (10)	−1 (1)	−1 (1)	−1 (5)	+1 (5)	69.04
R10	+1 (10)	+1 (10)	+1 (3)	−1 (1)	−1 (5)	−1 (1)	68.51
R11	−1 (5)	+1 (10)	+1 (3)	+1 (3)	−1 (5)	−1 (1)	68.72
R12	+1 (10)	−1 (5)	+1 (3)	+1 (3)	+1 (9)	−1 (1)	41.33

R1–R12 represents twelve different fermentations

with pellet form. The yielded pellets were applied onto thin-layer chromatography (TLC). Solvent mixture used in this study was chloroform, methanol, and water (65:25:4, V/V/V). Biosurfactant spots were detected by using orcinol reagent (Zhang Guo-liang et al. 2005).

Quantification of biosurfactant by orcinol method

The culture supernatant is acidified to pH 2 with 5 mol/l H_2SO_4 and kept at 4 °C overnight. Two hundred microliters of acidified culture supernatant was extracted three times with 1 ml of diethyl ether. Then, fractions were pooled, dried, and resuspended in 1 ml of 0.05 M sodium bicarbonate. Two hundred microliters of sample was treated with 1.8 ml of a orcinol solution (100 mg orcinol in 53 % H_2SO_4 and boiled for 20 min). After cooling at room temperature for 15 min, the readings are taken at 421 nm. Biosurfactant concentration was calculated from standard curves prepared with l-rhamnose and expressed as rhamnose equivalents (in mg/ml) (Chandrasekaran and BeMiller 1980; Koch et al. 1991).

Determination of surface tension

For surface tension measurements, 5 ml of broth supernatant was transferred to a glass tube that was submerged in a water bath at a constant temperature (28 °C). Surface tension was calculated by measuring the height reached by the liquid when freely ascended through a capillary tube (Munguia and Smith 2001). As control, non-inoculated broth was used, and the surface tension was calculated according to the following formula:

$$\gamma = \frac{rh\delta g}{2}$$

γ = surface tension (mN/m); d = density (g/ml); g = gravity (980 cm/s²); r = capillary radius (0.05 cm); h = height of the liquid column (cm).

Bacterial growth and dry cell weight

Cell growth was determined by monitoring the optical density of culture broth at 600 nm. The biomass was determined from the cells after centrifugation of the culture broth at 10,000 rpm (6,700g) 4 °C for 10 min. The dry cell weight (DCW) was obtained from the cell pellets by washing twice with distilled water and drying in hot air oven at 105 °C for 24 h (Suwansukho et al. 2008).

Kinetic model for growth and glycolipid product

Logistic equation: The sigmoidal shape of batch growth of *Pseudomonas aeruginosa* can be given as

$$\frac{dX}{dt} = \mu X \left(1 - \frac{X}{X_\infty}\right) \qquad (2)$$

where μ: specific growth rate, X_∞: maximum biomass concentration.

Logistic equations are a set of equations that characterize growth in terms of specific growth rate (Shuler and Kargi 2003). The integration of Eq. (2) with the boundary condition $X(0) = X_0$ yields the logistic curve.

$$X = \frac{X_0 e^{\mu t}}{1 - \frac{X_0}{X_\infty}(1 - e^{\mu t})} \qquad (3)$$

The above equation can be written as follows:

$$\mu = \frac{1}{\overline{X}}\frac{\Delta X}{\Delta t}\bigg/\left(1 - \frac{\overline{X}}{X_\infty}\right) \qquad (4)$$

$\overline{X}$: Average biomass concentration.

Leudeking–Piret model: Bio-products are classified as growth associated, non-growth associated, and mixed growth associated metabolites based on their production during log or stationary phases of cell growth. The kinetics of rate of product formation (r_p), in batch culture was described by Leudeking–Piret model and was given as follows:

$$r_p = \frac{dp}{dt} = \alpha(r_x) + \beta x \qquad (5)$$

where α—growth associated term; β—non-growth associated term.

The bio-product formation associated with exponential growth of *Pseudomonas aeruginosa* was modeled as

$$r_p = \frac{dp}{dt} = \alpha(r_x) \qquad (6)$$

A mixed growth associated product formation was modeled by the Eq. (8), and the product produced during stationary phase was modeled as follows:

$$r_p = \frac{dp}{dt} = \beta x \qquad (7)$$

The above equation is rearranged to calculate α and β by plotting a graph between $r_{p/x}$ and r_x/x.

The product yield on biomass growth,

$$Y_{P/X} = \frac{X - X_0}{P - P_0}. \qquad (8)$$

Results

In the present study, the screening of *Pseudomonas aeruginosa* 2297 was carried out using oil collapse and oil spreading techniques. It gave positive results for both oil collapse and oil spreading tests. The emulsification activities of the biosurfactant produced by *Pseudomonas aeruginosa* 2297 from different fermentation substrates and synthetic surfactants were tested with diesel, petrol, olive oil, and groundnut oil. When petrol and olive oil tested, maximum emulsification activity of 72.25 ± 2.47 and 59.23 ± 0.19 % was shown by biosurfactant produced from groundnut husk and coconut oil, respectively, used as a substrate and was comparable to all synthetic surfactants (Fig. 1). In diesel and groundnut oil, the synthetic surfactants (SDS) had the highest activity of 70.41 ± 0.56 and 62.31 ± 0.28 %, respectively, compared to all biosurfactants (Fig. 1). From the quantification experiment, the highest biosurfactant production was observed in sawdust

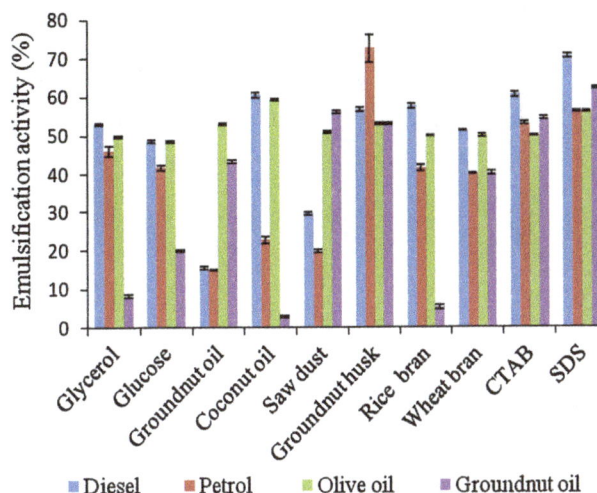

Fig. 1 Emulsification activity of different carbon and renewable sources from *P. aeruginosa*. *Values are represented as mean ±SD

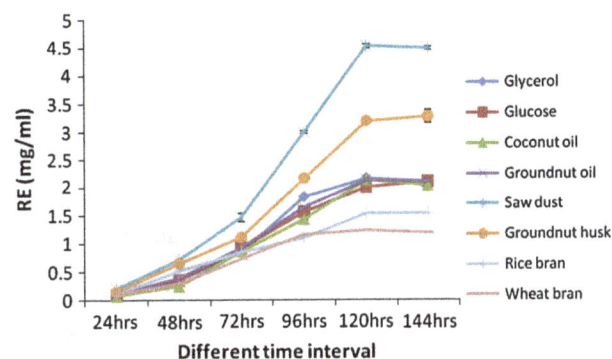

Fig. 2 Rhamnose equivalents of different carbon and renewable sources from *P. aeruginosa*. *Values are represented as mean ±SD

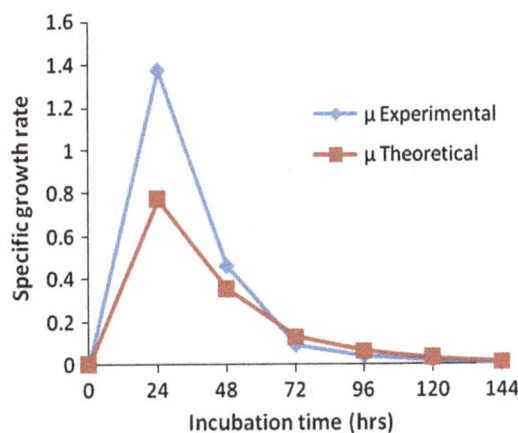

Fig. 3 Comparison of experimental and predicted specific growth rate of *P. aeruginosa*

with 4.53 ± 0.03 mg/ml (Fig. 2) followed by groundnut husk and glycerol as substrates.

The biomass concentration on sawdust was used in the calculation of specific growth rate, and it was obtained as

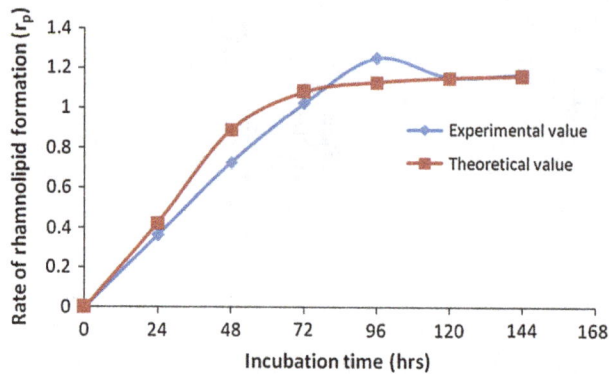

Fig. 4 Comparison of experimental and predicted rate of rhamnolipid formation (r_p) of *P. aeruginosa*

Table 2 Identification of significant substrates using PBD

Factor	Main effect	t stat
Sawdust	0.91	0.66
Groundnut husk	−0.91	−0.66
Glycerol	0.18	0.13
Groundnut oil	0.56	0.41
pH	−12.21	−8.86
Inoculum	−2.36	−1.71

Table 3 Three levels of substrates with actual and coded values and Box–Behnken experimental design matrix with experimental and predicted values of biosurfactant production

Trial	pH (X_1)	Glycerol (X_2) (ml)	Sawdust (X_3) (g)	Surface tension (mN/m), experimental	Surface tension (mN/m), predicted
1	−1 (5)	−1 (1)	0 (7.5)	0	5.28
2	−1 (5)	+1 (3)	0 (7.5)	0	5.43
3	+1 (9)	−1 (1)	0 (7.5)	69.32	74.52
4	+1 (9)	+1 (3)	0 (7.5)	58.91	52.63
5	−1 (5)	0 (2)	−1 (5)	0	0.65
6	−1 (5)	0 (2)	+1 (10)	0	0.25
7	+1 (9)	0 (2)	−1 (5)	62.33	62.55
8	+1 (9)	0 (2)	+1 (10)	62.33	61.65
9	0 (7)	−1 (1)	−1 (5)	69.04	63.78
10	0 (7)	−1 (1)	+1 (10)	68.54	62.45
11	0 (7)	+1 (3)	−1 (5)	39.23	62.08
12	0 (7)	+1 (3)	+1 (10)	41.31	63.76
13	0 (7)	0 (2)	0 (7.5)	62.33	62.33
14	0 (7)	0 (2)	0 (7.5)	62.33	62.33
15	0 (7)	0 (2)	0 (7.5)	62.33	62.33

R^2 (adj) = 93.36 %; R^2 = 97.63 %

1.12 day^{-1}, which would describe the most of the data. The lower value of specific growth rate (0.04 h^{-1}) was observed on cheap substrate (sawdust). The variation in experimental and predicted specific growth rate with time was depicted in Fig. 3. Higher value of F(11.14) and a very low value of P (<0.01) indicate that the logistic equation was the best-fit growth model for the *Pseudomonas aeruginosa* 2297 growth on sawdust consumption (Supplementary Table 1).

Based on the data obtained on the biomass and biosurfactant concentrations, the growth rate (r_x) and product formation (r_p) were calculated. The value of β was obtained as 0.186 with the negligible α which was an indication of the bio-product was non-growth associated product. From the above results, it was noticed that the more amount of product was formed during the stationary phase of *Pseudomonas aeruginosa* 2297. The comparison of experimental and predicted rate of biosurfactant production was depicted (Fig. 4). The estimated value of product yield on biomass growth ($Y_{p/x}$) was 1.02 g/g. The significance of model was verified with analysis of variance given in Supplementary Table 2. The higher value of F(13.19) and a very low value of P (<0.001) indicated the best fit of model to the experimental data.

In PBD approach, six parameters including carbon sources (groundnut oil and glycerol), renewable sources (groundnut husk and sawdust), and other factors (pH and inoculum level) were screened for biosurfactant production in 12 combinations with two test levels, and experiments were performed according to design matrix detailed in Table 1. From the PBD surface tension results, main effects were calculated at a confidence level of 95 % and are summarized in Table 2. Box–Behnken design (BBD) was adopted to optimize the levels of three identified substrates. The individual and interactive effects were noticed at three levels (−1, 0, and +1) of variables, and detailed design was given in Table 3. The response surface plot was represented as 3D plot with axes pH, sawdust, and

glycerol at three levels of −1, 0, and 1 (Fig. 5). The contour plots depicted interactions of two variables with fixed value of variable at its control level (Fig. 6). The outcomes of Table 3 revealed that the factors X_1, X_2, and X_1^2 (*P < 0.05) were significant, but the remaining terms were not significant (Supplementary Table 3). Computed F value (22.89) and P < 0.001 was an indication of fitness of polynomial model (Table 4).

Discussion

Selection of oil collapse and oil spreading methods was due to their strong advantages including simplicity, low cost, quick implementation, and use of relatively common equipment that is accessible in almost every microbiological laboratory. However, as expected, these methods are

Fig. 5 Quadratic surface model
for biosurfactant production by
P. aeruginosa

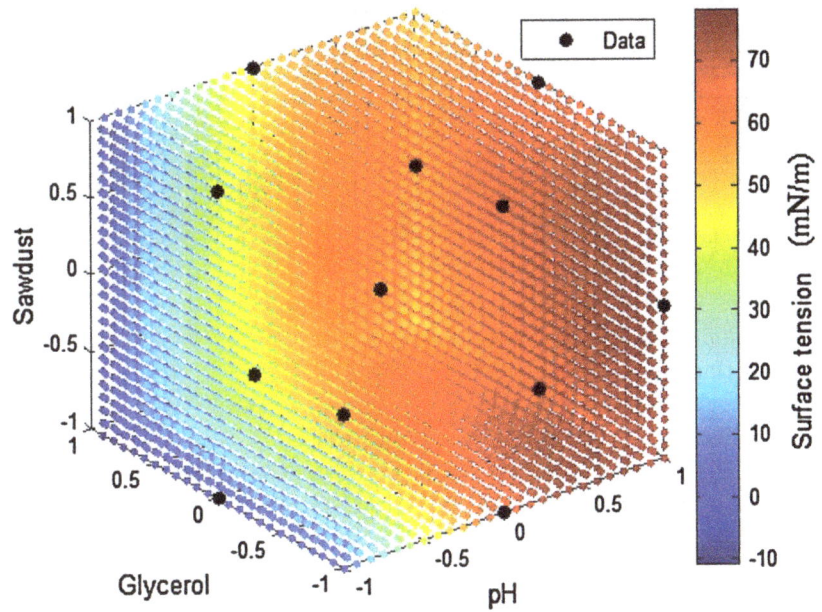

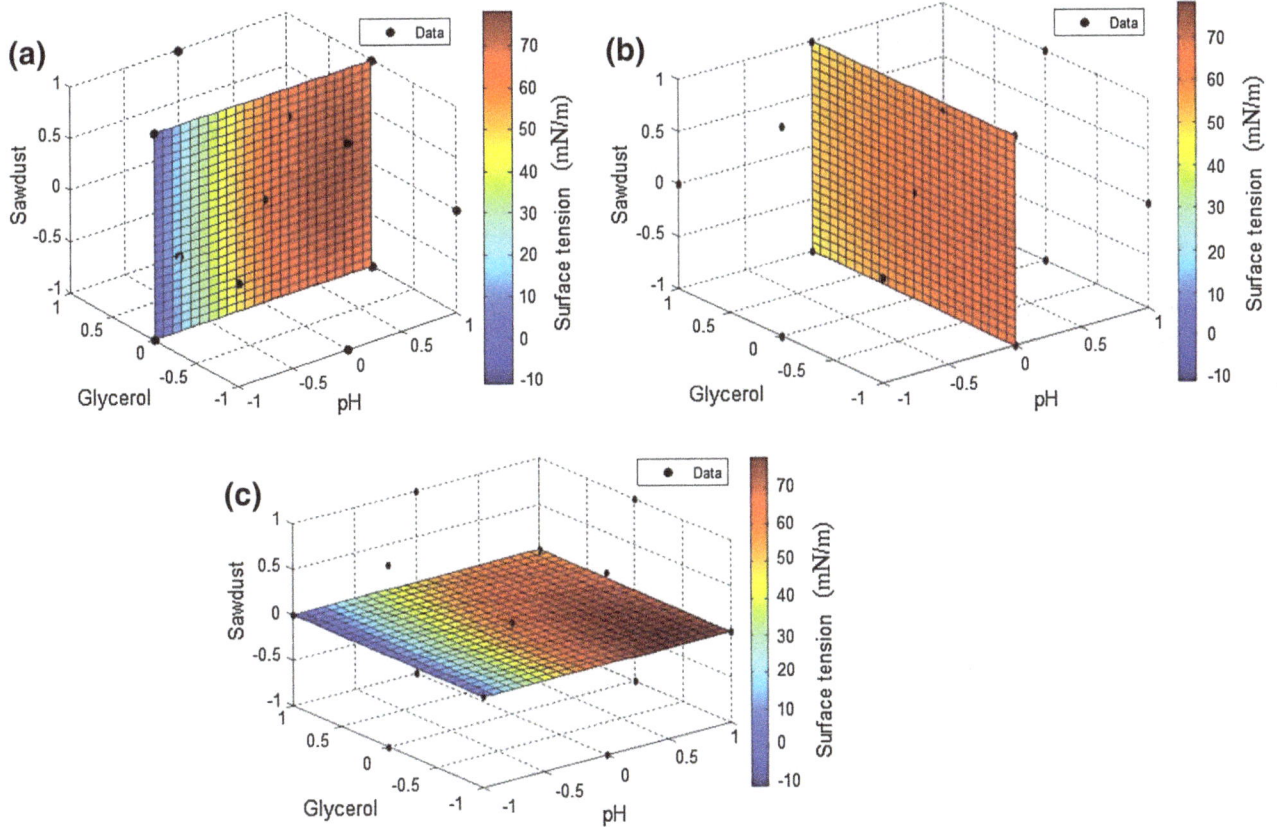

Fig. 6 **a** Interaction of pH and sawdust while glycerol at its control value, **b** interaction of glycerol and sawdust while pH at its control value, **c** interaction of glycerol and pH while sawdust at its control value

not perfect or flawless. The drop collapse method depends on the principle that a drop of liquid containing a biosurfactant collapses and spreads over the oily surface. There is a direct relationship between the diameter of the sample and concentration of the biosurfactant, and in contrast, the drop lacking biosurfactant remains beaded due to the hydrophobicity of the oil surface that cause aggregation of droplets (Youssef et al. 2004); Bodour and Miller-Maier

Table 4 Analysis of variance for the fitted second-order regression model for surface tension

Source	Df	SS	MS	F	P value
Regression	9	11,262.39	1,251.38	22.89	<0.001
Residual	5	273.39	54.68		
Total	14	11,535.78			

Df degrees of freedom, *SS* sum of squares, *MS* mean of squares, *F* Fischer's test value, *P value* probability value

1998; Christofi and Ivshina 2002; Bodour et al. 2003; Tugrul and Cansunar 2005; Krepsky et al. 2007), but this method is not sensitive in detecting low levels of biosurfactant production. The emulsification index range of 7.8–63.3 % EA from biosurfactant was reported on kerosene (Techaosi et al. 2007). In *Rhodococcus* strain, emulsification activity of 63 % was reported using sunflower frying oil as substrate (Sadouk et al. 2008).

The *Pseudomonas aeruginosa* 2297 was extensively utilizes the renewable sources. During the growth of *Pseudomonas aeruginosa,* it produces surface-active compounds, which were measured as rhamnose equivalents (RE). Similar type of result was reported by *Pseudomonas* strain (Deziel et al. 1996). Biosurfactant production, like cell growth, depends on the availability of the substrate. Agro-industrial wastes are considered as the promising substrate for biosurfactant production and can alleviate many processing industrial waste management problems (Makkar et al. 2011). In the present TLC studies, we found that the *Pseudomonas aeruginosa* have shown two spots, which resembles the presence of two types of biosurfactants produced by used organisms, and similar results were observed by Koch et al. (1991) and Matsufuji et al. (1997) with *Pseudomonas aeruginosa.*

According to Sastoque-Cala et al. (2010), the specific rate of 0.109 h^{-1} was obtained by using MMS broth from *Pseudomonas fluorescence.* Regarding the production rate, the Leudeking and Piret model was used to describe this parameter, as it is versatile for fitting product formation data obtained from several fermentation processes (Bailey and Ollis 1986). According to the Rashedi et al. (2005), the maximum yield of rhamnolipid ($Y_{p/x}$) from *Pseudomonas aeruginosa* was 0.21 g/g.

A suitable medium was also formulated through statistical optimization methodology since it has various advantages of being rapid and reliable in short listing of nutrients at varying concentrations leading to significant reduction in the total number of experiments. Two sequential steps of statistical approach such as Plackett–Burman design (PBD) and response surface methodology (RSM) were performed to design the optimized production medium for biosurfactant from *Pseudomonas aeruginosa.*

When additional factors are to be investigated, Plackett–Burman method may be adopted to find the variables influencing the metabolite production (Plackett and Burman 1944). This technique allows for the evaluation of '$n - 1$' variables in 'n' experiments at high and low levels, and 'n' must be the multiple of 4. Each row in the design matrix represents one experimental run, and each column represents the high and low values of one factor. It requires that the frequency of each level of a variable in a given column should be equal. Once the data are obtained for each trial, statistical analysis can be performed to evaluate and rank factors by their degree of impact on the fermentation process. This ranking begins with the calculation of main effect and probability value of each factor. The main effect of the factor refers to the change in response over the entire range (-1 to $+1$).

For a variable X,

$$\text{Main effect} = \frac{\sum X(H)}{n/2} - \frac{\sum X(L)}{n/2} \qquad (9)$$

where $\sum X(H)$: sum of high values of variable 'X' in one experimental run $\sum X(L)$: sum of low values of variable 'X' in one experimental run n: number of experimental runs

Software programs such as Microsoft Excel and MATLAB are used to calculate the main effects and P (probability) values of variables. A large estimate, either positive or negative, indicates that a factor has a large impact on metabolite productivity, while an estimate close to zero means that a factor has little or no effect. Generally, probability value <0.01 for a factor is considered as significant factor for the response of fermentation.

This statistical design is described by a first-order linear model as follows:

$$Y_S = \beta_0 + \sum_{i=1}^{n} \beta_i X_i \qquad (10)$$

Y_S: surface tension, β_i: linear coefficients, X_i: factors.

The confidence level of components below 95 % in biosurfactant production was considered insignificant. Here, positive effect means reduction in surface tension, and negative effect means increase in surface tension. Hence, the effect for each component was considered as opposite from calculated values, i.e., lower surface tension means positive effect, and higher value means negative effect.

Of the tested variables, the positive effects on biosurfactant production have shown by sawdust, groundnut oil, and glycerol, while the negative impacts had given by pH, inoculums size, and groundnut husk. Initial pH of fermentation medium has got the highest impact on the surface tension. Significance of the present experimental could be validated through analysis of variance (P value <0.001). Proposed linear model was as follows:

$$Y_s = 56.841 + 0.908\,X_1 - 0.908\,X_2 + 0.175\,X_3$$
$$+ 0.558\,X_4 - 12.802\,X_5 - 2.358X_6 \qquad (11)$$

Based on the main effects pH, sawdust and glycerol have profound effect on the biosurfactant production, and the exact optimal values of the individual factors are still unknown but can be optimized by RSM.

RSM is the process of adjusting variables toward the optimum response of fermentation. A second-order polynomial equation, fitted to data by multiple regression procedure, resulted in quadratic model, and it was given by Eq. 1.

$$Y = \beta_0 + \beta_1 X_1 + \beta_2 X_2 + \beta_3 X_3 + \beta_{12} X_1 X_2 + \beta_{13} X_1 X_3$$
$$+ \beta_{23} X_2 X_3 + \beta_{11} X_1^2 + \beta_{22} X_2^2 + \beta_{33} X_3^2 \qquad (1)$$

where Y—predicted response of fermentation X_1, X_2 and X_3 are the coded settings for three factors β_0—value of fitted response at the center point of the design β_1, β_2, and β_3—linear coefficients β_{12}, β_{13}, and β_{23}—interaction coefficients β_{11}, β_{22}, and β_{33}—quadratic coefficients.

An optimum combination of variables (pH: 7.37, sawdust: 7.656 g, and glycerol: 1.5 ml) was achieved through predicted plot of full quadratic model with surface tension of 70 ± 40 mN/m. To confirm the coefficients of second-order polynomial model by regression coefficient and analysis of variance (ANOVA), biosurfactant production was performed and the proposed model was written as follows:

$$Y = 62.3 + 31.6\,X_1 - 8.425\,X_2 + 0.2\,X_3 - 2.6\,X_1 X_2$$
$$+ 0.65\,X_2 X_3 - 26.8\,X_1^2 - 3.45\,X_2^2 - 4.35\,X_3^2 \qquad (12)$$

Results obtained from Table 3 illustrated that the factors X_1, X_2, and X_1^2 ($*P < 0.05$) were significant and the rest were not significant (Supplementary Table 3). Computed F value (22.89) and $P < 0.001$ was an indication of fitness of polynomial model (Table 4). The multiple correlation coefficient (R^2) was calculated as 97.63 %. This indicates that the second-order polynomial model could explain 97.63 % of variability in the response, and only 2.37 % of the total variations were not explained by the model. The adjusted R^2 (93.36 %) and predicted R^2 (97.63 %) were suggesting a high significance model used for analyzing the data.

Abalos et al. (2002) reported the utilization of response surface methodology to optimize the culture media for the production of rhamnolipids by *Pseudomonas aeruginosa* AT10. Similarly, Joshi et al. (2007) also reported the statistical optimization of medium components for the production of biosurfactant by *Bacillus licheniformis* K51. According to Joshi et al. (2007), Plackett–Burman and

Box–Behnken designs are very effective statistical tools to improving biosurfactant production.

Conclusion

In the present study, *P. aeruginosa* 2297 has showed good screening, emulsification, and surface-active properties. In the initial screening optimum process for carbon and renewable sources, sawdust, groundnut husk, groundnut, and glycerol were the best optimized substrates for biosurfactant production. The logistic equation and Leudeking–Piret models would fit to the growth of *Pseudomonas aeruginosa* on sawdust consumption and biosurfactant production. The statistical analysis of coefficient in Plackett–Burman design experiments demonstrates that pH, sawdust, and glycerol showed profound effect on the biosurfactant production. Optimization of these three selected variables while keeping the rest of the factors at their low levels through a Box–Behnken design shows maximum predicted biosurfactant production using pH: 7.37, sawdust: 7.656 g, and glycerol: 1.5 ml.

Acknowledgments The authors are thankful to university authorities of Sri Venkateswara University, Tirupati, Andhra Pradesh, India, for providing laboratory facilities.

Conflict of interest We are declaring that there is no conflict of interests regarding the publication of this article.

References

Abalos A, Maximo F, Manresa MA, Bastida J (2002) Utilization of response surface methodology to optimize the culture media for the production of rhamnolipids by *Pseudomonas aeruginosa* AT 10. J Chem Technol Biotechnol 77(7):777–784

Al-Araji L, Rahman R, Basri M, Salleh A (2007) Optimisation of rhamnolipids produced by *Pseudomonas aeruginosa* 181 using response surface modeling. Ann Microbiol 57(4):571–575

Bailey JE, Ollis DF (1986) Biochemical engineering fundamentals. McGraw-Hill International, London

Bodour AA, Miller-Maier RM (1998) Application of a modified drop-collapse technique for surfactant quantification and screening of biosurfactant producing microorganisms. J Microbiol Meth 32(3):273–280

Bodour AA, Drees KP, Maier RM (2003) Distribution of biosurfactant-producing bacteria in undisturbed and contaminated arid southwestern soils. Appl Environ Microbiol 69(6):3280–3287

Box GEP (1952) Multi-factor designs of first order. Biometrika 39(1–2):49–57

Camilios D, Meira JA, De Araujo JM, Mitchell DA, Krieger N (2008) Optimization of the production of rhamnolipids by *Pseudomonas aeruginosa* UFPEDA 614 in solid-state culture. Appl Microbiol Biotech 81(3):441–448

Chandrasekaran EV, BeMiller JN (1980) Constituent analysis of glycosaminoglycans. In: Whistler RL (ed) Methods in carbohydrate chemistry. Academic press Inc., New York, pp 89–96

Christofi N, Ivshina IB (2002) Microbial surfactants and their use in field studies of soil remediation. J Appl Microbiol 93(6):915–929

Deziel E, Paquette G, Villemur R, Lepine F, Bisaillion JG (1996) Biosurfactant production by a soil *Pseudomonas* strain growing on polycyclic aromatic hydrocarbons. Appl Environ Microbiol 62(6):1908–1912

Joshi S, Yadav S, Nerurkar A, Desai AJ (2007) Statistical optimization of medium components for the production of biosurfactant by *Bacillus licheniformis* K51. J Microbiol Biotechnol 17(2):313–319

Koch AK, Kappeli O, Feichter A, Reiser J (1991) Hydrocarbon assimilation and biosurfactant production in *Pseudomonas aeruginosa* mutants. J Bacteriol 173(13):4212–4219

Krepsky N, Da Silva FS, Fontana LF, Crapez M (2007) Alternative methodology for isolation of biosurfactant-producing bacteria. Braz J Biol 67(1):117–124

Lotfy WA, Ghanem KM, El-Helow ER (2007) Citric acid production by a novel *Aspergillus niger* isolate: II. Optimization of process parameters through statistical experimental designs. Bioresour Technol 98(18):3470–3477

Maier RM, Soberon-Chavez G (2000) *Pseudomonas aeruginosa* rhamnolipids: biosynthesis and potential applications. Appl Microbiol Biotech 54(5):625–633

Makkar RS, Cameotra SS, Banat IM (2011) Advances in utilization of renewable substrates for biosurfactant production. AMB Express 1:1–19

Matsufuji M, Nakata K, Yoshimoto A (1997) High production of rhamnolipids by *Pseudomonas aeruginosa* growing on ethanol. Biotechnol Lett 19(12):1213–1215

Montgomery DC (1997) Response surface methods and other approaches to process optimization. In: Montgomery DC (ed) Design and analysis of experiments. Wiley, New York, pp 427–510

Mukherjee S, Das P, Sen R (2006) Towards commercial production of microbial surfactants. Trends Biotech 24(11):509–515

Mulligan CN (2005) Environmental application for biosurfactants. Environ Pollut 133(2):183–198

Munguia T, Smith CA (2001) Surface tension determination through capillary rise and laser diffraction patterns. J Chem Educ 78(3):343–344

Perfumo A, Rancich I, Banat IM (2010) Possibilities and challenges for biosurfactants use in petroleum industry. Adv Exp Med Biol 672:135–145

Plackett RL, Burman JP (1944) The design of optimum multifactorial experiments. Biometrika 33(4):305–325

Rashedi H, Mazaheri Assadi M, Bonakdarpour B, Jamshidi E (2005) Environmental importance of rhamnolipid production from molasses as a carbon source. Int J Enviorn Sci Technol 2(1):59–62

Sadouk G, Hacene H, Tazerouti A (2008) Biosurfactants production from low cost substrate and degradation of diesel oil by a *Rhodococcus* strain. Oil Gas Sci Technol 63(6):747–753

Sarubbo LA, de Luna JM, de Campos-Takaki GM (2006) Production and stability studies of the bioemulsifiers obtained from a strain of *Candida glabrata* UCP 1002. Electron J Biotechol 9(4):400–406

Sastoque-Cala L, Cotes-Prado AM, Rodríguez-Vázquez R, Pedroza-Rodríguez AM (2010) Effect of nutrients and fermentation conditions on the production of biosurfactants using rhizobacteria isolated from fique plants. Universitas Scientiarum 15(3):251–264

Shuler ML, Kargi F (2003) A text book of bioprocess engineering-basic concepts, chapter 6, 2nd edn. Pearson Education, pp 175–177

Suwansukho P, Rukachisirikul V, Kawai F, Aran H-K (2008) Production and applications of biosurfactant from *Bacillus subtilis* MUV4. Songklanakarin J Sci Technol 30:87–93

Tabatabaee A, Assadi MM, Noohi AA, Sajadian VA (2005) Isolation of biosurfactant producing bacteria from oil reservoirs. Iran J Environ Health Sci Eng 2(1):6–12

Tanyildizi MS, Ozer D, Elibol M (2005) Optimization of α-amylase production by *Bacillus* sp. using response surface methodology. Process Biochem 40(7):2291–2296

Techaosi S, Leelapornpisid P, Santiarwarn D, Lumyong S (2007) Preliminary screening of biosurfactant-producing microorganisms isolated from hot spring and garages in Northern Thailand. KMITL Sci Technol J 7(S1):38–43

Tugrul T, Cansunar E (2005) Detecting surfactant-producing microorganisms by the drop-collapse test. World J Microbiol Biotechnol 21(6–7):851–853

Wei YH, Lai HC, Chen SY, Yeh MS, Chang JS (2004) Biosurfactant production by *Serratia marcescens* SS-1 and its isogenic strain SMdeltaR defective in SpnR, a quorum-sensing LuxR family protein. Biotech Lett 26(10):799–802

Youssef NH, Duncan KE, Nagle DP, Savage KN, Knapp RM, McInerney MJ (2004) Comparison of methods to detect biosurfactant production by diverse microorganism. J Microbiol Method 56(3):339–347

Zhang G-L, Wu Y-T, Qian X-P, Meng Q (2005) Biodegradation of crude oil by *Pseudomonas aeruginosa* in the presence of rhamnolipids. J Zhejiang Univ Sci B 6(8):725–730

14

Kinetics of rapamycin production by *Streptomyces hygroscopicus* MTCC 4003

Subhasish Dutta · Bikram Basak · Biswanath Bhunia · Samayita Chakraborty · Apurba Dey

Abstract Research work was carried out to describe the kinetics of cell growth, substrate consumption and product formation in batch fermentation of rapamycin using shake flask as well as laboratory-scale fermentor. Fructose was used as the sole carbon source in the fermentation media. Optimization of fermentation parameters and reliable mathematical models were used for the maximum production of rapamycin from *Streptomyces hygroscopicus* MTCC 4003. The experimental data for microbial production of rapamycin fitted well with the proposed mathematical models. Kinetic parameters were evaluated using best fit unstructured models, viz. Andrew's model, Monod model, Yano model, Aiba model. Andrew's model showed a comparatively better R^2 value (0.9849) among all tested models. The values of maximum specific growth rate (μ_{max}), saturation constant (K_S), inhibition constant (K_i), and growth yield coefficient ($Y_{X/S}$) were found to be 0.008 (h^{-1}), 2.835 (g/L), 0.0738 (g/L), and 0.1708 (g g^{-1}), respectively. The optimum production of rapamycin was obtained at 300 rpm agitation and 1 vvm aeration rate in the fermentor. The final production of rapamycin in shake flask was 539 mg/L. Rapamycin titer found in bioreactor was 1,316 mg/L which is 52 % higher than the latest maximum value reported in the literature.

Keywords Rapamycin · Growth kinetics · *Streptomyces hygroscopicus* · Antibiotic · Substrate inhibition

S. Dutta · B. Basak · S. Chakraborty · A. Dey (✉)
Department of Biotechnology, National Institute of Technology Durgapur, Mahatma Gandhi Avenue, Durgapur 713209, India
e-mail: apurbadey.bt@gmail.com; apurba.dey@bt.nitdgp.ac.in

B. Bhunia
Department of Bio Engineering, National Institute of Technology Agartala, Barjala, Tripura 799055, India

Abbreviations

dP/dt	Production rate (mg/L h^{-1})
K_S	Saturation coefficient (g/L)
K_i	Inhibition coefficient (g/L)
m	Cell maintenance coefficient
P_0, P	Product concentration at 0th h of fermentation (mg/L), product concentration at particular time of fermentation (mg/L)
R^2	Regression coefficient
R_t	HPLC retention time (min)
S_0, S	Initial substrate concentration (g/L), limited substrate concentration (g/L)
t	Incubation time (h)
vvm	Volume of air per volume of fermentation media per minute (m^3)
X_0, X	Initial cell mass concentration (g/L), cell mass concentration at any time of fermentation (g/L)
$Y_{X/S}$	Growth yield coefficient per unit substrate consumed (g g^{-1})

Greek symbols

μ, μ_{max}	Specific growth rate (h^{-1}), maximum specific growth rate (h^{-1})

Introduction

Rapamycin, also known as sirolimus, is an antibiotic commonly used as a potent antifungal and immunosuppressant drug produced by the soilborne actinomycete *Streptomyces hygroscopicus* (Fang and Demain 1995). It has also been reported to have antitumor, neuroprotective and anti-aging properties (Zou and Li 2013). In recent time, because of its exceptional biological and pharmaceutical potential it has generated interest for its therapeutic use

(Graziani 2009; Sehgal 2003). Its immunosuppressive activity due to inhibition of T-cell activation and proliferation has led to its potential use in clinical treatment of graft rejection in organ transplant (Weber et al. 2005) and autoimmune diseases such as rheumatoid arthritis (Foroncewicz et al. 2005). In addition to its therapeutic usefulness, semisynthetic derivatives of rapamycin has also been shown to display activities against cancer (Park et al. 2010), Parkinson's disease (Tain et al. 2009), and AIDS (Nicoletti et al. 2009). Its anti-aging activity has been published recently when scientists observed its ability to extend the life span of mice (Zou and Li 2013).

The mechanism of action of rapamycin is distinct from that of cyclosporine A and FK-506. The latter drugs inhibit the first phase of T-cell activation by blocking calcineurin, serine/threonine phosphatase (transcriptional activator of IL-2 gene); whereas, rapamycin interferes with the second phase of T-cell activation by blocking the IL-2 dependent signal transduction (Morelon et al. 2001). The anti-proliferative effects of rapamycin are mediated through the formation of an active complex with a cytosolic protein FK-506 and binding protein (FKBP12) allowing this drug receptor complex to interact with a putative lipid kinase (Wiederrecht et al. 1995), and inhibits the 289 kDa RAFT/FRAP proteins called mTOR (mammalian target of rapamycin) (Sabatini et al. 1995).

Despite the versatile activities and resultant demand of this drug, its use may be trimmed because of the low titer of rapamycin produced by S. hygroscopicus. The low titer of this drug produced by the organism has now become a rate-limiting factor in further development and industrialization of this natural product (Zhu et al. 2010). In the past decades, most efforts have focussed on rapamycin biosynthesis (Park et al. 2010; Graziani 2009), its pharmaceutical activities (Prapagdee et al. 2008; Park et al. 2010; Weber et al. 2005; Foroncewicz et al. 2005; Nicoletti et al. 2009), strain improvement (Zhu et al. 2010; Xu et al. 2005; Chen et al. 2009) and optimization of physiological factors for better production of rapamycin (Zou and Li 2013; Chen et al. 2008). In the literature, although there are many reports describing optimization of the media (Refaat and Abdel-Fatah 2008) and improvement of strain for better production of rapamycin (Xu et al. 2005; Jung et al. 2011), kinetic studies of growth and rapamycin production by S. hygroscopicus have not been satisfactorily done yet. Though an attempt was made by Schuhmann and Bergter investigating the branch formation, and cytological properties of mycelial growth of Streptomyces hygroscopicus on solid media (Schuhmann and Bergter 1976), the actual growth and rapamycin production kinetic parameters were not determined.

The metabolism and product formation pattern of a microorganism depend mainly on their fermentative,

nutritional, physiological, and genetic nature (Prakasham et al. 2007; Bhunia et al. 2012). Exploitation of such microbial metabolism for the desired product formation by regulating the critical fermentation parameters helps in commercialization (Subba Rao et al. 2008). Hence, careful kinetic studies are required to monitor the growth of microorganisms and product formation pattern in presence of various substrates. Kinetic studies provide good quantitative information regarding the behavior of a system, which is essential for study of growth of the organism and consequent product formation (Bhunia et al. 2012).

In the present study, we used various unstructured kinetic models to characterize the growth and rapamycin production by S. hygroscopicus. The kinetic parameters obtained by fitting the experimental data of growth and product formation with the unstructured models can be effectively used to explain the relationship between microbial growth and substrate utilization.

Materials and methods

Chemicals

All chemicals used were of analytical and HPLC grade and purchased from Sigma Aldrich (USA), Himedia (India), and Merck (India). HPLC-grade rapamycin standard was purchased from Merck, Germany. Deionized water used for HPLC analysis was prepared by Ultrapure Water System (Arium®, 611UF, Sartorius, Germany).

Microorganisms

Streptomyces hygroscopicus MTCC 4003 and Candida albicans MTCC 227 (test organism) were procured in lyophilized form from microbial type culture collection (MTCC) Chandigarh, India. S. hygroscopicus was grown and maintained in a medium consisting of (in g/L), glucose 4; yeast extract 4; malt extract 10; $CaCO_3$ 2. The pH of the medium was maintained at 7.2. C. albicans was grown on MYGP medium having the following composition (g/L): malt extract 3; yeast extract 3; peptone 5; glucose 10 (pH 7).

Preparation of inoculum

Inoculum (seed culture) was prepared by inoculating thawed S. hygroscopicus spores into 250-mL Erlenmeyer flask containing 100 mL growth medium with the help of a sterile inoculating loop under aseptic condition. The flasks were then incubated at 25 °C and 120 rpm for 7 days. S. hygroscopicus and C. albicans were maintained by bimonthly transfer to fresh medium and stored at 4 °C after incubation at 25 °C for 5 days.

Fermentation in shake flask

Two percent (v/v) seed culture was transferred into the production media (previously optimized in our lab) having the following composition (g/L): fructose 22; mannose 5; malt extract 10; casein 0.3; $(NH_4)_2SO_4$ 5.3; NaCl 5; K_2HPO_4 4; $ZnSO_4 \cdot 7H_2O$ 0.06; $MgSO_4 \cdot 7H_2O$ 0.0025; $MnSO_4 \cdot H_2O$ 0.012; $FeSO_4 \cdot 7H_2O$ 0.1; $CoCl_2 \cdot 6H_2O$ 0.010; Na_2SO_4 0.3; $CaCO_3$ 3 (pH 7.2). In another experiment, seed culture was inoculated into the same production media devoid of mannose where the amount of mannose was supplemented by fructose, as both have same empirical formula and molecular mass. Similarly, another two sets of production media were used which incorporated same amount of glucose, replacing fructose and mannose. Fructose, glucose and mannose were autoclaved separately and added to the production media aseptically so as to achieve the desired concentration in the media. Fermentation was carried out in 250-mL Erlenmeyer flasks, each contained 50 mL production medium and was incubated at 25 °C and 120 rpm for 7 days. All the experiments were performed in triplicate.

Fermentation in stirred tank reactor (STR)

Further work was done in the 2.2-L stirred tank reactor (STR) with 2-L working volume and Biocommand Plus fermentation supervising software (New Brunswick Scientific Co. Inc. USA) for advanced online control which result in higher cell density and rapamycin productivity. The reactor vessel containing production medium was sterilized by autoclaving along with the silicon tubes of 0.5 cm diameter and reagent bottles containing NaOH, HCl, antifoam agent (silicon oil) and one empty reagent bottle used for transferring the inoculum at 15 psi and 121.5 °C for 15 min. The pH probe was calibrated before autoclaving the reactor and set to 7.24 while the dissolved oxygen (DO) probe was calibrated after autoclaving and set to 100 %. pH was maintained at the set point 7.24 by supplying 0.1 (N) NaOH and 0.1 (N) HCl automatically to the fermentor with the help of peristaltic pumps. DO level was controlled by agitator speed and aeration rate during fermentation. Agitation was set at 300 rpm with six-bladed turbine impellers. Compressed sterile air was sparged at 1 vvm cultivating for 7 days at 25 °C.

Analytical procedures

During fermentation in shake flasks, the microbial growth under the submerged conditions appeared as spherical pellets. Hence, after inoculation, 50 mL of samples were taken from shake flask in 50 mL centrifuge tube at every 24 h interval starting from 0th h and centrifuged at 3,500 rpm for 15 min (Cheng 1995). The sample from STR was collected in every 24 h interval in a 50-mL centrifuge tube and centrifuged. After centrifugation, the supernatant obtained was taken in separate centrifuge tube and the cell pellet was washed twice with 3 mL of methanol by centrifuging at 200 rpm for 15 min. The methanolic extract was mixed with the supernatant and used for rapamycin concentration (Sallam et al. 2010). The cell pellets were dried at 80 °C to a constant weight for 48 h in a hot air oven. The biomass was expressed as dry weight in g/L.

Residual fructose concentration was determined by 3,5-dinitrosalicylic acid (DNS) method (Miller 1959). A standard calibration curve of optical density versus known concentration of fructose was prepared and the unknown concentration of residual fructose in the media was determined from this curve.

Bioassay determination of rapamycin was performed by "paper-disc agar diffusion method" as described by Kojima et al. (1995). The assay was conducted in agar plates of assay medium ("Microorganisms") seeded with *C. albicans* MTCC 227 as test organism. From the mean diameter of the inhibition zones the concentration of rapamycin was empirically determined. Rapamycin concentration was determined using high performance liquid chromatography (HPLC). A calibration curve of different concentrations of rapamycin standard versus area of HPLC peak was plotted and unknown concentration of rapamycin in the supernatant was calculated using the linear equation obtained from the calibration curve. All experiments were done in triplicate. For sample preparation, the supernatant was filtered through 0.22-μm membrane filters for organic solvents (Pall Corporation, India). After appropriate dilutions with HPLC-grade methanol, samples were analyzed by HPLC system (Waters™ 600) equipped with UV/visible detector and a C_{18} hypersil column (4.6 × 250 mm; 5 μm particle size; Waters, USA). Mobile phase used was methanol:acetonitrile (80:20 v/v), at a flow rate of 1 mL/min. An aliquot of 20 μL of filtrate supernatant was injected and analyzed at 272 nm using the UV/visible detector.

Mathematical background

In Monod's model, the growth rate is related to the concentration of a single growth-limiting substrate through the parameters μ_{max} and K_S. Monod model also relates the yield coefficient ($Y_{X/S}$) to the μ (Okpokwasili and Nweke 2005). The specific growth rate in the exponential phase was calculated using the following equation:

$$\frac{dX}{dt} = \mu X \qquad (1)$$

GraphPad Prism 5 software was used to calculate the kinetics parameters from the Monod equation:

$$\mu = \frac{\mu_{\max} S}{K_S + S} \quad (2)$$

When higher substrate concentration inhibits cell growth, the original Monod model becomes unsatisfactory. In this case, Monod derivatives that provided corrections for substrate inhibition (by incorporating the inhibition coefficient K_i) can be used to describe the kinetics study. Among the substrate inhibition models, the Andrew's equation is most widely used (Okpokwasili and Nweke 2005). After evaluating several kinetic models (Yano, Aiba) the Andrew's model was best fitted and gave the highest R^2 value.

$$\mu = \frac{\mu_{\max} S}{S^2 K_i + S + K_S} \quad (3)$$

The substrate utilization kinetics is given by Eq. (4). A carbon substrate is used to form cell material and metabolic products as well as for maintenance of the cell.

$$\frac{dS}{dt} = -\frac{1}{Y_{X/S}}\frac{dX}{dt} - \frac{1}{Y_{P/S}}\frac{dP}{dt} - mX \quad (4)$$

However, if the substrate used for product formation and cell maintenance is assumed to be negligible, Eq (4) can be written as:

$$\frac{dS}{dt} = -\frac{1}{Y_{X/S}}\frac{dX}{dt} \quad (5)$$

Now, $Y_{X/S}$ is the ratio of cell mass growth and substrate concentration used for cell growth. $Y_{X/S}$ can be expressed as:

$$Y_{X/S} = -\frac{dX}{dS} \quad (6)$$

$Y_{X/S}$ was calculated from experimental data using the Eq. (7).

$$Y_{X/S} = \frac{X - X_0}{S_0 - S} \quad (7)$$

Results and discussion

For rapamycin production, we evaluated four different combinations (1. glucose, 2. fructose, 3. fructose + mannose and 4. mannose) of carbon sources for better production. It can be seen in Fig. 1, among the carbohydrates used fructose yielded highest rapamycin titer. Catabolite repression may be the most likely reason for this lagging effect (Priest 1977; Kumar et al. 1999). It was previously established that a catabolite control protein (CcpA) was responsible for this regulatory mechanisms which transduced signal for the repression in rapamycin. Synthesis (Tobisch et al. 1999). The result is as accordance with Kojima et al. (1995).Therefore, we selected fructose in

further studies replacing all other carbon sources by the same amount as they have same molecular mass (180.16 g/mol) as well as same empirical formula ($C_6H_{12}O_6$). Although carbohydrates are also present in other media constituents, viz. malt extract, it is hard to analyze the amount of sugar present in the constituents and the amount utilized. Hence, analysis of substrate utilization was done by assuming supplemented carbohydrates as sole source of carbon and energy, thereby neglecting the amount of carbohydrate present in other media constituents. This result contradicts the findings reported by Lee et al., where they showed that a combination of fructose and mannose resulted in good growth of the organisms and consequent higher rapamycin titer (Lee et al. 1997). However, our result suggested that fructose together with nitrogen source casein and $(NH_4)_2SO_4$ supported good growth and rapamycin production by the employed strain (Fig. 1).

To study the effect of S on μ, μ for different S was calculated using Eq. (1) while X was obtained at different intervals of the incubation period. The values of μ for different S were fitted to Monod's model using GraphPad Prism 5 software. The growth kinetic parameters, viz. $\mu_{\max}$ and K_S were found to be 0.003869 h^{-1} and 0.8271 g/L. The lower correlation coefficient (R^2) value of 0.8713 found in Monod's model suggested that there was reasonable substrate or product inhibition on growth of the organism (Bhunia et al. 2012). Analysis of the Monod's model under different substrate concentration conditions suggested that fructose concentration regulates the microbial growth pattern. Since, there was no report of product inhibition for rapamycin production, the effect of substrate inhibition was only considered for modeling. Several substrate inhibition kinetic models were examined and compared in this work (Table 1). Andrew's model for substrate inhibition on microbial growth was found to fit the experimental values well, since an R^2 value of 0.9849 was obtained (Fig. 2a). The values of $\mu_{\max}$, K_S and K_i were found to be 0.0083 h^{-1}, 2.835 and 0.073 g/L, respectively. The higher R^2 value found with Andrew's model indicates that it is comparatively better fitted model for the experimental data than Monod and other substrate inhibition models (Table 1). In rapamycin production, the increase in biomass concentration was accompanied by a decrease of fructose concentration. It is assumed that fructose is consumed for cell growth and cell maintenance. $Y_{X/S}$ value was determined by averaging $Y_{X/S}$ values calculated using Eq. (7) at different data points (Fig. 2b). Its values ranged from 0.0598 to 0.1708 and were found to be maximal at lower S and minimal at higher S because of the substrate inhibition on $Y_{X/S}$. The average $Y_{X/S}$ value was calculated to be 0.107 g g^{-1} and was fairly constant up to fructose concentration of 0.548 g/L and then decreased to minimal at fructose concentration of 27 g/L.

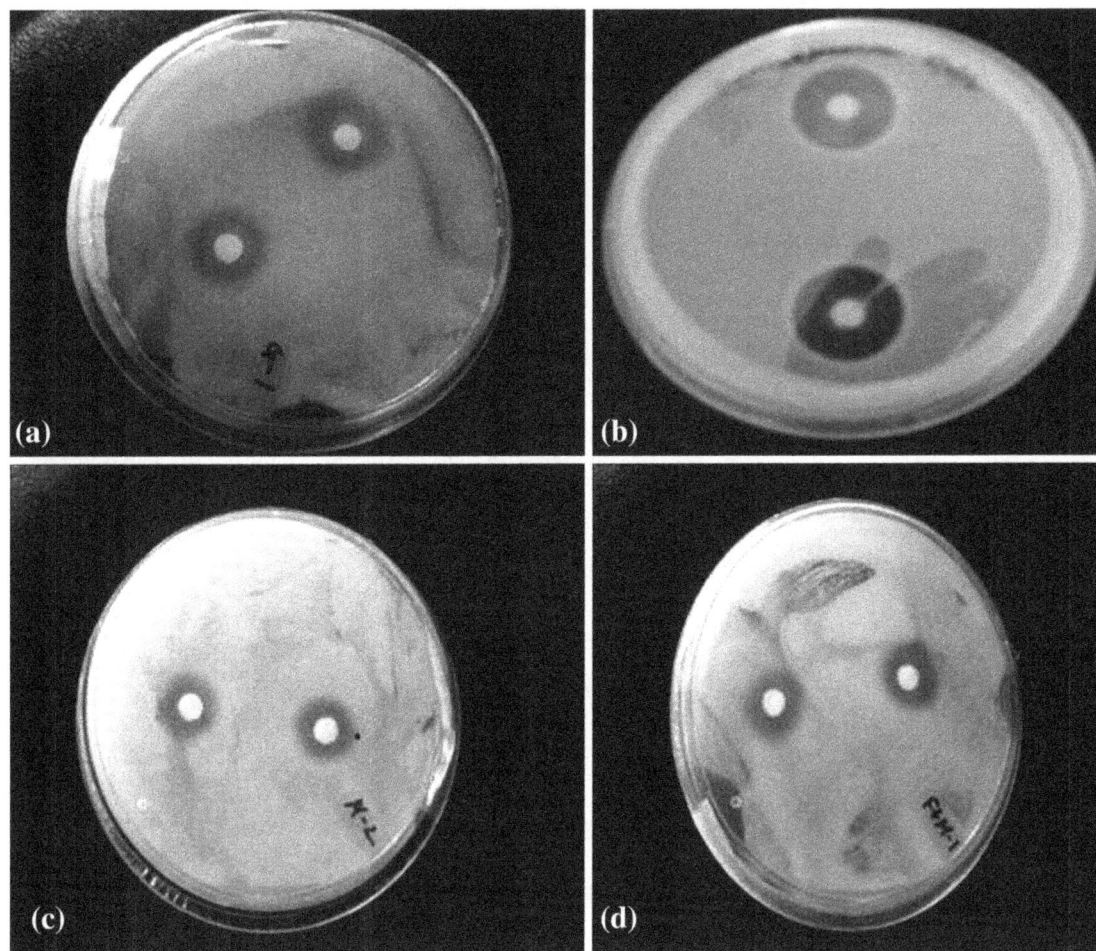

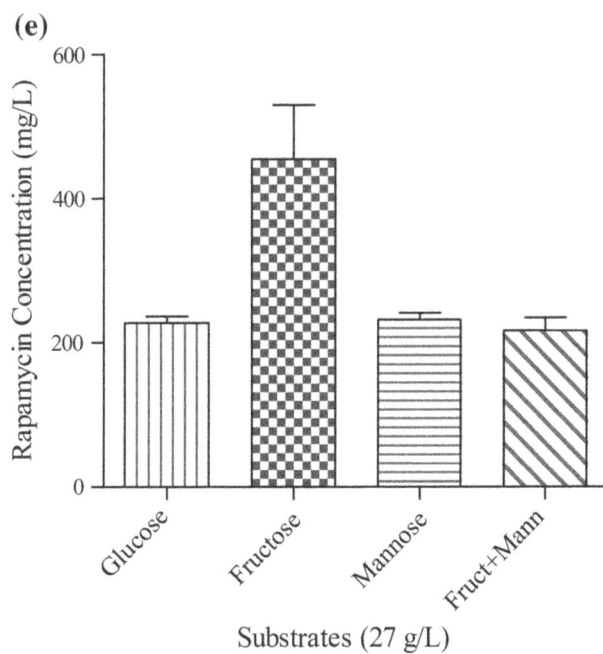

Fig. 1 Inhibition zones of rapamycin on agar plates of *Candida albicans* obtained using different carbon sources: **a** glucose, **b** fructose, **c** mannose, **d** fructose + mannose. **e** Concentrations of rapamycin found in shake flask using above-mentioned carbon sources

Table 1 Growth kinetic parameters for *S. hygroscopicus* MTCC 4003 obtained by different models

Mathematical models	μ_{max} (h^{-1})	K_S (g/L)	K_i (g/L)	R^2
Andrew's model $\mu = \frac{\mu_{max}S}{S^2 K_i + S + K_S}$	0.0083	2.835	0.073	0.9849
Yano model $\mu = \frac{\mu_{max}S}{S+K_S+\left(\frac{S^2}{K_i}+\frac{S^3}{K_i^2}\right)}$	0.0071	2.324	29.21	0.9818
Aiba model $\mu = \frac{\mu_{max}S}{K_S+S}e^{(-S+K_i)}$	0.0078	2.540	0.03963	0.9813
Monod model $\mu = \frac{\mu_{max}S}{K_S+S}$	0.003869	0.8271	–	0.8713

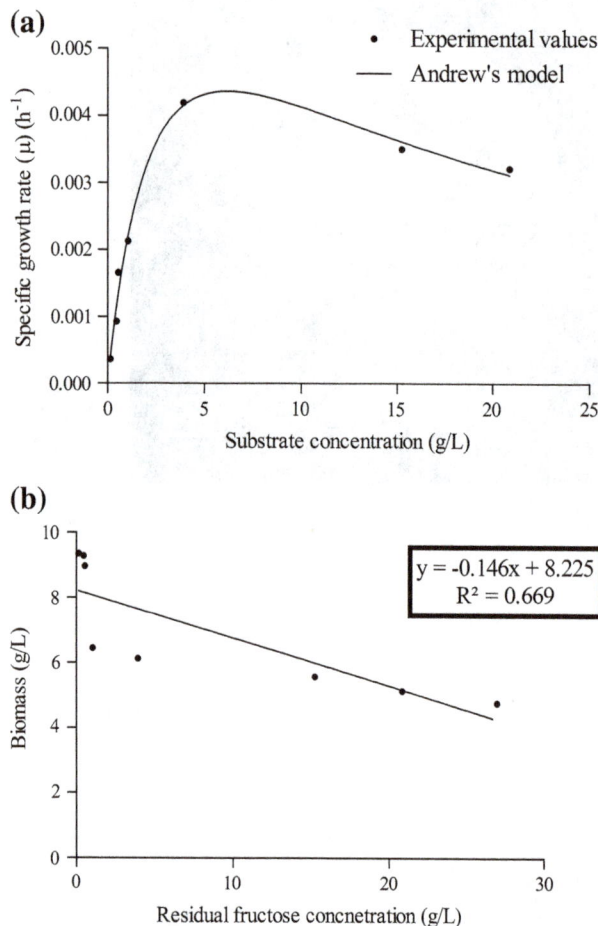

(a)

(b)

Fig. 2 a Relationship between specific growth rate (μ) and substrate concentrations (S). **b** Determination of growth yield coefficient ($Y_{X/S}$)

The fermentor was operated with constant impeller speed of 300 rpm for 8 consecutive days. The dissolved oxygen (DO) plays a vital role in the rapamycin production. The air flow rate was kept constant at 1 vvm throughout the fermentation process. Initially DO was set at 100 % at the time of inoculation, while during the lag phase microorganisms did not start to utilize the carbon of the media. However, when the fermentation enters the log phase, dissolved oxygen rapidly decreases as a result of increase in oxygen demand by the cells. Therefore, the dissolved oxygen level decreases and finally reached a certain stationary value. Finally, the DO level was maintained at 10 % of saturation level all the time (Zhu et al. 2010).

An antibiotic is a secondary metabolite which is synthesized mainly at the stationary phase of growth of a microorganism. However, the obtained data clearly show that exponential phase of growth is very important for rapamycin production (Zhu et al. 2010; Sanchez and Brana 1996). *Streptomyces hygroscopicus* MTCC 4003 showed a conventional growth pattern during the batch fermentation in the bioreactor (Fig. 3a). The production of rapamycin was started after 24 h of incubation and reached a maximal at 144 h (Fig. 3a). Similar trends were also reported in the literature (Xu et al. 2005; Lee et al. 1997). Rapamycin production was observed to increase rapidly during end of the exponential phase and early stationary phase (Xu et al. 2005). The productivity (mg/L h) is defined as the amount of product formation per unit of time. In the present study, we compared rapamycin productivity (d*P*/d*t*) in shake flask with that of bioreactor. Figure 3b represents the difference of rapamycin productivity in shake flask and bioreactor. Comparison of rapamycin production by different strains under various fermentation processes has been described in Table 2. Zhu et al., reported maximum productivity of rapamycin in fed-batch operate mode of 20,000-L fermentor, which is maximum productivity in fermentor reported till date (Zhu et al. 2010). In the present study, maximum rapamycin productivity of 9.13 mg/L h^{-1} was obtained in bioreactor (Table 2).

The production of rapamycin was detected using HPLC analysis. The calibration curve of rapamycin standard was prepared by plotting HPLC peak area against known rapamycin standard concentrations ($R^2 = 0.9954$) (Fig. 4a). For each concentration of rapamycin standard HPLC retention time (RT) was found to be 2.884 min (Fig. 4b). The methanolic extract of the supernatant was analyzed for presence of rapamycin. Its presence was confirmed by its retention time (2.814 min) which is about same as that of rapamycin standard (Fig. 4c). Figure 3c represents the HPLC chromatogram of methanolic extract of supernatant after 6 days of fermentation. Concentrations of rapamycin in the production medium were obtained by measuring HPLC peak areas (Refaat and Abdel-Fatah 2008). From the equation given in the Fig. 3a, the concentration of the 6th day's sample was calculated to be 1,316.02 mg/L which is the highest rapamycin produced in the bioreactor till date. It is 52 % higher than that reported previously (860.6 mg/L) by Zou and Li (2013). On the 12th day of fermentation, rapamycin

Fig. 3 Time-course profile of cell growth, rapamycin production, and substrate utilization

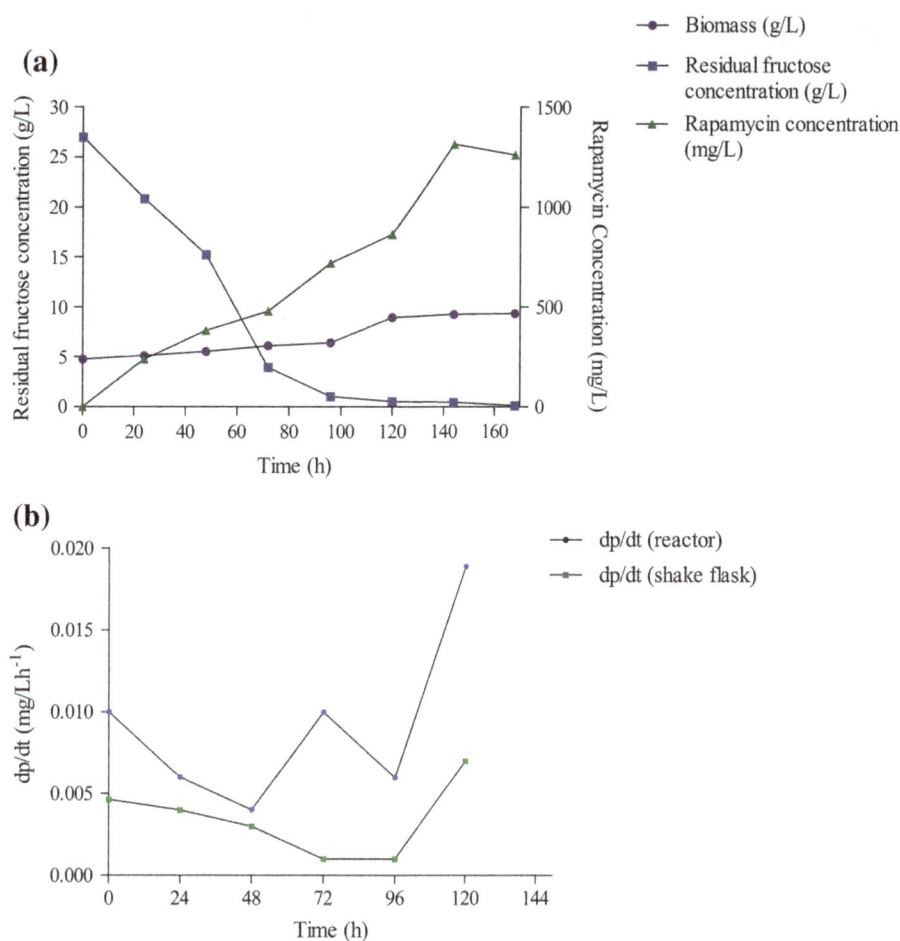

Table 2 Comparison of rapamycin production by different strains under various fermentation processes

Microorganism	Bioreactor	Operate mode	Time (h)	Rapamycin titer (mg/L)	Productivity (mg/L h)	References
S. hygroscopicus C9	Shake flask	Batch	144	186	1.29	Fang and Demain (1995)
S. hygroscopicus C9	Shake flask	Batch	144	130	0.90	Lee et al. (1997)
S. hygroscopicus NBS-9746	130L fermentor	Fed Batch	110	110	1.0	Cheng (1995)
S. hygroscopicus N5632	Shake flask	Batch	120	420	3.5	Xu et al. (2005)
S. hygroscopicus GS-1437	Shake flask	Batch	120	445	3.71	Chen et al. (2009)
S. hygroscopicus R060107	5L fermentor	Fed batch	120	500	4.17	Chen et al. (2008)
S. hygroscopicus HD-04-S	20,000L fermentor	Fed batch	168	783	4.66	Zhu et al. (2010)
S. hygroscopicus ATCC29253	Tubes	Batch	120	42.8	0.36	Jung et al. (2011)
S. hygroscopicus FMT11	7L fermentor	Fed batch	204	860.6	4.22	Zou and Li (2013)
S. hygroscopicus MTCC4003	Shake flask	Batch	144	539	3.74	This study
S. hygroscopicus MTCC4003	2.2L fermentor	Batch	144	1,316.02	9.13	This study

concentration decreased significantly to 91.67 mg/L. This phenomenon was observed by Xu et al. (2005) and might be attributed to the fact that rapamycin starts degrading after 7 days of fermentation (Prapagdee et al. 2008). This is also evident from Fig. 4d that shows the lower rapamycin concentration noted in the 12th day's sample and formation of two adjacent peaks in the HPLC chromatogram might be attributed to degradation of rapamycin and consequent production of rapamycin degradation products (Prapagdee et al. 2008; Wang et al. 1994).

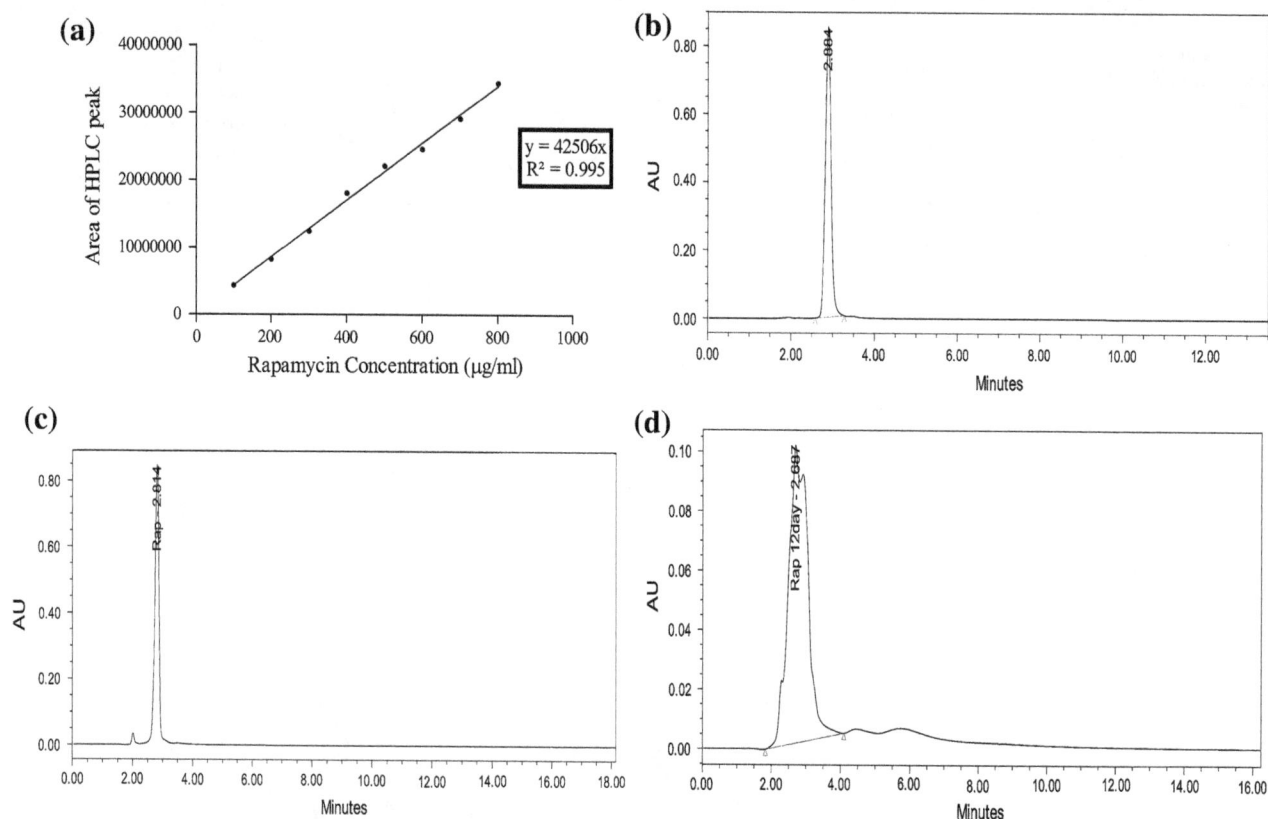

Fig. 4 a Calibration curve between the HPLC peaks areas and respective concentration of rapamycin standard; **b** HPLC peak of rapamycin standard; **c** HPLC analysis of rapamycin produced on 6th day of fermentation; **d** HPLC analysis of rapamycin on 12th day of fermentation

Conclusion

Fructose was found to be a better carbon source than mannose. The maximum production of rapamycin in the shake flask was found to be 539 mg/L using fructose as sole carbon source in combination with casein and $(NH_4)_2SO_4$. Extraction of rapamycin with methanol gave higher yield of rapamycin. Sixth day sample of fermentation gave the highest rapamycin titer of 1,316 mg/L in bioreactor quantified using HPLC technique.

Low titers of rapamycin produced by different strains of *S. hygroscopicus* limits the large-scale industrial production of this potent natural product. Therefore, different attempts, viz. strain improvement, process parameters optimization; precursor engineering studies have been employed to improve the production of this antibiotic. However, proper knowledge of nutritional requirements and kinetic behavior of the organism also need to be thoroughly studied if the maximum production of this immunosuppressant drug is intended. Therefore, the cell growth dynamic results obtained in the present study will help in the enhanced production of this drug.

Acknowledgments The authors would like to acknowledge Ministry of Human Research and Development, Govt. of India, for providing fund for this research.

Conflict of interest The authors would like to state that they have no potential conflict of interest regarding submission and publication of this manuscript.

References

Bhunia B, Basak B, Bhattacharya P, Dey A (2012) Kinetic studies of alkaline protease from *Bacillus licheniformis* NCIM-2042. J Microbiol Biotechnol 22:1758–1766

Chen Y, Krol J, Huang W, Cino JP, Vyas R, Mirro R, Vaillancourt B (2008) DCO2 on-line 5 measurement used in rapamycin fed-batch fermentation process. Process Biochem 43:351–355

Chen X, Wei P, Fan L, Yang D, Zhu X, Shen W, Xu Z, Cen P (2009) Generation of high-yield rapamycin-producing strains through protoplasts-related techniques. Appl Microbiol Biotechnol 83(3):507–512.

Cheng YR (1995) Phosphate, ammonium, magnesium and iron nutrition of *Streptomyces hygroscopicus* with respect to rapamycin biosynthesis. J Ind Microbiol Biotechnol 14:424–427

Fang A, Demain A (1995) Exogenous shikimic acid stimulates rapamycin biosynthesis in *Streptomyces hygroscopicus*. Folia Microbiol 40:607–610

Foroncewicz B, Mucha K, Paczek L, Chmura A, Rowinski W (2005) Efficacy of rapamycin in patient with juvenile rheumatoid arthritis. Transpl Int 18(3):366–368.

Graziani EI (2009) Recent advances in the chemistry, biosynthesis and pharmacology of rapamycin analogs. Nat Prod Rep 26(5):602–609.

Jung WS, Yoo YJ, Park JW, Park SR, Han AR, Ban YH, Kim EJ, Kim E, Yoon YJ (2011) A combined approach of classical mutagenesis and rational metabolic engineering improves rapamycin biosynthesis and provides insights into methylmalonyl-CoA precursor supply pathway in *Streptomyces hygroscopicus* ATCC 29253. Appl Microbiol Biotechnol 91(5):1389–1397.

Kojima I, Cheng YR, Mohan V, Demain AL (1995) Carbon source nutrition of rapamycin biosynthesis in *Streptomyces hygroscopicus*. J Ind Microbiol 14(6):436–439

Kumar CG, Malik RK, Tiwari MP, Jany KD (1999) Optimal production of *Bacillus* alkaline protease using a cheese whey medium. Microbiologie des Alimentes et Nutr 17:39–48

Lee MS, Kojima I, Demain AL (1997) Effect of nitrogen source on biosynthesis of rapamycin by *Streptomyces hygroscopicus*. J Ind Microbiol Biotechnol 19:83–86

Miller GL (1959) Use of dinitrosalicylic acid reagent for determination of reducing sugar. Anal Chem 31(3):426–428

Morelon E, Mamzer-Bruneel MF, Peraldi MN, Kreis H (2001) Sirolimus: a new promising immunosuppressive drug. Towards a rationale for its use in renal transplantation. Nephrol Dial Transplant 16:18–20

Nicoletti F, Lapenta C, Donati S, Spada M, Ranazzi A, Cacopardo B, Mangano K, Belardelli F, Perno C, Aquaro S (2009) Inhibition of human immunodeficiency virus (HIV-1) infection in human peripheral blood leucocytes-SCID reconstituted mice by rapamycin. Clin Exp Immunol 155(1):28–34.

Okpokwasili GC, Nweke CO (2005) Microbial growth and substrate utilization kinetics. Afr J Biotechnol 5:305–317

Park SR, Yoo YJ, Ban YH, Yoon YJ (2010) Biosynthesis of rapamycin and its regulation: past achievements and recent progress. J Antibiot (Tokyo) 63(8):434–441.

Prakasham RS, Subba Rao C, Sreenivas Rao R, Sarma PN (2007) Enhancement of acid amylase production by an isolated *Aspergillus awamori*. J Appl Microbiol 102(1):204–211.

Prapagdee B, Kuekulvong C, Mongkolsuk S (2008) Antifungal potential of extracellular metabolites produced by *Streptomyces hygroscopicus* against phytopathogenic fungi. Int J Biol Sci 4:330–337

Priest FG (1977) Extracellular enzyme synthesis in the genus *Bacillus*. Bacteriol Rev 41(3):711–753

Refaat Y, Abdel-Fatah E (2008) Non conventional method for evaluation and optimization of medium components for rapamycin production by *Streptomyces hygroscopicus*. Res J Microbiol 3:405–413

Sabatini DM, Pierchala BA, Barrow RK, Schell MJ, Snyder SH (1995) The rapamycin and FKBP12 target (RAFT) displays phosphatidylinositol 4-kinase activity. J Biol Chem 270: 20875–20878

Sallam L, El-Refai A, Osman M (2010) Some physiological factors affecting rapamycin production by *Streptomyces hygroscopicus* ATCC 29253. J Am Sci 6(6):188–194

Sanchez L, Brana AF (1996) Cell density influences antibiotic biosynthesis in *Streptomyces clavuligerus*. Microbiol 142(Pt 5): 1209–1220

Schuhmann E, Bergter F (1976) Microscopic studies of *Streptomyces hygroscopicus* growth kinetics. Z Allg Mikrobiol 16(3):201–205

Sehgal SN (2003) Sirolimus: its discovery, biological properties, and mechanism of action. Transplant Proc 35(3 Suppl):7S–14S (pii:S0041134503002112)

Subba Rao C, Madhavendra SS, Sreenivas Rao R, Hobbs PJ, Prakasham RS (2008) Studies on improving the immobilized bead reusability and alkaline protease production by isolated immobilized *Bacillus circulans* (MTCC 6811) using overall evaluation criteria. Appl Biochem Biotechnol 150(1):65–83.

Tain LS, Mortiboys H, Tao RN, Ziviani E, Bandmann O, Whitworth AJ (2009) Rapamycin activation of 4E-BP prevents parkinsonian dopaminergic neuron loss. Nat Neurosci 12(9):1129–1135.

Tobisch S, Zuhlke D, Bernhardt J, Stulke J, Hecker M (1999) Role of CcpA in regulation of the central pathways of carbon catabolism in *Bacillus subtilis*. J Bacteriol 181(22):6996–7004

Wang CP, Chan KW, Schiksnis RA, Scatina J, Sisenwine SF (1994) High performance liquid chromatographic isolation, spectroscopic characterization, and immunosuppressive activities of two rapamycin degradation products. J Liq Chromatogr Relat Technol 17:3383–3392

Weber T, Abendroth D, Schelzig H (2005) Rapamycin rescue therapy in patients after kidney transplantation: first clinical experience. Transpl Int 18(2):151–156.

Wiederrecht GJ, Sabers CJ, Brunn GJ, Martin MM, Dumont FJ, Abraham RT (1995) Mechanism of action of rapamycin: new insights into the regulation of G1-phase progression in eukaryotic cells. Prog Cell Cycle Res 1:53–71

Xu Z-N, Shen W-H, Chen X-Y, Lin J-P, Cen P-L (2005) A high throughput method for screening of rapamycin-producing strains of *Streptomyces hygroscopicus* by cultivation in 96-well microtiter plate. Biotechnol Lett 27:1135–1140

Zhu X, Zhang W, Chen X, Wu H (2010) Generation of high rapamycin producing strain via rational metabolic pathway-based mutagenesis and further titer improvement with fed -batch bioprocess optimization. Biotechnol Bioeng 107:506–514.

Zou X, Li J (2013) Precursor engineering and cell physiological regulation for high level rapamycin production by *Streptomyces hygroscopicus*. Ann Microbiol.

15

An approach to low-density polyethylene biodegradation by *Bacillus amyloliquefaciens*

Merina Paul Das · Santosh Kumar

Abstract Low-density polyethylene (LDPE) is a major cause of persistent and long-term environmental pollution. In this paper, two bacterial isolates *Bacillus amyloliquefaciens* (BSM-1) and *Bacillus amyloliquefaciens* (BSM-2) were isolated from municipal solid soil and used for polymer degradation studies. The microbial degradation LDPE was analyzed by dry weight reduction of LDPE film, change in pH of culture media, CO_2 estimation, scanning electron microscopy (SEM), and fourier transform infrared FTIR spectroscopy of the film surface. SEM analysis revealed that both the strains were exhibiting adherence and growth with LDPE which used as a sole carbon source while FTIR images showed various surface chemical changes after 60 days of incubation. Bacterial isolates showed the depolymerization of biodegraded products in the extracellular media indicating the biodegradation process. BSM-2 exhibited better degradation than BSM-1 which proves the potentiality of these strains to degrade LDPE films in a short span of time.

Keywords Biodegradation · LDPE · *Bacillus amyloliquefaciens* · Degradation analysis

Introduction

Plastic is a man-made hazardous long-chain synthetic polymer. The annual global demand for plastics has consistently increased over the recent years and presently stands at about 245 million tones. Being a versatile, lightweight, strong and potentially transparent material, plastics are ideally suited for a variety of applications. Their low cost, excellent oxygen/moisture barrier properties, bio-inertness and light weight make them excellent packaging materials (Andrady 2011). However, large amount of accumulation of plastic waste is leading a harmful effect on nature, causing environmental pollution as it is resistant to biodegradation. There are several chemicals within plastic material itself that have been added to give it certain properties such as Bisphenol A, phthalates and flame retardants. These all have known negative effects on human and animal health, mainly affecting the endocrine system. There are also toxic monomers, which have been linked to cancer and reproductive problems. Awareness of the waste problem and its impact on the environment has awakened new interest in the area of degradable polymers. The interest in environmental issues is growing and there are increasing demands to develop material which do not burden the environment significantly (Shah et al. 2008).

Low-density polyethylene (LDPE) is a widely used non-biodegradable thermoplastic. To deal with this environmental problem related to non-biodegradable thermoplastics, research to modify non-biodegradable thermoplastics to biodegradable materials is of great interest (Zheng et al. 2005). Furthermore, these synthetic polymers are normally not biodegradable until they are degraded into low molecular mass fragments that can be assimilated by microorganisms (Francis et al. 2010).

The biodegradable polymers are designed to degrade it quickly by microbes since microorganisms are capable of degrading most of the organic and inorganic materials, including lignin, starch, cellulose and hemicelluloses (Sadocco et al. 1997). In addition, biodegradable plastics offer a lot of advantages such as increased soil fertility, low

M. P. Das (✉) · S. Kumar
Department of Industrial Biotechnology, Bharath University, Chennai 600073, Tamil Nadu, India
e-mail: merinadas@gmail.com

accumulation of bulky plastic materials in the environment (which invariably will minimize injuries to wild animals), and reduction in the cost of waste management (Tokiwa et al. 2009). The microbial species associated with the degrading polymers were identified as bacteria (*Pseudomonas, Streptococcus, Staphylococcus, Micrococcus, Moraxella*), fungi (*Aspergillus niger, Aspergillus glaucus*), Actinomycetes sp. and *Saccharomonospora* genus (Swift 1997). Biodegradation has been considered as a natural process in the microbial world where polymers can be used as carbon and energy sources for their growth and plays a key role in the recycling of these materials in the natural ecosystem (Albertsson et al. 1987). The microbial degradation of plastics is caused by certain enzymatic activities that lead to a chain cleavage of the polymer into oligomers and monomers. These water soluble enzymatically cleaved products are further absorbed by the microbial cells where they are metabolized. Aerobic metabolism results in carbon dioxide and water (Starnecker and Menner 1996), whereas anaerobic metabolism results in carbon dioxide, water and methane as the end products, respectively (Gu et al. 2000). The aim of this research was to study the biodegradation of low-density polyethylene using various techniques in vitro by selected and potent microorganism isolated from municipal solid waste.

Materials and methods

Pre-treatment and preparation of LDPE powder

Low-density polyethylene (LDPE) was obtained from B.N. Polymers, Bangalore, India. LDPE films were cut into small pieces, immersed into xylene and boiled for 15 min, followed by crushing with blender at 3,000 rpm. As obtained LDPE powder was further washed with ethanol, dried overnight in hot air oven at 60 °C and stored at room temperature for further use.

Polyethylene degrading bacteria and culture conditions

The bacteria used in this study, *B. amyloliquefaciens* (BSM-1) (GenBank accession no. KC924446) and *B. amyloliquefaciens* (BSM-2) (GenBank accession no. KC924447) (Das and Kumar 2013), were isolated from the municipal solid waste landfill area, Pallikaranai (12.9377N/80.2153E, 7 m above sea level), Chennai, India and maintained on nutrient agar at 4 °C. The polymer degrading bacteria were identified using synthetic media supplemented with 0.3 % LDPE powder. The synthetic media composition was as follows: (g/L: K_2HPO_4, 1; KH_2PO_4, 0.2; $(NH_4)_2SO_4$, 1; $MgSO_4 \cdot 7H_2O$, 0.5; NaCl, 1; $FeSO_4 \cdot 7H_2O$, 0.01; $CaCl_2 \cdot 2H_2O$, 0.002; $MnSO_4 \cdot H_2O$, 0.001; $CuSO_4 \cdot 5H_2O$, 0.001; $ZnSO_4 \cdot 7H_2O$, 0.001; Agar 15; pH 7.0).

Biodegradation studies

Biodegradation tests were performed with 3 g samples of LDPE films (1.5 × 1.5 cm) that had been dried overnight at 60 °C, weighed, disinfested (30 min in 70 % ethanol), air-dried for 15 min in Laminar air flow chamber and added to Erlenmeyer flasks containing 300 ml of synthetic medium. LDPE degradation study was carried out using both the bacterial strains individually. Each flask inoculated with 3 ml of 24 h old culture (BSM-1 and BSM-2) grown on LDPE-supplemented medium was used as inoculums to avoid any associated long lag phase. Then cultures were incubated on a rotary shaker (Neolab Instruments) at 33.3 °C and 130 rpm for 60 days. Each test consisted of three replicates.

Measurement of biodegradation

Determination of pH change

Study of pH change was adopted to make sure about any metabolic activity of the microbial strain in supplemented medium, as metabolism shown by microbial cells may greatly support the evidence of degradation. The pH of the each bacterial suspension was measured at an interval of 10 days during the study. The pH probe was inserted in the broth to measure the pH. Initial value of the medium was ensured to be 7 ± 0.3 for both strains using phosphate buffer.

Determination of dry weight of residual polymer

To facilitate accurate measurement of the weight of the residual polyethylene, the polyethylene sheets were recovered after the 60 days of incubation and washed off the bacterial biofilm from the polymer surface with a 2 % (v/v) aqueous sodium dodecyl sulfate solution for 4 h (using shaker), followed by distilled water and finally with 70 % ethanol to ensure maximum possible removal of cells and debris. The washed polymer pieces were placed on a filter paper and dried overnight at room temperature before weighing.

CO₂ evolution test

The self-modified simple apparatus was designed which consists of control and test vessels and sterile air was supplied to the system for aeration. Here, the polymer incubated with microbes served as the test and polymer without microbes served as control. After incubation, both the metabolic and atmospheric CO_2 from the test vessel and atmospheric CO_2 from the control vessel were trapped and assessed using "Sturm test" (Sturm 1973) for each isolate.

Scanning electron microscopy (SEM)

The untreated and treated samples after 60 days of duration were subjected to SEM analysis (after washing with 2 % (v/v) aqueous SDS and distilled water repeatedly through mild shaking for few minutes and additionally flushed with 70 % ethanol with the objective of removal of cells so as to get maximum surface to be exposed for visualization. The samples were pasted onto the SEM Sample Stub using a carbon tape and sample was coated with gold for 40 s and analyzed under high-resolution scanning electron microscope (JEOL, Model JSM-6390LV).

FTIR analysis

Fourier transform infrared spectroscopic studies were carried out for control and bacteria-treated LDPE films. The analysis was performed using Perkin-Elmer Spectrum-One FTIR spectroscopy in the horizontal mode with thallium bromide disks.

Results and discussion

Biodegradable plastic is a favorable solution of plastic disposal or accumulation problem in nature. As domestic and industrial waste containing huge amount of low-density polyethylene, in this work municipal landfill solid waste sample was collected to isolate the microorganisms which showed potent biodegradation. The bacterial isolates can grow in a LDPE-supplemented synthetic medium utilizing LDPE as sole carbon and energy sources. These observations indicate the formation and attachment of a biofilm on LDPE film. The microbial colonization on polymer surface is the first requirement for its biodegradation (Yabannavar and Bartha 1993).

Biodegradation studies

After incubation period of 60 days, the degrading capability of the strains *Bacillus amyloliquefaciens* (BSM-1) and *Bacillus amyloliquefaciens* (BSM-2) was analyzed and interpreted using various parameters.

pH Change

Figure 1 shows the variation in pH of both bacterial suspensions during and after the biodegradation. Microorganisms secrete a variety of enzymes into the soil water, which begin the breakdown of the polymers. Two types of enzymes are involved in the process, namely intracellular and extracellular depolymerases. Exoenzymes from the microorganisms first breakdown the complex polymers

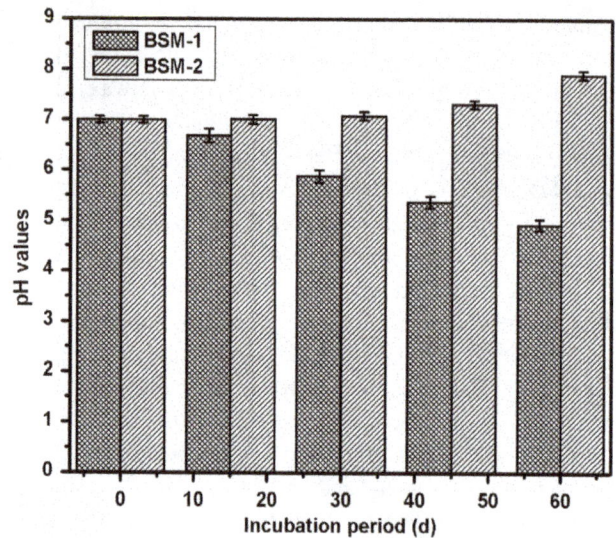

Fig. 1 Variation in pH level during biodegradation due to microbial activity

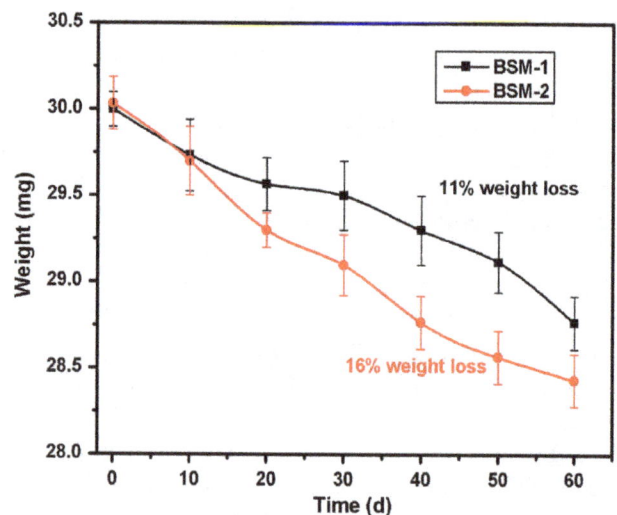

Fig. 2 Degradation of LDPE films (initial weight: 30 mg with 1.5 × 1.5 cm) on synthetic medium inoculated with strain BSM-1 and BSM-2 incubated at 33.3 °C for 60 days

giving short chains or monomers that are small enough to permeate through the cell walls to be utilized as carbon and energy sources by a process of depolymerization (Dey et al. 2012). Bacterial isolates, BSM-1 and BSM-2 showed the production of some enzymes and metabolites with the indication of pH change supporting the metabolic activity of strains on the LDPE substrate and also its degradation.

Weight reduction

A simple and quick way to measure the biodegradation of polymers is by determining the weight loss. Microorganisms that grow within the polymer lead to an increase in weight due to accumulation, whereas a loss of polymer integrity leads to weight loss. Weight loss is proportional to the surface area since biodegradation usually is initiated at the surface of the polymer. After the degradation period, the LDPE films were treated with SDS as surfactant which denatured the cells and completely washed off from the surface. The reduction in weight was observed after the biodegradation of LDPE (Fig. 2).

Assessment of mineralization level

Sturm test is the method where the degradation was attributed to the amount of metabolic carbon dioxide evolved during the growth period. The polymers are made up of carbon chain and when it degrades through the microbes CO_2 and H_2O are obtained as byproducts, the process is called mineralization in which polymer is first converted to monomers by breaking the links and then to more simpler compounds to be assimilated into the living cells. The level of CO_2 was calculated from the control (atmospheric CO_2) and reaction chamber (atmospheric and metabolic CO_2) after 60 days of biodegradation study. Theoretical carbon dioxide evolution for 3 % LDPE was estimated to be 11 g/L for complete biodegradation. Here, the percentage of biomineralization level of LDPE through evolved carbon dioxide from reaction chambers was calculated for strain BSM-1 and BSM-2 by comparing with the corresponding values of control chambers (Table 1). The result shows the potentiality of *Bacillus amyloliquefaciens* and supports the fact of biodegradation and biomineralization of this hazardous polymer.

Table 1 Biomineralization level of microbial isolates

Isolates	Carbon dioxide evolution (g/L)			Mineralization level (%)
	Control (atmospheric) (C)	Test (metabolic + atmospheric) (T)	T − C (metabolic)	
BSM-1	16.26	17.58	1.32	12
BSM-2	18.31	19.92	1.61	14.7

Fig. 3 SEM micrograph of LDPE film before treatment as Control (**a**), LDPE film after treatment with BSM-1 (**b**), and LDPE film after treatment with BSM-2 (**c**)

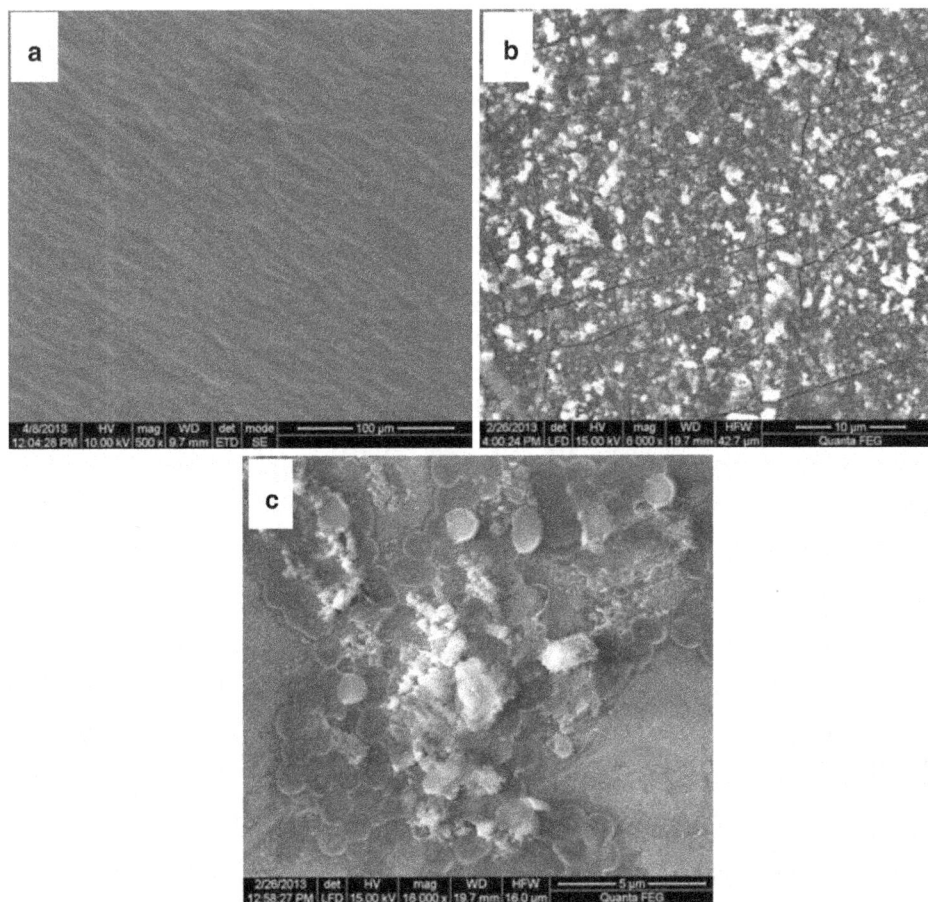

SEM analysis of LDPE film

While the pH change, weight reduction, mineralization level and absorption spectra provide solid evidence of polymer biodegradation, the changes of surface of LDPE films were elucidated by SEM. Control sample has an appearance of smooth surface having no pits, cracks or any particles attached on the surface (Fig. 3a). In the case of LDPE film treated with the bacterial isolate BSM-1, it was found that several cracks on the surface developed after 60 days of treatment. Simultaneously, microbes were also noticed on the film surface indicate its strong adhering capabilities as well as LDPE utilization capacities (Fig. 3b). The film treated with the bacterial isolate BSM-2 found to have bacterial attachment on higher rate as compared to BSM-1. Clear mark of degradation can be seen at places where initially microbes were attached along with the pockets and pits around (Fig. 3c). For both the strains, at different places on the surface several colonies forming biofilm can be observed.

FTIR analysis

FTIR spectroscopy is used as analytical technique in many biodegradation studies (Kiatkamjornwong et al. 1999; Klrbas et al. 1999; Arboleda et al. 2004; Drímal et al. 2007). It is a useful tool to determine the formation of new or disappearance of functional groups. So degradation products, chemical moieties incorporated into the polymer molecules such as branches, co-monomers, unsaturation and presence of additives such as antioxidants can be determined by this technique. Control spectra of polymer film (not treated with microbes) displayed a number of peaks reflecting the complex nature of the LDPE (Fig. 4a). There was a variation in the intensity of bands in different regions when test samples (after incubation with microbes, BSM-1 and BSM-2) were analyzed (Fig. 4b, c). For control spectrum, the characteristic absorption bands were assigned at 719 cm^{-1} (C–H bend-mono), 1,472 cm^{-1} (C=C stretch), 2,660 cm^{-1} (CHO stretch), and 2,919, 2,850 cm^{-1} (both due to C–H stretch). Significant and similar changes were found for both microbial strains. The peak at 2,660 cm^{-1} corresponds to CHO stretching vibration that has been disappeared in case of BSM-1 and 2 while new band has been observed at 939 cm^{-1} (O–H bend) which supports the depolymerization activity of the microbial isolates. The strong absorption peaks at 719 and 1,472 cm^{-1} became weaker after microbial treatment. In addition, the intensity of those peaks reduced more in case of BSM-2 than BSM-1 whereas peaks at 2,919 and 2,850 cm^{-1} became sharper in the treated sample than the control one, here also the same microbial activity pattern was seen. The change in the peak values of almost all functional groups supporting the conformational change on polymer surface.

Conclusion

The problem of plastic pollution is now really a mess for mankind. There is no part of the world untouched from its impact. In the present era of globalization some stress must be given to plan safe disposal of products before making it commercial. Making science to the leap and forgetting the other side of coin lead to such conditions. In the present study, two isolated strains of *Bacillus amyloliquefaciens* were found to be useful for the biodegradation which is first time reported with applicable evidences. This biodegradation approach is safe and eco-friendly. The results showed a promising hope to degrade LDPE faster rate than to be degraded naturally.

Acknowledgments The authors convey their thanks to Department of Industrial Biotechnology, Bharath University, Chennai, for providing laboratory facilities.

Conflict of interest The authors of this paper declare that they have no conflict of interest.

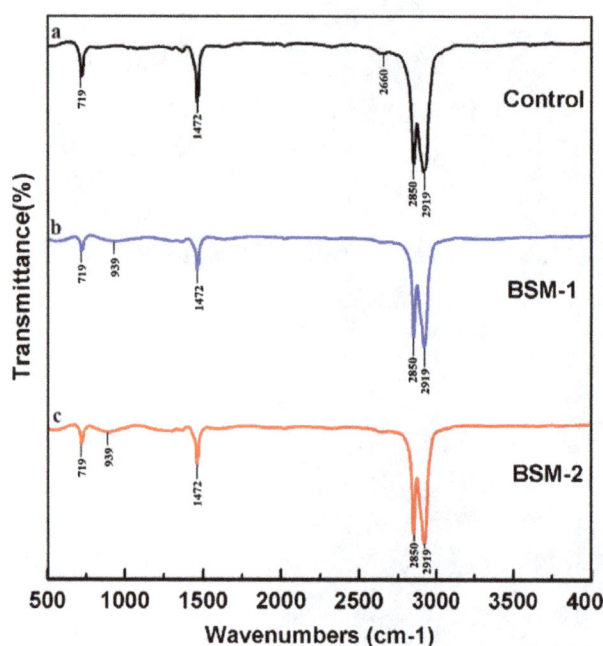

Fig. 4 FTIR spectra of Control (*a*), treated with BSM-1 (*b*), and treated with BSM-2 (*c*)

References

Albertsson AC, Andersson SO, Karlsson S (1987) The mechanism of biodegradation of polyethylene. Polym Degrad Stab 18:73–87

Andrady AL (2011) Microplastics in the marine environment. Mar Pollut Bull 62:1596–1605

Arboleda CE, Mejía AIG, López BLO (2004) Poly (vinylalcohol-co-ethylene) biodegradation on semi solid fermentation by *Phanerochaete chrysosporium*. Acta Farm Bonaer 23:123–128

Das MP, Kumar S (2013) Influence of cell surface hydrophobicity in colonization and biofilm formation on LDPE biodegradation. Int J Pharm Pharm Sci 5:690–694

Dey U, Mondal NK, Das K, Dutta S (2012) An approach to polymer degradation through microbes. IOSRPHR 2:385–388

Drímal P, Hoffmann J, Družbík M (2007) Evaluating the aerobic biodegradability of plastics in soil environments through GC and IR analysis of gaseous phase. Polym Test 26:729–741

Francis V, Raghul SS, Sarita GB, Eby TT (2010) Microbial degradation studies on linear low-density poly(ethylene)-poly (vinyl alcohol) blends using *Vibrio* sp. International Conference on advances in polymer Technology, pp 26–27

Gu JD, Ford TE, Mitton DB, Mitchell R (2000) Microbial corrosion of metals. In: Revie W (ed) The Uhlig Corrosion Handbook, 2nd edn. Wiley, New York, pp 915–927

Kiatkamjornwong S, Sonsuk M, Wittayapichet S, Prasassarakich P, Vejjanukroh PC (1999) Degradation of styrene-g-cassava starch filled polystyrene plastics. Polym Degrad Stab 66:323–335

Klrbas Z, Keskin N, Güner A (1999) Biodegradation of polyvinylchloride (PVC) by white rot fungi. Bull Environ Contam Toxicol 63:335–342

Sadocco P, Nocerino S, Dubini-Paglia E, Seves A, Elegir G (1997) Characterization of a poly (3-hydroxybutyrate) depolymerase from *Aureobacterium saperdae*: active site and kinetics of hydrolysis studies. J Environ Polym Degrad 5:57–65

Shah AA, Hasan F, Hameed A, Ahmed S (2008) Biological degradation of plastics: a comprehensive review. Biotech Adv 26:246–265

Starnecker A, Menner M (1996) Assessment of biodegradability of plastics under stimulated composting conditions in a laboratory test system. Int Biodeter Biodegr 37:85–92

Sturm RN (1973) Biodegradability of nonionic surfactants: screening test for predicting rate and ultimate biodegradation. J Oil Chem Soc 50:159–167

Swift G (1997) Non-medical biodegradable polymers: environmentally degradable polymers. In: Handbook of biodegradable polymers. Hardwood Academic, Amsterdam, pp 473–511

Tokiwa Y, Calabia BP, Ugwu CU, Aiba S (2009) Biodegradability of plastics. Int J Mol Sci 10:3722–3742

Yabannavar A, Bartha R (1993) Biodegradability of some food packaging materials in soil. Soil Biol Biochem 25:1469–1475

Zheng Y, Yanful EK, Bassi AS (2005) A review of plastic waste biodegradation. Crit Rev Biotechnol 25:243–250

16

Potentiality of *Bacillus subtilis* as biocontrol agent for management of anthracnose disease of chilli caused by *Colletotrichum gloeosporioides* OGC1

N. Ashwini · S. Srividya

Abstract A soil bacterium, *Bacillus subtilis*, isolated from the rhizosphere of Chilli, showed high antagonistic activity against *Colletotrichum gloeosporioides* OGC1. A clear inhibition zone of 0.5–1 cm was observed in dual plate assay. Microscopic observations showed a clear hyphal lysis and degradation of fungal cell wall. In dual liquid cultures, the *B. subtilis* strain inhibited the *C. gloeosporioides* up to 100 % in terms of dry weight. This strain also produced a clear halo region on chitin agar medium plates containing 0.5 % colloidal chitin, indicating that it excretes chitinase. The strain also produced other mycolytic enzymes—glucanase and cellulase, demonstrated by a clear zone of hydrolysis of yeast cell wall glucan (YCW 0.1 % v/v) and carboxymethylcellulose (CMC 0.1 % v/v). In liquid cultures, the strain showed appreciable levels of chitinase, glucanase and cellulase activities and hydrolytic activity with *C. gloeosporioides* OGC1 mycelia as the substrate. The role of the *B. subtilis* strain in suppressing the fungal growth in vitro was studied in comparison with a UV mutant of that strain, which lacked both antagonistic and hydrolytic activity. The mycolytic enzyme mediated antagonism of *B. subtilis* was further demonstrated by heat inactivation (70–100 °C), treatment with trypsin and TCA of the crude enzyme extract which lacked antifungal property also. Treatment of the chilli seeds with *Bacillus* sp. culture showed 100 % germination index similar to the untreated seeds. The treatment of the seed with co-inoculation of the pathogen with *Bacillus* sp. culture showed 65 % reduction in disease incidence by the treatment as compared to the seed treated with pathogen alone (77.5 %).

Keywords *B. subtilis* · Mycolytic enzymes · Antagonism · *C. gloeosporioides* · UV mutagenesis

Introduction

Rhizosphere bacteria are excellent agents to control soil-borne plant pathogens. Bacterial species such as *Bacillus*, *Pseudomonas*, *Serratia* and *Arthrobacter* have been proved in controlling the fungal diseases (Joseph et al. 2007) Earlier reports showed that microorganisms capable of lysing chitin, which is a major constituent of the fungal cell wall, play an important role in biological control of fungal pathogens (Yu et al. 2002; Zhang and Fernando 2004; Abdullah et al. 2008). Fungi such as *Trichoderma* and bacteria such as *Bacillus*, *Serratia* and *Alteromonas* were reported to have chitinolytic activity (Mabuchi et al. 2000; Someya et al. 2001; Wen et al. 2002; Huang et al. 2005; Viterbo et al. 2001). Non-pathogenic soil *Bacillus* species offer several advantages over other organisms as they form endospores and hence can tolerate extreme pH, temperature and osmotic conditions. *Bacillus* species were found to colonize the root surface, increase the plant growth and cause the lysis of fungal mycelia (Turner and Backman 1991; Podile and Prakash 1996; Takayanagi et al. 1991).

Chilli, *Capsicum annum* L. cultivation has existed for several 100 years as a sustainable form of agriculture in India and in many other countries. It is an annual herbaceous vegetable and spice grown in both tropical and subtropical regions. The crop is grown in almost all states of India, such as Andhra Pradesh, Maharashtra, Karnataka, Gujarat, Tamil Nadu and Orissa. India accounts for 25 %

N. Ashwini · S. Srividya (✉)
Department of Microbiology, Centre for PG Studies,
Jain University, 18/3, 9th Main, Jayanagar 3rd Block,
Bangalore 560011, India
e-mail: sk2410@yahoo.co.uk; sk.srividya@jainuniversity.ac.in

of the world's total production of chilli (http://agropedia. iitk.ac.in/content/area-and-production-chilli-world-2008-09).

The crop is a significant source of income making India the world's single largest producer and exporter to the USA, Canada, UK, Saudi Arabia, Singapore, Malaysia, Germany and many more countries across the world. The sustainability of chilli-based agriculture is threatened by a number of factors. Main biotic stresses such as bacterial wilt, anthracnose, viruses and several insect pests have been reported to impair the crop productivity (Isaac 1992).

The genus *Colletotrichum* and its teleomorph *Glomerella* contain an extremely diverse number of fungi including both plant pathogens and saprophytes. Plant pathogenic species are important worldwide, causing pre- and post-harvest losses of crops (Bosland and Votava 2003). These fungi cause diseases commonly known as anthracnose of grasses, legumes, vegetables, fruits and ornamentals. The disease can occur on leaves, stems and fruit of host plants (Sutton 1992). Anthracnose disease is a major problem in India and one of the more significant economic constraints to chilli production worldwide, especially in tropical and subtropical regions (Than et al. 2008).

Economic losses caused by the disease are mainly attributed to lower fruit quality and marketability. Although infected fruits are not toxic to humans or animals, severely affected fruits showing blemishes are generally considered unfit for human consumption. This is because the anthracnose causes an unpleasant colour and taste in chilli products. Management of the disease under the prevailing farming systems in India has become a recurrent problem to chilli growers (Thind and Jhooty 1985).

Effective control of anthracnose disease involves the use of one of, or a combination of, the following: resistant cultivars, cultural control and chemical control. The intensive use of fungicides has resulted in the accumulation of toxic compounds potentially hazardous to humans and the environment, and also in the build-up of resistance of the pathogens. In view of this, investigation and the application of biological control agents (BCAs) seems to be one of the promising approaches (Cook 1985). Biocontrol involves the use of naturally occurring non-pathogenic microorganisms that are able to reduce the activity of plant pathogens and thereby suppress diseases. Hence, controlling this pathogen using biocontrol agents will help in enhancing the yield of the crop.

B. subtilis JN032305, isolated from chilli rhizosphere produced appreciable levels of three mycolytic enzymes— chitinase, glucanase and cellulase and showed broad spectrum antagonism against potent bacterial and fungal phytopathogens (Srividya et al. 2012). The production of all three enzymes of this strain have been optimised using Plackett-Burman approach (Ashwini and Srividya 2012a, b). The objective of the present study was to ascertain the concerted role of these three mycolytic enzymes (chitinase, β1,3-glucanase and cellulase) mediated antagonism of *B. subtilis* against *C. gloeosporioides*.

Materials and methods

Isolation of rhizospheric *Bacillus* sp.

Chilli rhizosphere soils were collected from in and around Bangalore and *Bacillus* sp. were isolated by soil dilution method. The dilutions were heat treated at 80 °C for 20 min to ensure that only heat resistant strains remained. The different isolates obtained were screened for chitinase production based on the halo produced on plates with minimal salts medium amended with chitin (1 % chitin) (Cook 1985). The *Bacillus* sp., thus, obtained were maintained on Nutrient agar (NA) amended with chitin (1 %).

Morphological and phenotypic characterization of *Bacillus* sp.

This strain was characterised morphologically and biochemically by following Bergey's Manual of Systematic Bacteriology (Sneath 1986) and was found to be a *Bacillus* sp. It was grown and maintained on NA at 30 °C. Polymerase chain reaction (PCR) was performed to amplify a partial 16S rRNA gene of the bacteria, and partial 16S rDNA sequencing was used to assist in the identification of the isolate. Isolation of genomic DNA, PCR amplification and sequencing of PCR product for analysis of 16S rRNA were conducted according to Marchesi et al. (1998). A similarity search for the nucleotide sequence of 16S rRNA of the test isolate was carried out using a blast search at NCBI (Altschul et al. 1990).

Preparation of colloidal chitin

Colloidal Chitin was prepared from crab shell chitin powder (Roberts and Selitrennikoff 1988). A few modifications were made as described: 10 g of chitin powder was added slowly into 100 mL of concentrated HCl under vigorous stirring for 2 h. The mixture was added to 1,000 mL of ice-cold 95 % ethanol with rapid stirring and kept overnight at 25 °C and then stored at −20 °C until use. When needed, the filtrate was collected and washed with 0.1 M phosphate buffer (pH 7) until the colloidal chitin became neutral (pH 7) and used for further applications.

Phytopathogens and chilli seeds

Colletotrichum gloeosporioides (OGC1) and chilli seeds
(Arka Shweta variety) were obtained as a kind gift from
IIHR, Hessarghatta, Bangalore.

Dual plate assay

The fungal growth inhibition capacity of *Bacillus* sp.
strains was determined (Huang and Hoes 1976). A few
modifications were made to suit the need. One 5-mm disc
of a pure culture of the pathogen was placed at the centre of
a Petri dish containing PDA. The *Bacillus* sp. was inoculated at two opposing corners. Plates were incubated for
72 h, at 28 °C, and growth diameter of the pathogen was
measured and compared to control growth, where the
bacterial suspension was replaced by sterile distilled water.
Each experiment using a single pathogen isolate was run in
triplicate. Results were expressed as the means of the
percentage of inhibition of growth of the corresponding
pathogen isolate in the presence of any of the strain of
Bacillus sp.

   Percent inhibition was calculated using the following
formula:

$$\% \text{ inhibition} = [1 - (\text{Fungal growth/Control growth})] \times 100.$$

Detection of extracellular hydrolytic activity

Plates with minimum salts medium (MSM) containing
(1 % w/v) carboxy methyl cellulose (CMC) was prepared.
The *Bacillus* sp. was spot inoculated in the centre of the
plate. After an appropriate incubation period at 30 °C for
48 h, the agar medium was flooded with an aqueous
solution of Congo red for 15 min. The Congo red solution
was then poured off and plates containing CMC were
visualised for zones of hydrolysis (Shanmugaiah et al.
2008; Moataza 2006; Teather and Wood 1982) detecting β-
1,4 cellulase. Yeast Glucan containing plates (Chen et al.
1995) was used to detect β-1,3 cellulase activity. MSM
with (1 % v/v) yeast cell glucan was prepared and spot
inoculated with the isolate. Development of a clear zone
surrounding the colony indicated enzyme production.

Assay of hydrolytic enzymes

Chitinase enzyme assay

Chitinase activity was measured with colloidal chitin as a
substrate. The culture broth was centrifuged and enzyme
solution of 1 ml was added to 1.0 ml of substrate solution,
which was made by suspending 1 % of colloidal chitin in

Phosphate buffer (pH 7.0). The mixture was incubated at
50 °C for 30 min. 1 ml of DNS was added and incubated at
100 °C in boiling water bath. The amount of reducing
sugar produced in the supernatant was determined by
dinitrosalicylic acid method (DNS) (Miller 1959). One unit
of chitinase activity was defined as the amount of enzyme
that produced 1 μmol of reducing sugars per min (Annamalai et al. 2008).

Cellulase β-1,3 assay

The specific activity of β-1,3-cellulase was determined by
measuring the amount of reducing sugars liberated using
dinitrosalicylic acid solution (DNS) (Annamalai et al.
2008). The culture broth was centrifuged and enzyme
solution of 1 ml was added to 1.0 ml of substrate solution
which contained 1 ml of yeast cell wall extract (YCW 1 %
v/v). The mixture was incubated in a water bath at 50 °C
for 30 min and the reaction was terminated by adding 1 ml
of DNS solution and incubated in boiling water bath for
10–15 min till the development of the colour of the end
product. Reducing sugar concentration was determined by
optical density at 540 nm (Gadelhak et al. 2005).

Cellulase β-1,4 assay

The specific activity of β-1,4-glucanase was determined
by measuring the amount of reducing sugars liberated
using dinitrosalicylic acid solution (DNS) (Annamalai
et al. 2008). The culture broth was centrifuged and
enzyme solution of 1 ml was added to 1.0 ml of substrate
solution which contained 1 ml of carboxy methyl cellulose
solution (CMC 1 % v/v). The mixture was incubated in a
water bath at 50 °C for 30 min and the reaction was terminated by adding 1 ml of DNS solution and incubated in
boiling water bath for 10–15 min till the development of
the colour of the end product. Reducing sugar concentration was determined by optical density at 540 nm (Moataza 2006).

Induction with fungal mycelium

The *Bacillus* sp. was grown on Nutrient broth supplemented with dead fungal mycelium (*Colletotrichum gloeosporioides* OGC1) as inducer for enzymes production at
a concentration of 1.0 % and dispensed in Erlenmeyer
flasks (250 ml), each flask containing 50 ml of medium.
The flasks were autoclaved and inoculated with 1.0 ml of a
pre-cultured *Bacillus* sp. culture. The culture was incubated
in a shaker (120 rpm.), at 28 ± 2 °C. Aliquots from the
flask were analyzed daily for chitinase and cellulases (β-1,3
and β-1,4) for a period of 5 days (Moataza 2006).

Hydrolytic assay

Preparation of hyphal wall

The pathogenic fungal culture (*C. gloeosporioides*) was inoculated into 50 ml of PDB broth and incubated at 30 °C for 5 days under shaking conditions. After incubation, the mycelia were collected by filtration. The mycelia were thoroughly washed with autoclaved distilled water and homogenised on ice, with a homogenizer for 5 min. The mycelia suspension was centrifuged at 10,000 rpm for 20 min at 4 °C (Remi: C 24). The pellet was resuspended in sodium phosphate buffer (0.1 M, pH 7.0). This preparation was used as substrate for the hydrolytic assay (Moataza 2006).

Hydrolytic activity of B. subtilis (wild type and mutants) culture filtrate

For assessing the hydrolytic activity, the reaction mixture (1 ml) containing 1 mg/ml of fungal mycelia with 1.0 ml of crude enzyme (from *B. subtilis* grown on NB + CMC) was incubated at 30 °C for 24 h. The released total reducing sugars (Miller 1959) in control and treated fungal cell wall was estimated using the DNS method.

Dual culture method

The differences in dry weights between the fungal cultures grown with *B. subtilis* strain or the control culture grown without any bacterium were recorded according to Broekaert et al. 1990. For this, 48 h grown dual cultures were passed through the pre-weighed Whatman No 1 filter paper. It was dried for 24 h at 70 °C and weights were measured (Saleem and Kandasamy 2002).

Microscopy

The fungal culture grown on NA agar plate without any bacterial culture served as control. The damage caused by the bacterium to the fungal mycelium by dual plate assay was studied microscopically. The mycelium along with the agar disc present in the inhibition zone and control mycelium was taken, stained with lactophenol cotton blue and observed under a Nikon Trinocular microscope (Saleem and Kandasamy 2002).

UV mutagenesis

To characterise the antagonistic mechanism by this bacterium, a mutant of this bacterium was developed, which lost its antagonistic activity. UV mutagenesis of *B. subtilis* was carried out following the procedure of Miller (1992). During UV mutagenesis, the log phase culture was exposed to short wavelength UV light (280 nm, Philips TUV 30 W, G3018, Holland) from a distance of 30 cm for various time intervals (for 0.1 % UV survivors). From the serial dilutions of the mutagenized culture, 0.1 ml was plated on NA plates for isolated mutant colonies. The isolated colonies were further screened for the loss of antifungal activity using dual plate assay. The mutants were also screened for mycolytic enzyme activity and hydrolytic activity with the pathogen mycelium.

Sensitivity of the culture supernatant of wild type B. subtilis to proteolytic enzymes, TCA and heat

The crude supernatant (1 mL) was subjected to treatments for 1 h at 37 °C (for enzymes) or room temperature (for TCA). The proteolytic enzymes (Sigma) were used at a final concentration of 1 mg ml^{-1} in 10 mmol^{-1} potassium phosphate buffer, pH 7.0. The crude supernatant in buffer without enzymes as well as the enzyme solutions was exposed to the same conditions. For the heat treatment, the preparations were incubated at 70, 80 and 90 °C and autoclaving for 20 min. Antifungal activities were checked before and after all treatments on a test plate made with *C. gloeosporioides* as mentioned earlier (Tendulkar et al. 2007).

Seed testing

Germination efficiency and antagonism of the *Bacillus* sp. against *C. gloeosporioides* was checked on chilli seeds in vitro. The water agar plates were seeded with the following: Set 1-Seeds control-plain seeds coated with carboxy methyl cellulose (CMC); Set 2- Seeds coated with CMC and *C. gloeosporioides* spores; Set 3- Seeds coated with CMC and *Bacillus* sp. culture and Set 4- Seeds coated with CMC and both *C. gloeosporioides* spores and *Bacillus* sp. culture. Chilli seeds were surface sterilised successively with sterile distilled water and 0.1 % HgCl$_2$. To remove the residual HgCl$_2$, the seeds were washed with sterile distilled water. The isolate was inoculated into NB medium and incubated for 24 h at 30 °C (Moataza 2006). *C. gloeosporioides* was inoculated onto PDA plates and incubated at 28 °C for 3–4 days. The above three sets of treated seeds were seeded onto 1 % water agar plates. Plain CMC coated seeds on water agar were used as control. The four sets were monitored regularly for germination and growth. After 1 week, the sets were observed for germination and biocontrol against *C. gloeosporioides* coated seeds by the isolate.

Reproducibility

All the experiments were done in triplicates and the values represented statistically are in ANOVA form.

Results and discussion

15 chilli rhizosphere soils samples were collected from in and around Bangalore, heat treated and plated on chitin amended plates. Out of 15 samples plated, 9 chitinolytic colonies were isolated based on the clearance zones that they formed.

Dual plate assay

These chitinolytic bacteria were subjected to dual plate assay with nine different chilli fungal pathogens, one particular isolate designated as isolate two showed inhibition of all the nine pathogens and this was chosen for further work. The selected isolate showed broad spectrum antagonism against *Alternaria* (3) spp. (55 %), *C. gloeosporioides* (57 %), *Phytophthora capsici* (55 %), *Rhizoctonia solani* (42 %), *Fusarium solani* (42 %), *Fusarium oxysporum* (40 %) and *Verticillium* sp. (36 %), the range of percentage inhibition varied from 40 to 62 (Table 1). The isolate was checked for other hydrolytic enzyme production and was found to produce both Cellulase β-1,3 and β-1,4 on the basis of clearance zones produced on yeast glucan plates as well as CMC plates, respectively (Fig. 1).

Strain identification

This strain was characterised morphologically and biochemically by following Bergey's Manual of Systematic Bacteriology and was found to be a *Bacillus* sp. It was

Table 1 Results of the dual plate assay

Pathogens	Inhibition (%)
C. gloeosporioides (OGC1)	57 ± 0.901^d
A. brassicae (OCA1)	57 ± 2.685^a
A. brassicicola (OCA3)	53 ± 1.443^a
A. alternata (OTA36)	45 ± 1.01^c
P. capsici (98-01)	62 ± 0.935^c
Verticillium sp.	52 ± 2.049^c
F. solani	41 ± 1.527^b
F. oxysporum	40 ± 1.607^e
R. solani	56 ± 5.204^f

Results are the mean of three replicates ± SD. In columns, values with the same letters are not significantly different (*P* < 0.05 Duncan test)

grown and maintained on NA at 30 °C. The isolate upon Gram's staining was identified as a gram positive, spore forming rod. The isolate was motile and answered positive for catalase, nitrate reduction, Voges Proskauer, starch hydrolysis and growth on 6.5 % NaCl medium and negative for parasporal crystal formation and citrate utilisation, hence, was tentatively identified as *Bacillus* sp. Further, 16SrDNA sequencing identified the isolate to be *B. subtilis* and the Gene-Bank accession no. for the nucleotide sequence is JN032305 (Fig. 2) (Ashwini et al. 2012).

UV mutagenesis

To characterise the antagonistic mechanism by this bacterium, a mutant of this bacterium was developed, which lost its antagonistic activity. Out of 61 putative mutant colonies tested, six showed no antagonistic property against *C. gloeosporioides* (Fig. 3). These mutant isolates also did not grow over the mycelial mat and were named as M17, M22, M23, M27, M28 and M30.

Microscopy

The fungal mycelium grown with *Bacillus* BC2 culture showed damage, swelling and distortions as compared to the control (Fig. 4a, b). The mycelium which grew with *Bacillus* BCUVM cultures and the control mycelium which was not grown with any bacterial culture did not show these abnormal features (Fig. 4c). This clearly indicates the mycolytic activity of the *Bacillus* BC2 culture. A similar observation has been made in the antagonism of *Arthrobacter* sp. to *Fusarium* sp. (Barrows-Broaddus and Kerr 1981). Similarly, Podile and Prakash (1996) reported the lysis and dissolution of fungal mycelium of *Aspergillus niger* by *B. subtilis* AF1 strain.

Mycolytic enzyme activities

The induction profile of the *Bacillus* sp. was checked with autoclaved *C. gloeosporioides* mycelium used as the carbon source in the medium. Data represented in (Fig. 5) showed that the lysis of dead mycelia of *C. gloeosporioides* was very efficient by the *Bacillus* sp. Appreciable levels of all the three enzymes were observed in the presence of the autoclaved mycelia—chitinase peaked on day 1 (2.84 U/mL) and gradually decreased (2 U/mL by day 2 and 1.4 U/mL by day 3); β-1,3 cellulase was detected from day 1 (2.8 U/mL) which peaked on day 2 (3.2 U/mL) followed by a gradual decrease (2.7 U/mL by day 3 and 1 U/mL by day 4) and β-1,4 cellulase was also detected from day 1 (6 U/mL) and increased to a maximum of 13.21 U/mL by day 3 followed by a gradual decrease on day 4 (8 U/mL) suggesting the possible role of these enzymes in antibiosis

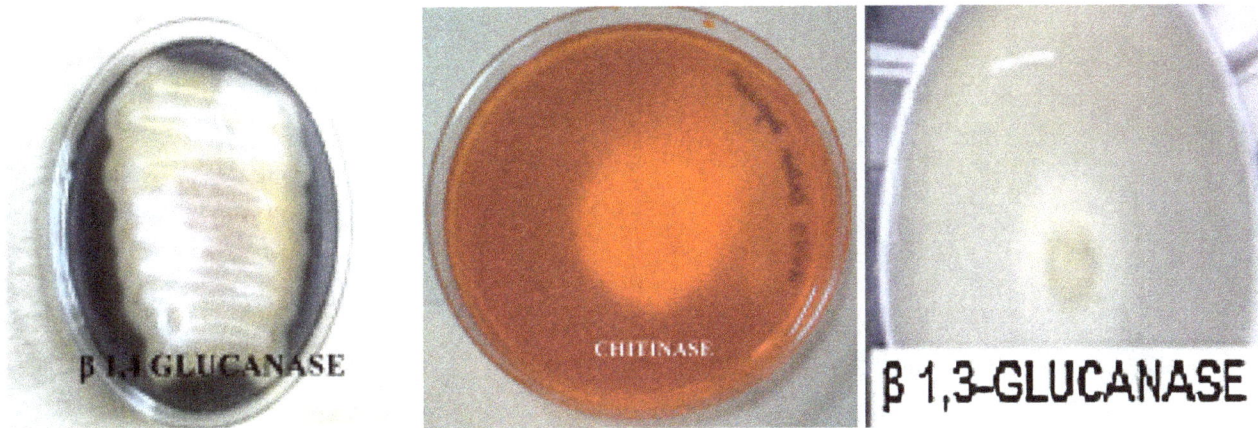

Fig. 1 Plates showing zone of hydrolysis for β1,4-glucanase; chitinase and β 1,3 glucanase by WT *B. subtilis*

Fig. 2 Phylogenetic tree constructed with the 16S rDNA sequences

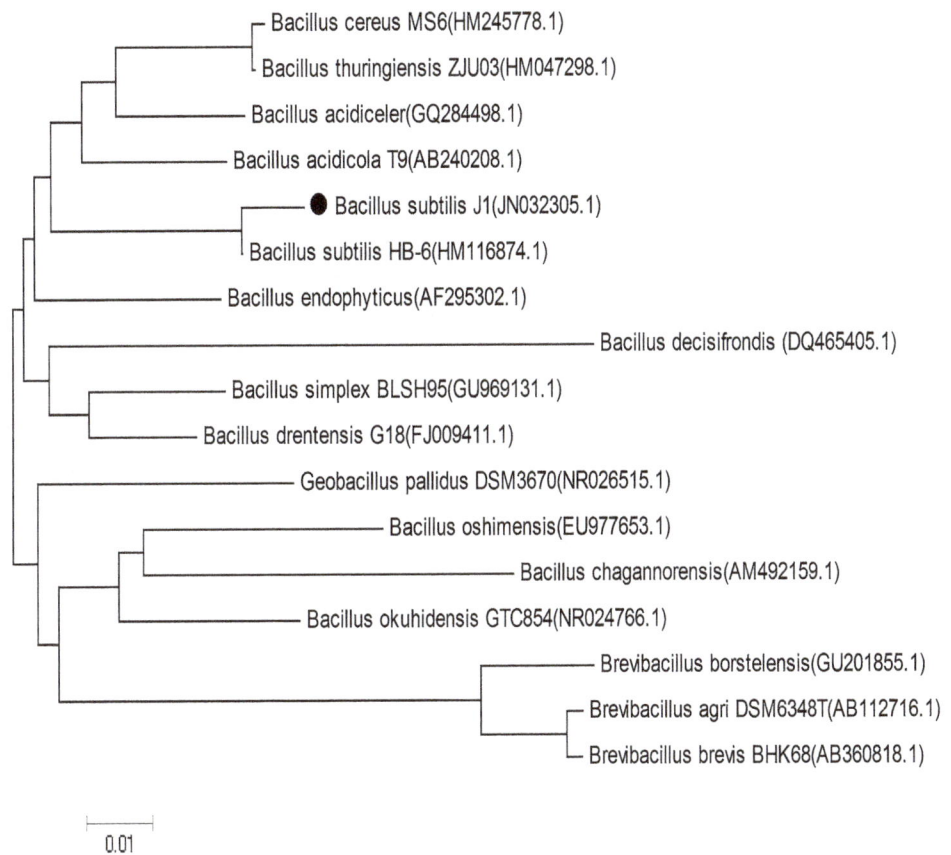

Bacillus cereus MS6(HM245778.1)

Bacillus thuringiensis ZJU03(HM047298.1)

Bacillus acidiceler(GQ284498.1)

Bacillus acidicola T9(AB240208.1)

● Bacillus subtilis J1(JN032305.1)

Bacillus subtilis HB-6(HM116874.1)

Bacillus endophyticus(AF295302.1)

Bacillus decisifrondis (DQ465405.1)

Bacillus simplex BLSH95(GU969131.1)

Bacillus drentensis G18(FJ009411.1)

Geobacillus pallidus DSM3670(NR026515.1)

Bacillus oshimensis(EU977653.1)

Bacillus chagannorensis(AM492159.1)

Bacillus okuhidensis GTC854(NR024766.1)

Brevibacillus borstelensis(GU201855.1)

Brevibacillus agri DSM6348T(AB112716.1)

Brevibacillus brevis BHK68(AB360818.1)

0.01

of the mycelia. Moataza 2006 also reported varied levels and types of mycolytic enzymes by different *Pseudomonas* strains with different pathogens such as *P. capsici* and *R. solani*. Further, the mutants were studied for their mycolytic enzyme activities under shake flask conditions. All the mutants showed significant loss of all three mycolytic enzyme activities (Fig. 5). The mutants also exhibited low levels of hydrolytic activity with *C. gloeosporioides* mycelia as compared to the wild type strain indicating

clearly the mycolytic enzyme mediated antagonism of this strain (Fig. 6).

Dual culture method

To test the antifungal activity of the *Bacillus* strain BC2, dual liquid culture method was employed. The differences in dry weights between the fungal cultures grown with BC2 strain or the mutant strains or the control culture grown

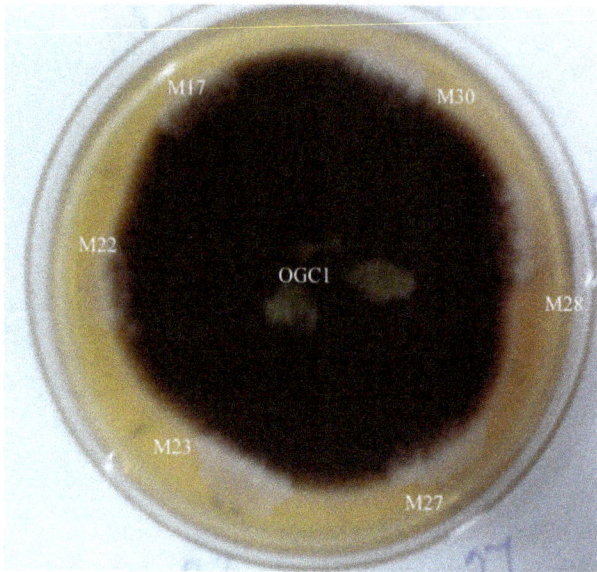

Fig. 3 *Bacillus* strain mutants M17, M22, M23, M27, M28 and M30 not showing inhibition to *Colletotrichum* on PDA

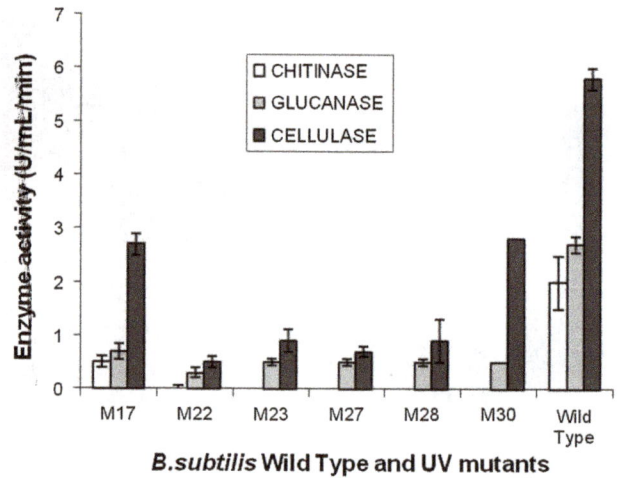

Fig. 5 Comparison of mycolytic enzyme activities of the mutants and the wild strain BC2

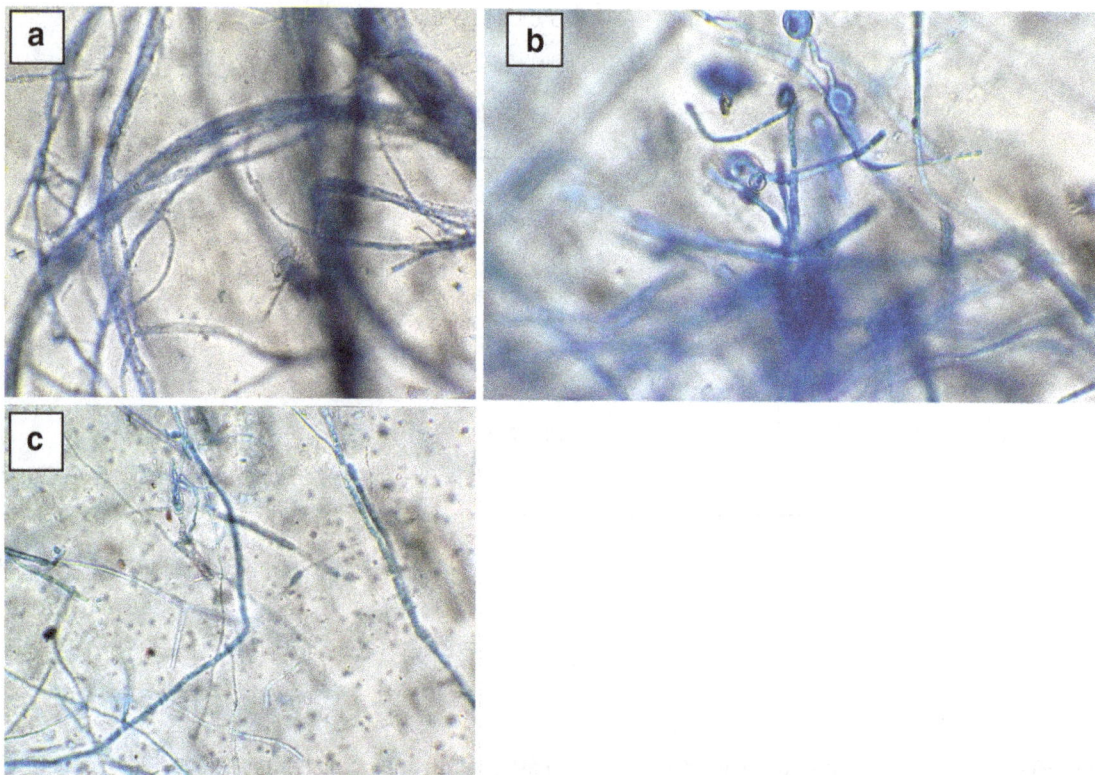

Fig. 4 Light microscopic observations of mycelium inhibited by BC2 strain (1,000×). Mycelium of *Colletotrichum* **a** grown on PDA (control), **b** present in the inhibition zone, when grown with *Bacillus* strain BC2 on PDA, **c** grown with mutant strains on PDA medium

without any bacterium were recorded according to Broekaert et al. (1990). There was almost 100 % reduction in dry weight of the culture grown with BC2 strain when compared to the control. There was very little reduction in dry weight of the culture when grown with the mutant strains. This clearly shows that the reduction in dry weight of the fungus when grown with the BC2 strain is due to the antifungal activity of this bacterium. The mutant strains which had lost the antifungal activity could not reduce the dry weight of the fungus.

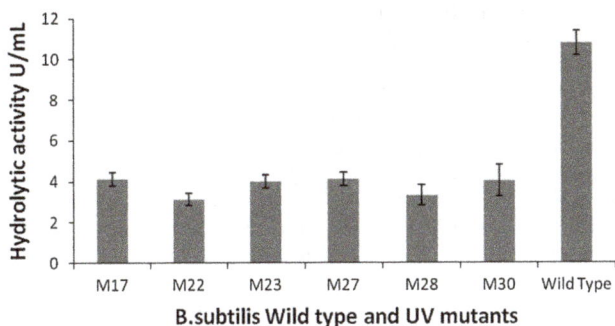

Fig. 6 Comparison of hydrolytic activities of the mutants and the wild strain BC2. Activities of all three enzymes on day 1 by WT *B. subtilis* is represented

Sensitivity of the culture supernatant of *B. subtilis* to proteolytic enzymes, TCA and heat

The sensitivity of the WT crude culture filtrate of *B. subtilis* BC2 was tested with TCA, heat and proteolytic enzymes. The results revealed that the activity was not preserved. Further, when such treated extracts were subjected to antagonism assay, the extract had lost its antagonistic property which supported the mycolytic enzyme mediated antagonism of the fungal pathogen.

Seed bacterization

Treatment of the chilli seeds (Arka variety, obtained as a kind gift from IIHR, Bangalore) with *Bacillus* sp. culture showed 100 % germination index similar to the untreated

Table 2 Effect of seed treatment with effective *B. subtilis* on *C. gloeosporioides*

Treatments	Infection %	Germination %
Control	0	90
Seeds with pathogen	77.50	1
Seeds with *B. subtilis*	0	100
Seeds with Pathogen + *B. subtilis*	12.50	85

seeds (Table 2). The treatment of the seed with co-inoculation of the pathogen with *Bacillus* sp. culture showed 65 % reduction in disease incidence by the treatment as compared to the seed treated with pathogen alone (77.5 %). Kamil et al. (2007) reported that the seed coat treatment of sunflower seeds with *B. licheniformis* induced high reduction in percentage of infection of *R. solani* damping off (from 60 to 25 %) as compared with the pathogen alone. Our observations also comply with these reports.

Statistical analysis

The analysis of variance (ANOVA) has been performed for all three mycolytic enzymes and hydrolytic activity by the wild type and mutants of *B. subtilis*. *P* value was found to be very low at both $P = 0.05$, which indicated that there is significant difference in mycolytic enzyme and hydrolytic activity between the strains (Table 3).

In a similar study, Saleem and Kandasamy (2002) showed the role of the *Bacillus* strain BC121 in suppressing the fungal growth in vitro when studied in comparison with a mutant of that strain, which lacks both antagonistic activity and chitinolytic activity. Another study by Balasubramanian et al. (2010) reported that on testing the biocontrol efficacy of the mutants and wild strain against phytopathogens such as *Fusarium oxysporum, Bipolaris oryzae, Rhizoctonia solani* and *Alternaria* sp. by dual culture assay on PDA medium, the UV H11 mutant and adapted mutant showed increased biocontrol activity when compared to wild strain. Balasubramanian et al. (2010) further reported that the antagonism of these two mutants with *F. oxysporum, R. solani, B. oryzae* and *Alternaria* sp. were varied and could be related with lytic enzyme production with fast growing ability. However, Lorito et al. (1993) reported chitinolytic enzymes contributing to the ability of *Trichoderma sp* to act as biocontrol agents. Graeme Cook and Faull (1991) reported that high antibiotic production by two *T. harzianum* mutant strains, BC10 and BC63, increased inhibition of hyphal growth of *R. solani* and *P. ultimum*; while Papavizas et al. (1982) have shown UV-induced benomyl resistant mutant to suppress the

Table 3 ANOVA for mycolytic enzyme production by wildtype and mutants of *B. subtilis*

Sources of variation	Chitinase		Glucanase		Glucanase		Hydrolytic assay	
	Between	Within	Between	Within	Between	Within	Between	Within
Degrees of freedom	5	3	5	3	5	3	5	3
Sample square	13.98	0.52	11.72	0.13	44.99	0.61	31.76	2.59
Mean square	2.79	0.17	2.34	0.04	8.99	0.20	6.352	0.86
F value	15.92		52.75		44.01		7.34	
P (5 %)	**2.9**		**2.37**		**2.45**		**3.97**	

Bold values indicate significant *P* values (5 %)

saprophytic activity of *R. solani* more effectively than the wild strain (Papavizas et al. 1982).

Conclusion

The selection of effective antagonistic organisms is the first and foremost step in biological control. On the basis of these studies, it is concluded that the *Bacillus* BC2 isolate is showing antagonistic property probably through the enzyme mediated lytic mechanism, which has been proved to be an effective mechanism in controlling the fungal pathogens (Chet et al. 1990). The in vitro seed bacterization studies also have revealed the success of biocontrol of the pathogen. These observations and further studies will help in developing the *Bacillus* BC2 isolates as a potential biological control agent against *C. gloeosporioides*.

Acknowledgments The authors thank the management of Jain University for providing the necessary facilities for carrying out this work.

Conflict of interest The authors declare that they have no conflict of interest in the publication.

References

Abdullah TM, Ali YN, Suleman P (2008) Biological control of *Sclerotinia sclerotiorum* (Lib.) de Bary with *Trichoderma harzianum* and *Bacillus amyloliquefaciens*. Crop Prot 27: 1354–1359

Altschul SF, Gish W, Miller W, Myers EW, Lipman DJ (1990) Basic local alignment search tool. J Mol Biol 215:403–410

Annamalai N, Giji S, Arumugam M, Balasubramanian T (2008) Purification and characterization of chitinase from *Micrococcus* sp.AG84 isolated from marine environment. Afr J Microbiol Res 4(24):2822–2827

Ashwini N, Srividya S (2012a) Optimization of chitinase produced by a biocontrol strain of *B. subtilis* using Plackett-Burman design. Eur J Exp Biol 2(4):861–865

Ashwini N, Srividya S (2012b) Optimization of fungal cell wall degrading glucanases produced by a biocontrol strain of *B. subtilis* using Plackett-Burman design. In: Proceeding Intl Conf Biol Active Mol, Excel India Publisher New Delhi, pp 430–433

Ashwini N, Samantha S, Deepak B, Srividya S (2012) Enhancement of mycolytic activity of an antagonistic *Bacillus subtilis* through ethyl methane sulfonate (EMS) mutagenesis. Turk J Biol.

Balasubramanian N, Thamil Priya V, Gomathinayagam S, Shanmugaiah V, Jashnie J, Lalithakumari D (2010) Effect of chitin adapted and ultra violet induced mutant of *trichoderma harzianum* enhancing biocontrol and chitinase activity. Aust J Basic Appl Sci 4(10):4701–4709

Barrows-Broaddus J, Kerr TJ (1981) Inhibition of *Fusarium moniliforme* var.*subglutinans* the causal agent of pine pitch canker, by the soil bacterium *Arthrobacter* sp. Can J Microbiol 27:20–27

Bosland PW, Votava EJ (2003) Peppers: vegetable and spice capsicums. CAB International, Oxford, p 233

Broekaert WF, Franky RG, Terras Bruno PA, Cammue J (1990) An automated quantitative assay for fungal growth inhibition. FEMS Microbiol Lett 69:55–60

Chen MH, Shen Bobin S, Kahn PC, Lipke PN (1995) Structure of *Saccharomyces cerevisiae* a-agglutinin. J Biol Chem 270: 26168–26177

Chet I, Ordentlich A, Shapira Oppenheim A (1990) Mechanism of biocontrol of soil borne plant pathogens by rhizobacteria. Plant Soil 129:85–92

Cook RJ (1985) Biological control of plant pathogens: theory to application. Phytopathology 75:25–29

Gadelhak G, Khaled A, El-Tarabily KA, Fatma K, Al-Kaabi FK (2005) Insect control using chitinolytic soil actinomycetes as biocontrol agents. Int J Agri Biol 7(4):627–633

Graeme Cook KA, Faull JL (1991) Effect of ultraviolet induced mutants of *T. harzianum* with altered antibiotic production of selected pathogens in vitro. Can J Microbiol 37:659–664

Huang HC, Hoes JA (1976) Penetration and infection of *Sclerotinia sclerotiorum* by *Coniothyrium minitans*. Can J Bot 54:406–410

Huang CJ, Wang TK, Chung SC, Chen CY (2005) Identification of an antifungal chitinase from a potential biocontrol agent, *Bacillus cereus*. J Biochem Mol Biol Sci 38:82–88

Isaac S (1992) Fungal plant interaction. Chapman and Hall Press, London, p 115

Joseph B, Patra RR, Lawrence R (2007) Characterization of plant growth promoting Rhizobacteria associated with chickpea (*Cicer arietinum* L). Intern J Plant Prod 1(Suppl 2):141–152

Kamil Z, Rizk M, Saleh M, Moustafa S (2007) Isolation and Identification of Rhizosphere Soil Chitinolytic bacteria and their Potential in Antifungal Biocontrol. Global J Mol Sci 2:57–66

Lorito M, Harman GE, Hayes CK, Broadway RM, Tronsmo A, Woo SL, Di Pietro A (1993) Chitinolytic enzymes produced by *Trichoderma harzianum*: Antifungal activity of purified endo-chitinase and chitobiosidase. Phytopathology 83:302–307

Mabuchi N, Hashizume I, Araki Y (2000) Characterization of chitinases exerted by *Bacillus cereus CH*. Can J Microbiol 46: 370–375

Marchesi JR, Sato T, Weightman AJ, Martin TA, Fry JC, Hiom SJ, Wade WJ (1998) Design and evaluation of useful bacterium-specific pcr primers that amplify genes coding for bacterial 16 s rrna. Appl Environ Microbiol 64(2):795–799

Miller GL (1959) Use of dinitrosalicylic acid reagent for determination of reducing sugar. Anal Chem 31:426–428

Miller H (1992) A short course in bacterial genetics. A laboratory manual and handbook for E.coli and related bacteria. Cold Spring Harbor Laboratory Press, New York

Moataza MS (2006) Destruction of *Rhizoctonia solani* and *Phytophthora capsici* causing tomato root-rot by *Pseudomonas fluorescences* lytic enzymes. Res J Agri Biol Sci 2(6):274–281

Papavizas GC, Lewis JA, Moity Abd- Ei (1982) Evaluation of new biotypes of *Trichoderma harzianum* for tolerance of Benomyl and enhanced biocontrol capabilities. Phytopathology 72(1):127–132

Podile AR, Prakash AP (1996) Lysis and biological control of *A. niger* by *Bacillus subtilis* AF1. Can J Microbiol 42:533–538

Roberts WK, Selitrennikoff CP (1988) Plant and bacterial chitinases differ in antifungal activity. J Gen Microbiol 134:169–176

Saleem B, Kandasamy U (2002) Antagonism of *Bacillus* species (strain BC121) towards *Curvularia lunata*. Curr Sci 82(12): 1457–1463

Shanmugaiah V, Mathivanan N, Balasubramanian N, Manoharan PT (2008) Optimization of Cultural conditions for production of

Chitinase by *Bacillus laterosporous* MMI2270 isolated from rice rhizosphere soil. Afr J of Biotechnol 7(15):2562–2568

Sneath PHA (1986) In Bergey's manual of systematic bacteriology. In: Sneath PHA et al (eds) Williams and Wilkins, vol 2. Baltimore, pp 1104–1207

Someya N, Nakajima M, Hirayae K, Hibi T, Akutsu K (2001) Synergistic antifungal activity of chitinolytic enzymes and prodigiosin produced by the biocontrol bacterium *Serratia marcescens* strain B2 against the gray mold pathogen, *Botrytis cinerea*. J Gen Plant Pathol 67:312–317

Srividya S, Sasirekha B, Ashwini N (2012) Multifarious antagonistic potentials of rhizosphere associated bacterial isolates against soil borne diseases of Tomato. Asian J Plant Sci Res 2(2):180–186

Sutton BC (1992) The genus *Glomerella* and its anamorph *Colletotrichum*, In: Bailey JA, Jeger MJ (eds) *Colletotrichum*—biology, pathology, and control, CAB International, Wallinngford, pp 1–26

Takayanagi T, Ajisaka K, Takiguchi Y, Shimahara K (1991) Isolation and characterization of thermostable chitinases from *Bacillus licheniformis* X-7u. Biochem Biophys Acta 12:404–410

Teather RM, Wood PJ (1982) Use of Congo red-polysaccharide interactions in enumeration and characterization of cellulolytic bacteria from the bovine rumen. Appl Environ Microbiol 43:777–780

Tendulkar SR, Saikumari YK, Patel V, Raghotama S, Munshi TK, Balaram P, Chattoo BB (2007) Isolation, purification and characterization of an antifungal molecule produced by *Bacillus licheniformis* BC98, and its effect on phytopathogen Magnaporthe grisea. J Appl Microbiol 103:2331–2339

Than PP, Jeewon R, Hyde KD, Pongsupasamit S, Mongkolporn O, Taylor PWJ (2008) Characterization and pathogenicity of *Colletotrichum* species associated with anthracnose on chilli (*Capsicum* spp.) in Thailand. Plant Pathol 57:562–572

Thind TS, Jhooty JS (1985) Relative prevalence of fungal disease of chilli fruit in Punjab. Ind J Mycol Plant Path 15:305–307

Turner JT, Backman PA (1991) Factors relating to peanut yield increases after seed treatment with *Bacillus subtilis*. Plant Dis 75:347–353

Viterbo A, Haran S, Friesem D, Ramot O, Chet I (2001) Antifungal activity of a novel endochitinase gene (chit36) from Trichoderma harzianum Rifai TM. FEMS Microbiol Lett 200:169–174

Wen C, Tseng MCS, Cheng CY, Li YK (2002) Purification, characterization and cloning of a chitinases from *Bacillus* sp. NCTU2. Biotechnol Appl Biochem 35:213–219

Yu GY, Sinclair GL, Hartman GL, Bertagnolli BL (2002) Production of iturin A by *Bacillus amyloliquefaciens* suppressing *Rhizoctonia solani*. Soil Biol Biochem 34:955–963

Zhang Y, Fernando WGD (2004) Zwittermicin A detection in *Bacillus* spp. controlling *Sclerotinia sclerotiorum* on canola. Phytopathology 94:S116

17

Biosynthesis of silver nanoparticles from *Schizophyllum radiatum* HE 863742.1: their characterization and antimicrobial activity

Ram Prasad Metuku · Shivakrishna Pabba ·
Samatha Burra · S. V. S. S. S. L. Hima Bindu N ·
Krishna Gudikandula · M. A. Singara Charya

Abstract Development of reliable and eco-friendly process for synthesis of silver nanoparticles is an important step in the field of application in nanotechnology. One of the options to achieve this objective is to use natural biological processes. They have an advantage over conventional methods involving chemical agents associated with environmental toxicity. This study demonstrates the extracellular synthesis of stable silver nanoparticles using the white rot fungus, *Schizophyllum radiatum* with GenBank Accession no HE 863742.1. The supernatant of the seed media obtained after separating the cells has been used for the synthesis of silver nanoparticles. The morphology and structure of synthesized silver nanoparticles were characterized using FT-IR, XRD, UV–visible spectrum of the aqueous medium containing silver ion showed a peak in the range of 420–430 nm corresponding to the Plasmon absorbance of silver nanoparticles. Scanning electron microscopy micrograph showed formation of well-dispersed silver nanoparticles in the range of 10–40 nm. The effect of different carbon sources and the time taken for formation particles and the anti-microbial activity of synthesized nanoparticles were carried and compared with silver nitrate solution and with standard streptomycin. The process of reduction being extra-cellular and fast may lead to the development of an easy bioprocess for synthesis of silver nanoparticles.

Keywords White rot fungi · Silver nanoparticles · Antimicrobial activity · Scanning electron microscopy

Introduction

Nanobiotechnology attempts to utilize biological templates in the development of nano-scaled products for diverse and specialized applications. Nanoparticles are clusters of atoms in the size range of 1–100 nm. Increasing concern in green chemistry approaches for nanomaterial synthesis and process technology development has provided additional impetus for bioprocess studies using both prokaryotic bacterial cells and eukaryotic organisms (Bhattacharya and Gupta 2005). The size, shape and intercalation properties are the attributes of nanomaterials. The whole cell and cellular biomolecules have evolved as one of the best ways for generation and functionalization in the nanocomposite (Agag and Takeichi 2000). Biological organisms can be used as the environmental friendly techniques to create predictable nanoparticles. Silver and gold nanoparticles synthesized extra-cellularly that have the potentials in the opto electronic devices, thin films and non linear optics (Dahl et al. 2007). Macromolecules like enzymes and polysaccharides from the bioprocesses are increasingly in focus for nanomaterial production and utilization. Bacteriorhodopsin is one of the extensively studied nanomaterials for technical application, including in photoelectric and proton transport devices (Hampp 2004). Silver nanoparticles are in demand for the photo chemical applications, catalysis, and chemical analysis. The microbial silver bio-inorganics were studied intensively (Lengke et al. 2007; Shiying et al. 2007; Morones et al. 2005). The applications of AgNPs are of great concern in waste water treatment, pesticide degradation, killing human pathogenic

R. P. Metuku · S. Pabba · S. Burra ·
S. V. S. S. S. L. Hima Bindu N · K. Gudikandula ·
M. A. Singara Charya (✉)
Department of Microbiology, Kakatiya University,
Hanamkonda, Warangal 506009, India
e-mail: mascharya@gmail.com

bacteria (Kuber and Souza 2006). They exhibited cyto-protectivity toward HIV-1 infected cells (Elechiguerra et al. 2005). The study of biosynthesis of nanomaterials offers a valuable contribution as eco-friendly technologies into material chemistry. The ability of some microorganisms such as bacteria and fungi to control the synthesis of metallic nanoparticles should be employed in the search for new materials (Mandal et al. 2006).

This study was on the potentials of extra-cellular biosynthesis of silver nanoparticles by *Schizophyllum radiatum* and their characterization, antimicrobial activity on Gram-positive and Gram-negative bacteria. The influence of carbon sources on the silver nanoparticles biosynthesis was carried out. The *S. radiatum* was also studied for its antimicrobial and antioxidative properties under submerged fermentation. The supernatant of the seed media obtained after separating the mycelia has been used for the synthesis of silver nanoparticles.

Materials and methods

Chemicals

Dextrose, silver nitrate Merck (Germany) Malt extract was procured from Himedia (India). Sterile distilled water was used throughout the experiments.

Collection and molecular identification of *Schizophyllum radiatum*

Fungi in the form of fruit body were collected from the Eturnagaram forest of Warangal, Andhra Pradesh, India. The fruit body was cleaned with disinfectants and approximately 3 × 3 mm was placed on MEA agar medium in petri-dishes. When the mycelium had grown on the medium in the vicinity of the tissues, the sample was transferred to fresh agar media in tubes. This was repeatedly carried out until pure culture was obtained. Molecular-based characterization on ribotyping of 18S rRNA was performed at the Xcelris Genomics, Ahmedabad, India and sequence was deposited to EMBL database for accession number.

Production of extra-cellular silver nanoparticles

Schizophyllum radiatum was grown in yeast malt broth containing dextrose 10 g l^{-1}, malt extract 5 g l^{-1}. The final pH was adjusted to 6.0. The flasks were incubated in the orbital shaker at 200 rpm at 32 °C. After 5 days of incubation, the mycelium was separated by filtration and

supernatant was challenged with equal amount of with various concentrations (0.5, 1.0, 1.5, 2.0, 2.5 mM) of silver nitrate solution (prepared in deionized water) and incubated in shaker at 200 rpm in dark condition at 32 °C. Simultaneously, a positive control of silver nitrate solution and deionized water and a negative control containing only silver nitrate solution were maintained under same conditions.

Influence of carbon sources on silver nanoparticle synthesis

In order to investigate the influence of carbon sources on the silver nanoparticle synthesis, the time taken for the formation of particles (i.e. from colorless to brown) was studied. Sources like glucose, fructose, sucrose, lactose, and starch were added separately in place of dextrose at a concentration of 0.4 % to the basal medium containing malt extract (1 %). *S. radiatum* was inoculated into the basal medium containing particular carbon source and incubated at 32 °C for 5 days at 200 rpm in orbital shaker. After incubation, the culture broth was filtered and separated the mycelial biomass. Supernatant was challenged with equal volume of effective concentration of AgNO$_3$ solution. Control consisting of filtered broth of particular carbon source and deionized water was maintained under same conditions.

UV–visible spectral analysis

Change in color was observed in the silver nitrate solution incubated with the *S. radiatum*. The UV–visible spectra of this solution were recorded in ELICO SL-159 Spectrophotometer in the range of 350–470 nm.

Scanning electron microscope

After freeze drying of the purified silver particles, the size and shape were analyzed by scanning electron microscopy (JOEL-Model 6390).

Fourier-transform infrared (FT-IR) chemical analysis

Fourier-Transform Infra-Red spectroscopy measurements, the biotransformed products present in extracellular filtrate were freeze-dried and diluted with potassium bromide in the ratio of 1:100. The FT-IR spectrum of samples was recorded on a FT-IR instrument (Digital Excalibur 3000 series, Japan) with diffuse reflectance mode (DRS-800) attachment. All measurements were carried out in the range of 400–4,000 cm^{-1} at a resolution of 4 cm^{-1} (Saifuddin et al. 2009).

X-ray diffraction analysis

The fungal mycelium embedded with the silver nanoparticles was freeze-dried, powdered and used for XRD analysis. The spectra were recorded in Philips® automatic X-ray Diffractometer with Philips® PW 1830 X-ray generator. The diffracted intensities were recorded from 30° to 90° 2θ angles.

Antibacterial activity

Biosynthesized silver nanoparticles produced by the *S. radiatum* were tested for antimicrobial activity as method suggested by Srinivasulu et al. (2002) using various Gram-positive and Gram-negative bacteria by the agar well-diffusion method. Approximately, 20 ml of nutrient agar medium was poured into sterilized petri-dishes. The bacterial test organisms were grown in nutrient broth for 24 h. A 100 µl nutrient broth culture of each bacterial organism (1×10^5 CFU ml^{-1}) was used to prepare bacterial lawns. Agar wells of 8 mm diameter were prepared with the help of a sterilized stainless steel cork borer. The wells were loaded with 60 µl of Ag nanoparticles solution, 60 µl of 1.5 mM silver nitrate and 60 µl of culture broth from *S. radiatum* culture without AgNO$_3$ as a negative control, along with 60 µl of 30 µg ml^{-1} of streptomycin as a positive control. The plates were incubated at 37 °C for 24 h and then were examined for the presence of zones of inhibition. The diameter of such zones of inhibition was measured and the mean value for each organism was recorded and expressed in millimeter unit.

Results and discussions

The 18S rRNA gene sequencing analysis of the isolate yielded 1,112 base pairs, and NCBI BLAST search analysis based on the topology of phylogenetic analysis revealed that the sequence was 99 % related with *S. radiatum*. The obtained sequence was deposited in EMBL database with the Accession number HE 863742.1.

Biosynthesis of silver nanoparticles

Bioreduction of silver nitrate into nanosilver can be primarily characterized using UV–vis spectroscopy. In the present study, the color of the silver nitrate solution in the flask containing culture filtrate of *S. radiatum* changed from colorless to brown within 48 h. Various concentrations of silver nitrate (0.5, 1.0, 1.5, 2.0 and 2.5 mM) that were used, the formation of silver nanoparticles synthesis was observed at 1.5 mM within 48 h (Fig. 1) and further increased with increase in silver nitrate concentration. An appreciable amount of silver nanoparticle was obtained at 2.5 mM concentration. Kathiresan et al. (2009), working with silver nanoparticles in the range of 400–500 nm wave length, reported that there is a decrease in the value of optical density (OD) with a proportional increase in silver nitrate concentration. Extracts from organisms may act as reducing and capping agents in AgNPs synthesis. The reduction of silver ions by combinations of bio-molecules found in these extracts such as enzymes, proteins, amino acids, polysaccharides and vitamins is environmentally benign, yet chemically complex. But, the mechanism which is widely accepted for the synthesis of silver nanoparticles is the presence of enzyme "Nitrate reductase" (Anil Kumar et al. 2007; Kalimuthu et al. 2008). Cysteine a biomolecule present in the cell-free extract from *Trichoderma asperellum* acts as a potential reducing and capping agent in the synthesis of stable silver nanoparticles (Roy et al. 2011). From application point of view, the extracellular synthesis of nanoparticles is more important than the intracellular assimilations. Carbon sources such as glucose and fructose enhanced the production of enzyme which in turn reduced the silver nitrate hence the amounts of silver nanoparticles were improved. The time taken for the formation of silver nanoparticles appeared with in 48 h in glucose- and fructose-containing carbon sources. Lactose and galactose showed 72 h, whereas starch has no influence on silver nanoparticle synthesis (Fig. 2).

Fig. 1 **a** Showing the formation of silver nanoparticles after 48 h of incubation. **b** Effect of various concentrations of silver nitrate AgNO$_3$ on nanoparticle synthesis

Fig. 2 Influence of carbon sources on nanoparticles synthesis **a** glucose, **b** sucrose, **c** fructose, **d** lactose, **e** galactose, **f** starch with respective controls

Characterization of silver nanoparticles

The homogeneous spherical silver nanoparticles are known to produce the surface plasmon resonance band at 413 nm (Ahmad et al. 2003). In the present system absorption maxima (λ_{max}) of biosynthesized silver nanoparticles (Fig. 3) observed in the range of 420–430 nm. Brause et al. (2002), working with silver colloids in aqueous solution, reported that optical absorption spectra of metal nanoparticles are mainly dominated by surface plasmon resonance, and the absorption peak has relationship with particle size. Smaller AgNPs will have an absorbance maximum around 400 nm, which increases with size and disappears when particle size falls outside nanodimensions.

Scanning electron microscopy (SEM) analysis

This SEM image (Fig. 4) provided further insight into the morphology and size of the produced nanoparticles. It is evident from the figure that the biosynthesized silver nanoparticles are in different size and shapes and mostly observed as individual particles as well as a few aggregates. SEM-mediated characterization of biosynthesized nanomaterials has been performed by several investigators. Particle size analysis revealed that the silver nanoparticles are in the size range of 10–40 nm with a mean diameter of 14.5 nm suggesting the production of different-sized nanoparticles. Silver nanoparticles in the range of 35–46 nm by *Pseudomonas stutzeri* (Klaus et al. 1999) 20–50 nm particles by *Lactobacillus* sp. (Nair and Pradeep 2002) 2–20 nm particles by *Verticillium* sp. and 2–50 nm sized particles by *Fussarium oxysporum* (Ahmad et al. 2003) have been reported. Mukherjee et al. (2008) demonstrated the synthesis of low dispersive and highly stabilized nanocrystalline silver particles by a non-pathogenic and agriculturally important fungus, *Trichoderma asperellum*.

Fourier-transform infrared (FT-IR) chemical analysis

FT-IR spectroscopy, used to characterize the surface chemistry of silver nanoparticles produced by *S. radiatum*

Fig. 3 UV–visible absorption spectra of silver nanoparticles after 48 h of incubation

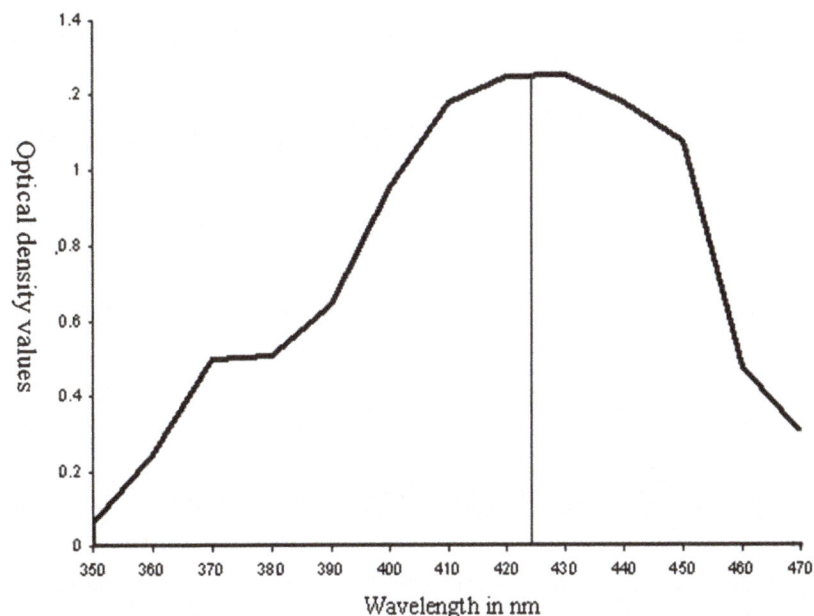

(Fig. 5), showed the FT-IR spectrum of the freeze-dried powder of silver nanoparticles formed after 48 h of reaction. The spectral data revealed two types of vibrations (i.e. stretching and bending) in the wavelength range of 4,000–500 cm^{-1}. It is evident from the figure that the presence of an amine vibration band at 3,400 cm^{-1} represents a primary amine (N–H) stretching, and amide (N–H) bending vibration bands at 1,650 and 1,644 cm^{-1}. Furthermore, the FT-IR spectra of biosynthesized silver nanoparticles also revealed peaks at 2,026 and 2,116 cm^{-1} stretching vibrations of aliphatic C–H bonds. A band presence at 1,412 cm^{-1} can be assigned to CH2-scissoring stretching vibration at the planar region. Several C–N stretching vibration peaks at 1,258, 1,143, 1,102, 1,027 and 908 cm^{-1} were also observed in the spectral range of 1,230–900 cm^{-1}. In addition, the presence of bands at 1,356 and 1,250 cm^{-1} in the FT-IR spectra suggested that the capping agent of biosynthesized nanoparticles possesses an aromatic amine groups with specific signatures of amide linkages between amino acid residues in the proteins in the infrared region of the electromagnetic spectrum (Shaligram et al. 2009). This type of FT-IR spectra supports the presence of a protein type of compound on the surface of biosynthesized nanoparticles, confirming that metabolically produced proteins acted as capping agents during production and prevented the reduced silver particles agglomeration.

X-ray diffraction analysis

Freeze-dried reaction mixture embedded with the silver nanoparticles was used for X-ray diffraction (XRD) analysis. Crystallinity of biosynthesized silver nanoparticles was assessed from their X-ray powder diffraction patterns. The diffractogram showed (Fig. 6) the phase purity of the material. The diffraction peaks are above 37°, XRD patterns were recorded the four prominent 111, 200, 220 and 311 reflections at $2\theta = 38.2, 44.4, 64.5$ and 77.7 indicating the face centered cubic (FCC) structure of silver nanoparticles.

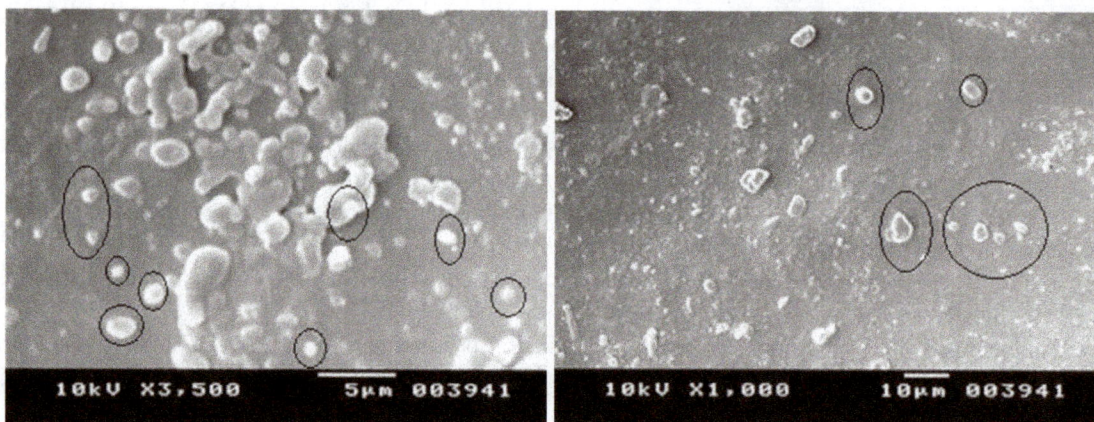

Fig. 4 Shows SEM micrographs of silver nanoparticles synthesized from fungal extracts

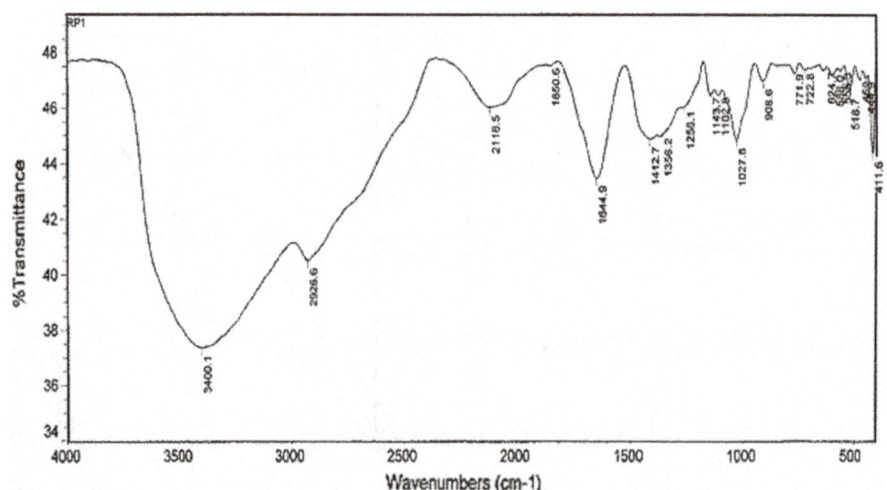

Fig. 5 FT-IR spectrum recorded with synthesized silver nanoparticles

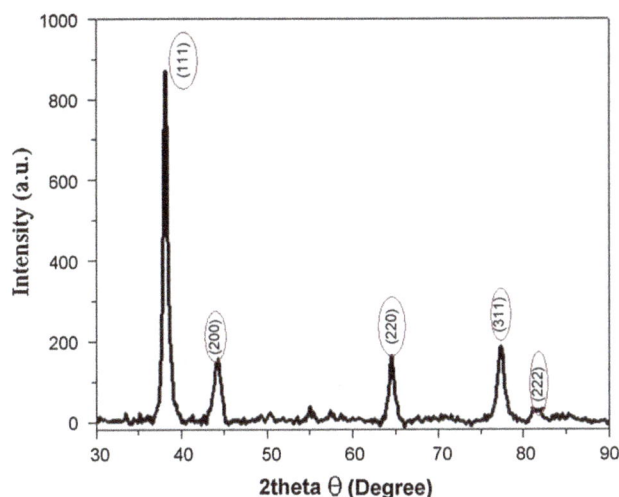

Fig. 6 XRD pattern of silver nanoparticles

Anti bacterial activity

Silver has been in use since time immemorial in the form of metallic silver, silver nitrate, silver sulfadiazine for the treatment of burns, wounds and several bacterial infections. But due to the emergence of several antibiotics the use of these silver compounds has been declined remarkably.

Silver in the form of silver nanoparticles has made a remarkable comeback as a potential antimicrobial agent and proved to be most effective as it has good antimicrobial efficacy against bacteria, viruses and other eukaryotic micro-organisms (Gong et al. 2007). The use of silver nanoparticles is also important, as several pathogenic bacteria have developed resistance against various antibiotics. The antibacterial activity of silver nanoparticles was investigated against various pathogenic Gram-positive and Gram-negative bacteria, like *Escherichia coli*, *Klebsiella pneumoniae*, *Enterobacter aerogenes*, *Pseudomonas aeroginosa*, *Staphylococcus aureus*, *Salmonella paratyphi*, *Bacillus sterothermophilus*, *Bacillus subtilus* using well-diffusion technique (Fig. 7). The diameter of inhibition zones around each well with AgNPs and AgNO$_3$ is recorded in Table 1. The highest antimicrobial activity was observed against, *B. subtilus* and *S. paratyphi*, followed by *Bacillus stearothermophilus*, *Staphylococcus aureus*, *Enterobacter aerogenes*, *Salmonella typhi* and the least was noticed against *Klebsiella pneumonia*. Feng et al. (2000) reported the mechanism of silver nanoparticles action and used *E. coli* and *S. aureus* as model organisms. The nanoparticles get attached to the cell membrane and also penetrate inside the bacteria. The bacterial membrane contains sulfur-containing proteins and the silver

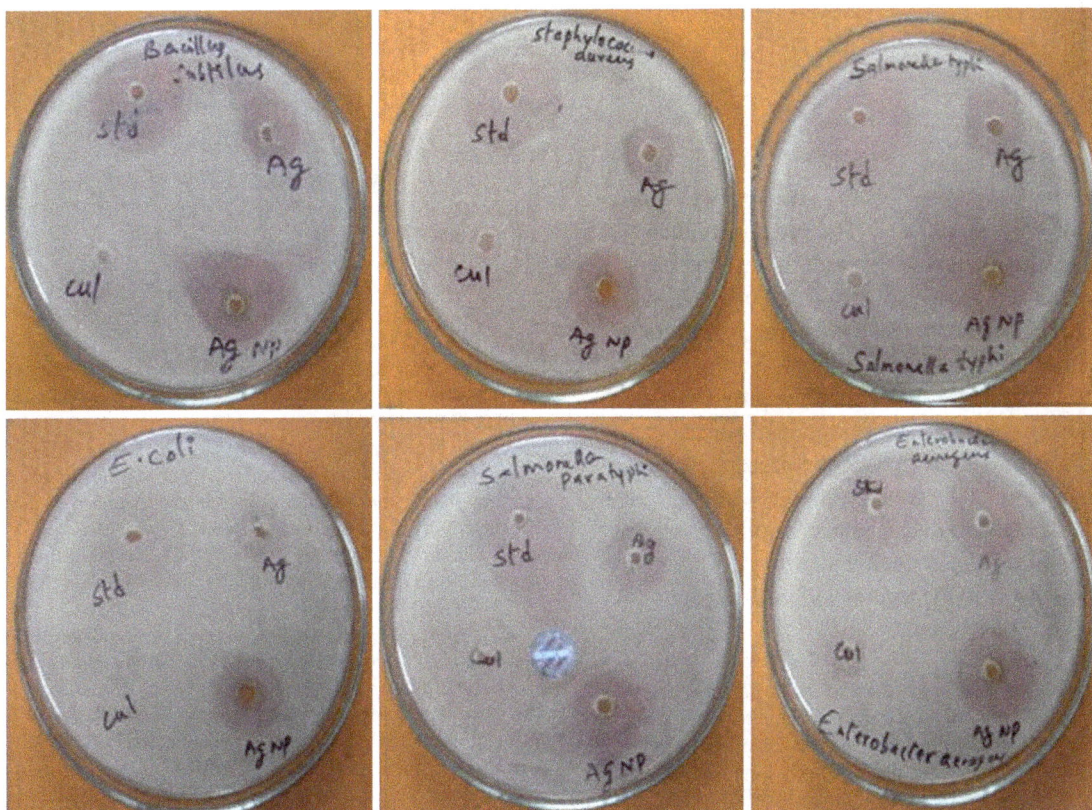

Fig. 7 Antibacterial activity of silver nanoparticles produced by *Schizophyllum radiatum* against bacterial species, where *std* standard (streptomycin), *cul* culture broth, *Ag* silver nitrate solution, *AgNP* silver nanoparticles solution

Table 1 Antibacterial activity of silver nanoparticles produced by *Schizophyllum radiatum*

Microorganism	Zone of inhibition (mm)		
	Streptomycin[a]	AgNO$_3$ sol[b]	Silver nanoparticle sol[c]
Bacillus subtilus	20	9	19
B. stearothermophilus	24	13	17
Salmonella paratyphi	24	8	19
S. typhi	23	9	14
Staphylococcus aureus	22	8	15
Pseudomonas aeroginosa	23	9	13
Enterobacter aerogenes	23	16	16
E. coli	24	6	14
Klebsiella pneumoniae	14	3	7

[a] Zone of growth inhibition diameter of streptomycin (positive control)

[b] Zone of growth inhibition diameter with silver nitrate sol

[c] Zone of growth inhibition diameter with silver nanoparticles

nanoparticles interact with these proteins in the cell as well as with the phosphorus-containing compounds like DNA. Nanoparticles preferably attack the respiratory chain, cell division finally leading to cell death. Silver nanoparticles have emerged up with diverse medical applications in silver-based dressings (Duran et al. 2007).

Conclusion

The nanoscale understanding of the bioprocesses and interventions are only in infancy. The areas particularly bioprocess product exploitations are only in current focus that in itself has opened vistas in a technology upsurge. The order beneath the chaos at the genomic level is recognizable but not yet been fully understood. The microorganisms such as bacteria and algae have already proved to be inorganic nanofactories of the enormous dimensions. Among the different methods for NP synthesis, the chemical reduction method and green synthesis method were widely studied due to their advantage in controlling particle size and morphology. White rot fungus, easy to produce biomass and non-pathogenic nature will add strength to silver nanoparticles biosynthesis.

Acknowledgments The authors are thankful to Prof. S. Girisham, Head of the department Microbiology, Kakatiya University Warangal for providing laboratory facilities. We acknowledge the support extended by Sophisticated Analytical Instrumentation Facility at Department of Physics for analyzing the samples by SEM and FTIR.

Conflict of interest The authors declare that they have no conflict of interest.

References

Agag T, Takeichi T (2000) Polybenzoxazine–montmorillonite hybrid nanocomposites synthesis and characterization. Polymer 41:7083–7090

Ahmad A, Mukherjee P, Senapati S, Mandal D, Khan MI, Kumar R, Sastry M (2003) Extracellular biosynthesis of silver nanoparticles using the fungus *Fusarium oxysporum*. Colloids Surf B 28:313–318

Anil Kumar S, Abyaneh MK, Gosavi Sulabha SW, Ahmad A, Khan MI (2007) Nitrate reductase mediated synthesis of silver nanoparticles from AgNO$_3$. Biotechnol Lett 29:439–445

Bhattacharya D, Gupta RK (2005) Nanotechnology and potential microorganisms. Crit Rev Biotechnol 25:199–204

Brause R, Moeltgen H, Kleinermanns K (2002) Characterization of laser-ablated and chemically reduced silver colloids in aqueous solution by UV–vis spectroscopy and STM/SEM microscopy. Appl Phys B 75:711–716

Dahl JA, Maddux BLS, Hutchison JE (2007) Toward greener nanosynthesis. Chem Rev 107:2228–2269

Duran N, Marcarto PD, De Souza GIH, Alves OL, Esposito E (2007) Antibacterial effect of silver nanoparticles produced by fungal process on textile fabrics and their effluent treatment. J Biomed Nanotechnol 3:203–208

Elechiguerra JL, Burt JL, Morones JR, CamachoBragado A, Gao X, Lara HH, Yaca Man MJ (2005) Interaction of silver nanoparticles with HIV-I. J Nanobiotechnol 3:6–16

Feng QL, Wa J, Chen GQ, Cui KZ, Kim TM, Kim JO (2000) A mechanistic study of the antibacterial effect of silver ions on *Escherichia coli and Staphylococcus aureus*. J Biomed Mater Res 52:662–668

Gong P, Li H, He X, Wang K, Hu J, Tan W (2007) Preparation and antibacterial activity of Fe$_3$O$_4$ Ag nanoparticles. Nanotechnology 18:604–611

Hampp N (2004) Bacteriorhodopsin and its potential in technical application. In: Niemeyer CM, Mirkin CA (eds) Nanobiotechnology concepts applications and perspectives. Wiley-VCH, New York, pp 146–167

Kalimuthu K, Babu RS, Venkataraman D, Mohd B, Gurunathan S (2008) Biosynthesis of silver nanocrystals by *Bacillus licheniformis*. Colloids Surf B 65:150–153

Kathiresan K, Manivannan S, Nabeel MA, Dhivya B (2009) Studies on silver nanoparticles synthesized by a marine fungus, *Penicillium fellutanum* isolated from coastal mangrove sediment. Colloids Surf B Biointerfaces 71:133–137

Klaus T, Joerger R, Olsson E, Granquist CG (1999) Silver based crystalline nanoparticles, microbially fabricated. Proc Natl Acad Sci USA 96:13611–13614

Kuber CB, Souza SF (2006) Extracellular biosynthesis of silver nanoparticles using the fungus *Aspergillus fumigatus*. Colloids Surf B 47:160–164

Lengke MF, Fleet ME, South G (2007) Biosynthesis of silver nanoparticles by filamentous cyanobacteria from a silver (I) nitrate complex. Langmuir 23:2694–2699

Mandal D, Bolander ME, Mukhopadhyay D, Sarkar G, Mukherjee P (2006) The use of microorganisms for the formation of metal nanoparticles and their application. Appl Microbial Biotechnol 69:485–492

Morones JR, Elechiguerra JL, Camacho A, Holt K, Kouri JB, Tapia J, Jose Yacaman M (2005) The bactericidal effect of silver nanoparticles. Nanotechnology 16:2346–2353

Mukherjee P, Roy M, Mandal B, Dey G, Mukherjee P, Ghatak J (2008) Green synthesis of highly Stabilized nanocrystalline silver particles by a non-pathogenic and agriculturally important fungus *T. asperellum*. Nanotechnology 19:75103–75110

Nair B, Pradeep T (2002) Coalescence of nanoclusters and formation of sub-micron crystallites assisted by *Lactobacillus* strains. Crystal growth Design 4:295–298

Roy M, Mukherjee P, Mandal P, Sharma K, Tyagi K, Kale P (2011) Biomimetic synthesis of nano crystalline silver sol using cysteine: stability aspects and antibacterial activities. RSC Advances 2:6496–6503

Saifuddin N, Wang WC, Nur Yasumira AA (2009) Rapid biosynthesis of silver nanoparticles using culture supernatant of bacteria with microwave irradiation. E J Chem 6:61–70

Shaligram SN, Bule M, Bhambure R, Singhal SR, Singh K, Szakacs S, Pandey A (2009) Biosynthesis of silver nanoparticles using aqueous extract from the compacting producing fungi. Process Biochem 44:939–943

Shiying H, Zhirui G, Yu Z, Song Z, Jing W, Ning G (2007) Biosynthesis of gold nanoparticles using the bacteria *Rhodo pseudomonas* capsulate. Mater Lett 61:3984–3987

Srinivasulu B, Prakasham RS, Jetty A, Srinivas S, Ellaiah P, Ramakrishna SV (2002) Neomycin production with free and immobilized cells of *Streptomyces marinensis* in an airlift reactor. Process Biochem 38:593–598

18

Isolation and characterization of plant growth promoting endophytic bacteria from the rhizome of *Zingiber officinale*

B. Jasim · Aswathy Agnes Joseph · C. Jimtha John ·
Jyothis Mathew · E. K. Radhakrishnan

Abstract Endophytes, by residing within the specific chemical environment of host plants, form unique group of microorganisms. Microbially unexplored medicinal plants can have diverse and potential microbial association. The rhizome of ginger is very remarkable because of its metabolite richness, but the physiological processes in these tissues and the functional role of associated microorganisms remain totally unexplored. Through the current study, the presence of four different endophytic bacterial strains were identified from ginger rhizome. Among the various isolates, ZoB2 which is identified as *Pseudomonas* sp. was found to have the ability to produce IAA, ACC deaminase and siderophore. By considering these plant growth promoting properties, ZoB5 can expect to have considerable effect on the growth of ginger.

Keywords Endophytic bacteria · Indole 3 acetic acid · HPLC · *Pseudomonas* sp. · 16S rDNA sequencing

Introduction

Plants are associated with a diverse community of microorganisms. The microorganisms residing within the plants or endophytes are unique in their adaptations to specific chemical environment of host plant. Even some of these microorganisms are shared genetically with the molecular machinery for the synthesis of plant specific compounds. This makes endophytes to be considered as

untapped source of natural products (Rosenblueth and Martínez-Romero 2006; Strobel 2003). Endophytes also provide advantages to the host plant by producing plant growth regulators, by providing resistance to diseases and also by assisting in phytoremediation (Lodewyckx et al. 2002).

The mechanisms by which endophytes deal with ever-changing environmental conditions may provide better survival advantages to host plants. The evolution of endophytic biochemical pathways for the production of plant growth hormones is very interestingly present in plants (Strobel 2003). Endophytic microorganisms can vary based on the plant source, age, type of tissue, season of sampling, and environment. Generally, the concentration of the endophytic bacteria is more at the root than at shoot tissue (Zinniel et al. 2002). A large number of plant species are shown to be associated with bacteria like *Pseudomonas, Bacillus, Azospirillum* etc. (Chanway 1996). At the same time various species of bacteria can be associated with specific plants as in the case of rice where bacteria like *Pantoea, Azospirillum, Methylobacterium, Rhizhobium, Herbaspirillum, Burkholderia* etc. are found to be endophytically associated. These bacterial species have been shown to have added contribution to the yield and growth of the rice plants (Mano and Morisaki 2008). Hard wooded trees are also shown to be endophytically associated with bacteria such as *Serratia* sp., *Rahnella* sp., *Pseudomonas* sp., *Stenotrophomonas* sp. etc. (Taghavi et al. 2009). So by considering the remarkable features of ginger rhizome, much diverse and even specific bacterial association can be well expected.

Endophytic bacteria have been reported from wide variety of plants but the functional role is known only with limited number of isolates. One of the major contributions of these microorganisms towards plant growth is the

B. Jasim · A. A. Joseph · C. J. John · J. Mathew ·
E. K. Radhakrishnan (✉)
School of Biosciences, Mahatma Gandhi University,
Priyadharshini Hills PO, Kottayam Dist, Kerala 686560, India
e-mail: radhakrishnanek@mgu.ac.in

production of auxin-like molecules (Spaepen et al. 2007). Indole 3 acetic acid (IAA) being an auxin can stimulate both rapid responses like cell elongation and long term responses like cell division and differentiation in plants (Taghavi et al. 2009). Indole 3 acetic acid (IAA) is shown to be produced by many root associated bacteria including *Enterobacter* sp., *Pseudomonas* sp., and *Azospirillium* sp. (El-Khawas and Adachi 1999). Due to its important role in plants, the level as well as distribution of IAA in plant tissue and endophytic production of IAA has gained a great deal of attention (Matsuda et al. 2005).

In addition to the IAA production, plant growth promoting bacteria (PGPB) are also shown to exhibit other properties like ACC deaminase, phosphate solubilization, siderophore production, etc. The enzyme ACC deaminase catalyzes degradation of 1-aminocyclopropane-1-carboxylic acid (ACC), the immediate precursor of ethylene, into α-ketobutyrate and ammonia and this inturn reduce the inhibitory effects of elevated level of ethylene. Plant associated bacteria can also have the capability to solubilise non-available phosphate to available form and there by enhance plant growth and yield (de Freitas et al. 1997). Siderophores are iron-chelating agents secreted by some microorganisms under iron-limiting conditions. Siderophore productions by some microorganisms make them successful in surviving several adverse environments and also make the iron limiting to plant pathogens (Miethke and Marahiel 2007).

Thus isolation and characterization of endophytes with diverse properties from unexplored sources will have much applications to manipulate plant growth promotion (Patten and Glick 2002; Sergeeva et al. 2007). In order to explore the promising potential of endophytes, diverse communities of endophytes should be isolated from various tissues of taxonomically diverse and metabolically distinct plants. Plants of Zingiberaceae, especially ginger are well known for the presence of structurally diverse bioactive metabolites including those of the gingerol group (Ramirez-Ahumada Mdel et al. 2006). Also ginger forms a model plant of the family where much interesting and unexplored rhizome specific metabolism is present. In addition to the complex chemical constituents, rhizome is also well known for its ability to survive under adverse conditions (Ramirez-Ahumada Mdel et al. 2006). So many interesting groups of microorganisms with diverse roles in plant physiology can be expected from the rhizome. However the growth promoting properties of endophytic bacteria from ginger has not yet been well studied. So studies on isolation and characterization of endophytic bacteria from ginger is very significant. In the current study four endophytic bacteria were isolated from ginger rhizome and one among the isolates was found to have the ability to produce IAA, ACC deaminase and siderophore.

Materials and methods

Isolation and characterization of endophytic bacteria

Rhizome of ginger (*Zingiber officinale*) was collected from Navajyothisree Karunakara Guru Research Centre for Ayurveda and Siddha, Uzhavoor, Kottayam and was used as the source material for the isolation of endophytic bacteria. The rhizome pieces of *Z. officinale* were washed with tap water to remove soil and were made to 1–2 cm long pieces. This was further treated with Tween 80 for 10 min with vigorous shaking. This was followed by wash with distilled water for several times to remove Tween 80. After the treatment with Tween 80, the samples were dipped in 70 % ethanol for 1 min and then treated with 1 % sodium hypochlorite for 10 min. The samples were then washed several times with sterilized distilled water and the final wash was spread plated onto nutrient agar plate (g/L; peptone 5, beef extract 2, yeast extract 3, sodium chloride 5 and agar 18, pH 7.0) as control. For the isolation of endophytic bacteria, the outer surface of the sterilized plant material was trimmed, the pieces were further macerated in Phosphate buffer saline (PBS) (g/L—sodium chloride 8, potassium chloride 0.2, disodium hydrogen phosphate 1.44 and potassium dihydrogen phosphate 0.24, pH 7.4) and was serially diluted up to 10^{-3} dilution. From this, 0.1 mL was plated onto nutrient agar plates. All plates including the control were incubated at room temperature for 5 days and observed periodically for bacterial growth. Those batches of experiments where the bacterial growth, if any present, in the control plate were completely discarded. Morphologically distinct colonies as identified by colony characters were selected, purified and used for further studies.

Identification of the isolates by 16S rDNA sequencing

Genomic DNA was isolated from all the bacterial isolates and was used as template for PCR. Primers used for the amplification of part of 16S rDNA were 16SF (5′-AgA gTT TgA TCM Tgg CTC-3′) and 16SR (5′-AAg gAg gTg WTC CAR CC-3′) and were selected based on the previous reports of Chun and Goodfellow (1995). PCR was carried out in a 50 μL reaction volume containing 50 ng of genomic DNA, 20 pmol of each primer, 1.25 units of Taq DNA polymerase (Bangalore Genei), 200 μM of each dNTPs and 1X PCR buffer. PCR was carried out for 35 cycles in a Mycycler™ (Bio-Rad, USA) with the initial denaturation at 94 °C for 3 min, cyclic denaturation at 94 °C for 30 s, annealing at 58 °C for 30 s and extension at 72 °C for 2 min with a final extension of 7 min at 72 °C. The PCR product was checked by agarose gel electrophoresis, purified and was further subjected to sequencing. The sequence data was checked by BLAST analysis (Zhang et al. 2000).

The phylogenetic analysis of the 16SrDNA sequences of the isolates obtained in the study was conducted with MEGA 5 using neighbor-joining method with 1,000 bootstrap replicates (Tamura et al. 2011).

Screening of isolates for plant growth promoting properties (PGP)

IAA production

The bacterial isolates were inoculated into 20 mL of nutrient broth supplemented with 0.2 % (v/v) of L-tryptophan and incubated for 10 days at 28 °C. After incubation, the culture was centrifuged at 3,000 rpm for 20 min and the supernatant was used for analysing indole 3 acetic acid production (Rahman et al. 2010). Initially one mL supernatant was mixed with 2 mL of Salkowski reagent and tubes were incubated in dark for 30 min. The development of the red color was observed as the indication for positive result. Uninoculated growth medium was used as negative control. The IAA positive isolates were further inoculated into 200 mL of nutrient broth supplemented with 0.2 % (v/v) of L-tryptophan and incubated for 10 days at 28 °C. After incubation the cell free extract was collected by centrifugation at 3,000 rpm for 20 min. The supernatant was then acidified to pH 2.5–3.0 with 1 N HCl and was extracted twice with ethyl acetate. The extracted ethyl acetate fraction was vacuum dried in a rotary evaporator at 40 °C. The dried powder was dissolved in 1 mL of methanol (MeOH) and stored at -20 °C. For confirmation of presence of IAA, the methanol extract of culture supernatant was subjected to reverse-phase HPLC analysis on a Supelcosil LC-18 column with a flow rate of 1 mL min^{-1} as described by Jensen et al. (1995). Elution was performed with mixture of H_2O and MeOH (60:40), both containing 0.5 % acetic acid. Elution was monitored at 280 nm by shimadzu UV–Vis Detector model SPD 10A.

ACC deaminase production

The ACC deaminase production of the endophytic bacterial isolates from ginger were screened using the methods described by Jasim et al. (2013). For this, the isolates were inoculated on to DF salts minimal medium (potassium dihydrogen phosphate 4 g/L, disodium hydrogen phosphate 6 g/L, magnesium sulfate heptahydrate 0.2 g/L, ferrous sulfate heptahydrate 0.1 g/L, boric acid 10 µg/L, manganese(II) sulfate 10 µg/L, zinc sulphate 70 µg/L, copper(II) sulfate 50 µg/L, molybdenum (VI) oxide 10 µg/L, glucose 2 g/L, gluconic acid 2 g/L, citric acid 2 g/L, agar 12 g/L) amended with 0.2 % ammonium sulphate (w/v). The bacterial growth in this media after 2 days of incubation was considered as positive result.

Phosphate solubilization

The endophytic bacterial isolates were screened for phosphate solubilization using the procedure described by Jasim et al. (2013). For this, Pikovskaya medium (g/L—glucose 10, tri-calcium phosphate 5, ammonium sulphate 0.5, sodium chloride 0.2, magnesium sulphate heptahydrate 0.1, potassium chloride 0.2, ferrous sulfate heptahydrate 0.002, yeast extract 0.5, manganese (II) sulfate dehydrate 0.002, agar 20, pH 7.0) containing 2.4 mg/mL bromophenol blue was used. The media inoculated with the isolates were incubated for 48 h and was observed for the formation yellow zone around the colony due to the utilization of tricalcium phosphate present in the medium.

Siderophore production

The isolates were checked for the production of siderophores on blue agar CAS medium containing chrome azurol S (CAS) and hexadecyltrimethylammonium bromide (HDTMA) as indicators (Schwyn and Neilands 1987). The blue agar CAS medium was prepared by adding 850 mL of autoclaved MM9 salt medium [added with 32.24 g piperazine-N, N′-bis 2- ethanesulfonic acid (PIPES) at pH 6], 100 mL of blue dye, 30 mL of filter sterilized 10 % Casaminoacid solution and 10 mL of 20 % glucose solution. The blue agar medium was aseptically poured on to sterile plates and allowed to solidify. All the bacterial isolates obtained were inoculated into the CAS medium and incubated at 28 °C for 24 h. Development of yellowish orange halo around the colonies was taken as the indication for the production of siderophore.

Results and Discussion

Fresh and cleaned ginger rhizomes were used for the isolation of endophytic bacteria. The rhizomes were surface sterilized to remove the epiphytic microorganisms. The surface sterilization procedure for the isolation of endophytic bacteria as standardized in the experiment was quite satisfactory as no growth appeared on the control plate. Also, adequate number of colonies obtained in the nutrient agar plates which were inoculated with plant samples macerated in PBS. Based on the distinct colony characteristics, the bacterial isolates obtained were grouped into four and were named as ZoB1–ZoB4. As no microbial growth was observed in control plate, the isolates ZoB1–ZoB4 obtained in the study can be considered as endophytic bacteria of ginger.

Molecular identification of the isolates was done by sequencing part of the 16S rDNA. The amplification of the 16S rDNA was confirmed by agarose gel electrophoresis.

The PCR product was gel eluted and sequenced. The 16S rDNA sequences of the bacterial isolates were submitted to NCBI under the accession numbers as explained in Table 1.

The sequence data of the 16S rDNA was subjected to BLAST analysis. As 16S rDNA gene sequence provide accurate grouping of organism even at subspecies level it is considered as a powerful tool for the rapid identification of bacterial species (Jill and Clarridge 2004). The sequence analysis of 16S rDNA sequences of ZoB1, ZoB2, ZoB3 and ZoB4 showed its maximum identity of 99 % to *Bacillus barbaricus* (JF727665), 99 % to *Pseudomonas putida* (JN596120), 99 % to *Stenotrophomonas maltophilia* (FN645734) and 98 % to *Staphylococcus pasteuri* (JF510533), respectively. Therefore, the isolates ZoB1–ZoB4 can be considered as strains of *Bacillus* sp., *Pseudomonas* sp., *Stenotrophomonas* sp, and *Staphylococcus* sp, respectively (Table 1). The presence of these organisms as endophytes has not been reported from ginger rhizome.

Strains of *P. putida* and *S. maltophilia* were previously reported as endophytes from other plants like *Populus* sp. (Taghavi et al. 2009). *S. pasteuri* strains were previously identified as endophyte from plants like Arabidopsis and Soyabean (Panchal and Ingle 2011). From the comparative analysis, it is very clear that bacteria like *P. putida*, *S. maltophilia*, and *S. pasteuri* are present as endophytes in wide variety of plants. However, *Bacillus barbaricus* was reported as an endo-lithosphere associated bacteria of *Musa* sp. (Thomas and Soly 2009). As information on the presence of *Bacillus barbaricus* as an endophyte is very limited, the presence of related strain as endophyte in ginger rhizome may be taken as an indication of its specific or limited association as an endophyte. Thus, it can be considered that endophytic bacteria present in ginger rhizome, include those species with wide range of host specificity and those with limited distribution. However, the presence these organisms in specific habitat of ginger make them much more interesting. This is because these stains can expect to have strain specific plant growth promoting potential. The phylogenetic analysis of 16S rDNA sequence of the isolates along with the sequences retrieved from the NCBI was carried out with MEGA 5 using the neighbor-joining method with 1,000 bootstrap replicates.

The result of phylogenetic analysis showed distinct clustering of the isolates (Fig. 1).

The culture supernatant of the bacterial isolates ZoB1–ZoB4 were checked for indole 3 acetic acid production calorimetrically. For this, the supernatant of the culture was treated with Salkowski reagent as explained by Rahman et al. (2010). Among the four endophytic isolates, ZoB2 gave positive result for the production of indole 3 acetic acid. The positive reaction was confirmed by comparing this with positive control, which had pure indole 3 acetic acid and a negative control which had uninoculated culture medium. The positive result appeared had a color similar to that of the positive control. Thus, ZoB2 (*Pseudomonas* sp.) was found to have the ability to produce IAA (Table 2). The production of the indole 3 acetic acid was also confirmed by HPLC. For this, the extracts were run on a C18-reversed phase column and absorbance was measured at 280 nm. The pure indole 3 acetic acid produced a peak at 8 min retention time and the crude extract had a predominant peak at the same retention time (Fig. 2). This confirmed the production of indole 3 acetic acid by the bacterial isolate ZoB2 (*Pseudomonas* sp.) from ginger. Even though ZoB1 also showed positive result when treated with Salkowski reagent, the results could not be confirmed by HPLC.

Some endophytic microorganisms have the potential to synthesize IAA. This may be a reason for the increased growth promotion of some plants when the plant is colonized with endophytes (Shi et al. 2009). For the microbial synthesis of IAA in tryptophan-dependent route, tryptophan is used as the precursor. There are different pathways that can lead to tryptophan-dependent microbial production of IAA. The various pathways for IAA biosynthesis include tryptophol, tryptamine, indole-3-pyruvic acid and indole-3-acetamide pathways (Gravel et al. 2007). The ability of the production of IAA different species of *Pseudomonas* was reported by many authors (Karnwal 2009). Even there are reports that suggest the ability of *P. putida* GR12-2, when inoculated on seeds can result in 2–3 fold increase in the length of seedling roots (Glick et al. 1986; Caron et al. 1995).

Even though the presence of *Pseudomonas* sp. as endophyte and its ability to produce IAA has already been

Table 1 16S rDNA sequence analysis of endophytic bacterial isolates from *Zingiber officinale*

Plant material	Endophytic isolate	NCBI accession number	Closest NCBI database match with accession number	Percentage of identity
Ginger rhizome	ZoB1	JN835212	*Bacillus barbaricus* JF727665	99
Ginger rhizome	ZoB2	JN835214	*Pseudomonas putida* JN596120	99
Ginger rhizome	ZoB3	JN835215	*Stenotrophomonas maltophilia* FN645734	99
Ginger rhizome	ZoB4	JN835216	*Staphylococcus pasteuri* JF510533	98

Fig. 1 Phylogenetic analysis of
16S rDNA sequences of the
bacterial isolates (ZoB1–ZoB4)
from ginger along with the
sequences from NCBI. The
analysis was conducted with
MEGA5 using neighbor-joining
method

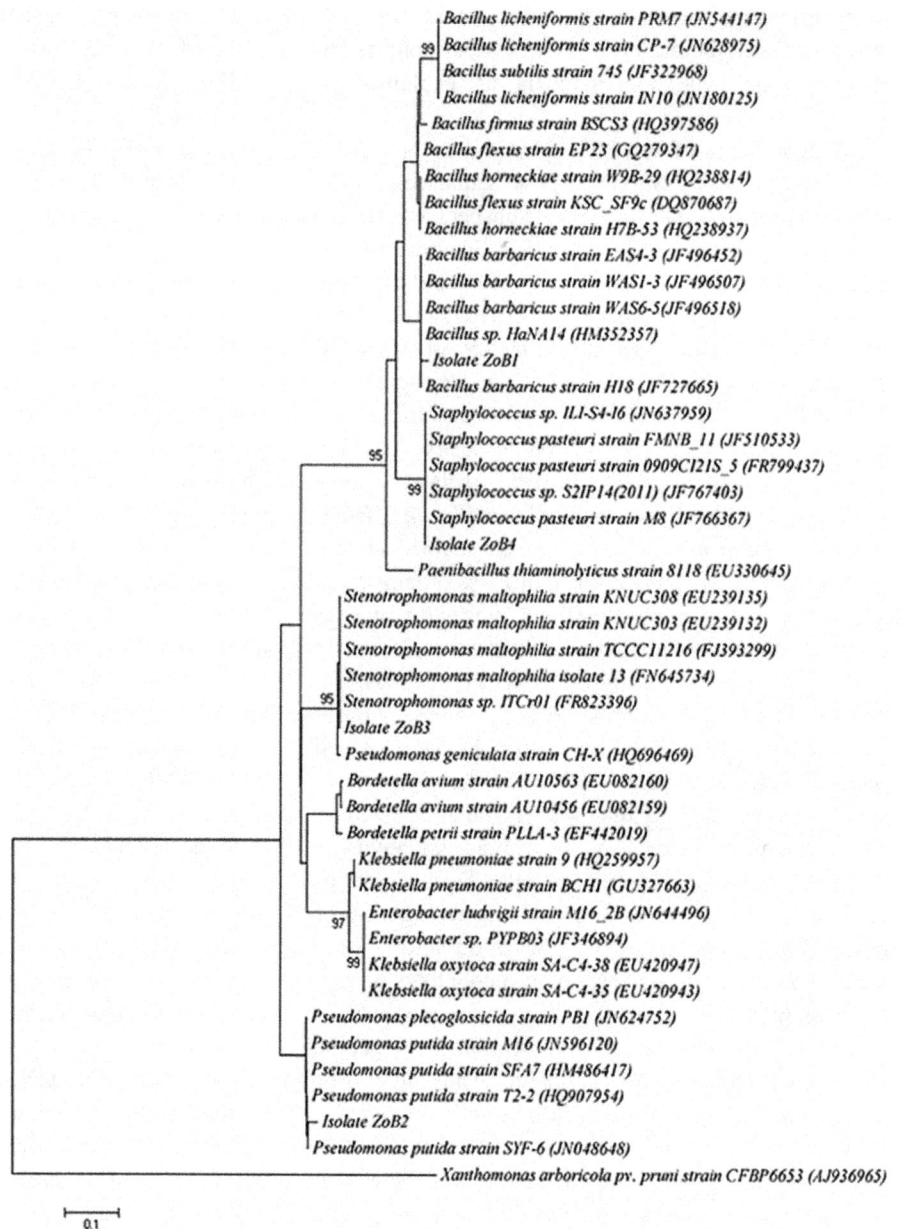

Bacillus licheniformis strain PRM7 (JN544147)
Bacillus licheniformis strain CP-7 (JN628975)
Bacillus subtilis strain 745 (JF322968)
Bacillus licheniformis strain IN10 (JN180125)
Bacillus firmus strain BSCS3 (HQ397586)
Bacillus flexus strain EP23 (GQ279347)
Bacillus horneckiae strain W9B-29 (HQ238814)
Bacillus flexus strain KSC_SF9c (DQ870687)
Bacillus horneckiae strain H7B-53 (HQ238937)
Bacillus barbaricus strain EAS4-3 (JF496452)
Bacillus barbaricus strain WAS1-3 (JF496507)
Bacillus barbaricus strain WAS6-5(JF496518)
Bacillus sp. HaNA14 (HM352357)
Isolate ZoB1
Bacillus barbaricus strain H18 (JF727665)
Staphylococcus sp. ILI-S4-16 (JN637959)
Staphylococcus pasteuri strain FMNB_11 (JF510533)
Staphylococcus pasteuri strain 0909CI21S_5 (FR799437)
Staphylococcus sp. S2IP14(2011) (JF767403)
Staphylococcus pasteuri strain M8 (JF766367)
Isolate ZoB4
Paenibacillus thiaminolyticus strain 8118 (EU330645)
Stenotrophomonas maltophilia strain KNUC308 (EU239135)
Stenotrophomonas maltophilia strain KNUC303 (EU239132)
Stenotrophomonas maltophilia strain TCCC11216 (FJ393299)
Stenotrophomonas maltophilia isolate 13 (FN645734)
Stenotrophomonas sp. ITCr01 (FR823396)
Isolate ZoB3
Pseudomonas geniculata strain CH-X (HQ696469)
Bordetella avium strain AU10563 (EU082160)
Bordetella avium strain AU10456 (EU082159)
Bordetella petrii strain PLLA-3 (EF442019)
Klebsiella pneumoniae strain 9 (HQ259957)
Klebsiella pneumoniae strain BCH1 (GU327663)
Enterobacter ludwigii strain M16_2B (JN644496)
Enterobacter sp. PYPB03 (JF346894)
Klebsiella oxytoca strain SA-C4-38 (EU420947)
Klebsiella oxytoca strain SA-C4-35 (EU420943)
Pseudomonas plecoglossicida strain PB1 (JN624752)
Pseudomonas putida strain M16 (JN596120)
Pseudomonas putida strain SFA7 (HM486417)
Pseudomonas putida strain T2-2 (HQ907954)
Isolate ZoB2
Pseudomonas putida strain SYF-6 (JN048648)
Xanthomonas arboricola pv. pruni strain CFBP6653 (AJ936965)

Table 2 Growth promotion
capability of the isolated
endophytic bacterial isolates

Name of the Isolates	Isolate Identified as	Plant Growth promoting properties			
		Phosphate solubilization	ACC Deaminase	Siderophore production	IAA production
ZoB1	*Bacillus sp.*	−	−	+	−
ZoB2	*Pseudomonas sp.*	−	+	+	+
ZoB3	*Stenotrophomonas sp.*	−	−	+	−
ZoB4	*Staphylococcus sp.*	−	−	−	−

reported from other plants including both monocot and dicots, its reports from ginger is limited. This makes the present finding much more interesting. The strain of *Pseudomonas* sp. identified in this study can have important growth regulating role in ginger rhizome. The strain may also have unique features to survive in the unique chemical environment of rhizome under various conditions. Confirmation of this by further experiments may pave the way for exploring the potential application of the isolate in ginger yield enhancement.

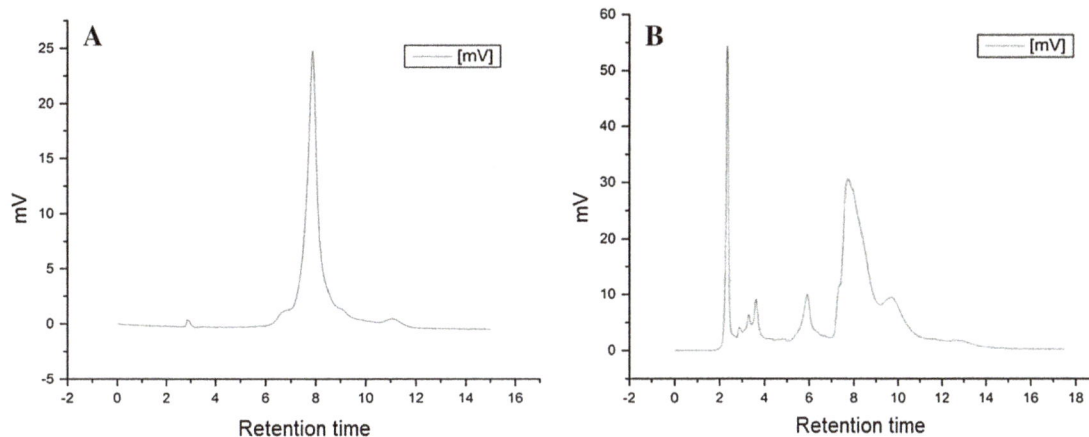

Fig. 2 HPLC analysis carried out using water and methanol, both containing 0.5 % acetic acid in the ratio 60:40 on reversed phase C18 column with a flow rate 1 mL min^{-1} and detected under UV at 280 nm. **a** Positive control (indole 3 acetic acid only), **b** methanolic extract from ZoB2 (Pseudomonas sp.)

All the four isolates (ZoB1-ZoB4) were screened for the production of ACC deaminase on DF salts minimal medium amended with 0.2 % ammonium sulphate. ZoB2 (*Pseudomonas sp.*) was found to be positive for ACC deaminase production as indicated by its growth in the media (Table 2). Glick et al. (2007) suggests that some microbes can utilize the ACC as nitrogen source from the exudates of roots or seeds. This decrease in the levels of ACC and ethylene may prevent the ethylene-mediated plant growth inhibition. Endophytic microbes with these capabilities residing inside the host plants can benefit the host by reducing the stress and increasing the plant growth (Hardoim et al. 2008). Alizadeh et al. (2012) has explained the application of the ACC deaminase which has been synthesised by different genera of *Pseudomonas* in increasing the senescence of the plants. The endophytic bacterial isolates were also screened for phosphate solubilization, but the results were negative for all of the isolates.

The endophytic bacterial isolates were screened for siderophore production using the chrome azurol S (CAS) agar. Among the four endophytic isolates, ZoB1 (*Bacillus sp.*), ZoB2 (*Pseudomonas sp.*) and ZoB3 (*Stenotrophomonas sp.*) were found to have the ability to produce siderophore (Table 2). The formation of orange halo around the colonies due to the chelation of iron was the indication for production of siderophore. The formation of orange halo is as a result of the production of siderophore, which removes the iron from the dye complex that changes the color of the medium from blue to orange (Schwyn and Neilands 1987). Siderophores producing bacteria can sequestrate the limited iron and thereby reduce its availability for growth of phytopathogens. Thus, they enable the plant growth promotion indirectly (Alexander and Zeeberi 1991). Different species of *Bacillus* have been reported to have the ability to produce of siderophores even that of petrobactin type (Gardner et al. 2004; Wilson et al. 2010).

Many reports reveals the ability of both gram negative bacterial isolates (*Pseudomonas* sp.) and bacterial genera of *Bacillus* and *Rhodococcus* that belongs to the gram-positive group with the capability to produce siderophores (Tian et al. 2009). Structural studies suggest that there are more than 50 structurally related siderophores like pyoverdins, that are produced by different species of *Pseudomonas* (Abdallah 1991; Budzikiewicz 1993).There are various reports that suggest the ability of *Stenotrophomonas maltophilia* to synthesize different types of siderophores using the universal CAS assay method. Chhibber et al. (2008) reported the production of ornibactin type siderophore by *Stenotrophomonas maltophilia* and Ryan et al. (2009) mentioned its ability to produce the catechol type siderophore compound enterobactin based on their recently sequenced genomes.

Conclusion

The results from the study demonstrated the diverse community of endophytic bacteria associated with ginger rhizome. Among these endophytic bacterial isolates obtained, *Pseudomonas* sp. (ZoB2) isolated from the rhizome was found to have the ability to form Indole 3 acetic acid as confirmed by HPLC analysis. The isolate was also found to have the capability to produce ACC deaminase and siderophore which have high impact on growth of the plant. Hence, this isolate can be considered to have growth promoting effect in ginger.

Acknowledgments This study was supported by Department of Biotechnology (DBT), Government of India under DBT-RGYI support scheme.

Conflict of interest The authors declare that they have no conflict of interest in the publication.

References

Abdallah MA (1991) Pyoverdines and pseudobactins. In: Handbook of Microbial Iron Chelates. Boca Raton, FL, CRC Press, pp 139–153

Alexander BD, Zeeberi DA (1991) Use of chromazurol S to evaluate siderophore production by rhizosphere bacteria. Biol Fertil Soils 2:39–54

Alizadeh O, Sharafzadeh S, Firoozabadi AH (2012) The effect of plant growth promoting rhizobacteria in saline condition. Asian J Plant Sci 11:1–8

Budzikiewicz H (1993) Secondary metabolites from fluorescent pseudomonads. FEMS Microbiol Rev 104:209–228

Caron M, Patten CL, Ghosh S (1995) Effects of plant growth promoting rhizobacteria Pseudomonas putida GR- 122 on the physiology of canola roots. Proceedings of the Plant Growth Regulation Society of America 7:18–20

Chanway CP (1996) Endophytes: they're not just fungi! Can J Bot 74:321–322

Chhibber S, Gupta A, Sharan R, Gautam V, Ray P (2008) Putative virulence characteristics of Stenotrophomonas maltophilia: a study on clinical isolates. World J Microbiol Biotechnol 24:2819–2825

Chun J, Goodfellow M (1995) A phylogenetic analysis of the genus Nocardia with 16 s rRNA gene sequences. Int J Syst Bacteriol 45:240–245

de Freitas JR, Banerjee MR, Germida JJ (1997) Phosphate solubilizing rhizobacteria enhance the growth and yield but not phosphorus uptake of canola (Brassica napus). Biol Fertil Soils 24:358–364

El-Khawas H, Adachi K (1999) Identification and quantification of auxins in culture media of Azospirillum and Klebsiella and their effect on rice roots. Biol Fertil Soils 28:377–381

Gardner RA, Kinkade R, Wang C, Phanstiel O (2004) Total synthesis of petrobactin and its homologues as potential growth stimuli for Marinobacter hydrocarbonoclasticus, an oil-degrading bacteria. J Org Chem 69:3530–3537

Glick BR, Brooks HE, Pasternak JJ (1986) Physiological effects of plasmid DNA transformation of Azotobacter vinelendi. Can J Microbiol 32:145–148

Glick BR, Cheng Z, Czarny J, Duan J (2007) Promotion of plant growth by ACC deaminase-producing soil bacteria. Eur J Plant Pathol 119:329–339

Gravel V, Antoun H, Tweddell RJ (2007) Growth stimulation and fruit yield improvement of greenhouse tomato plants by inoculation with Pseudomonas putida or Trichoderma atroviride: possible role of indole 3 acetic acid (IAA). Soil Biol Biochem 39:1968–1977

Hardoim PR, van Overbeek LS, van Elsas JD (2008) Properties of bacterial endophytes and their proposed role in plant growth. Trends Microbiol 16:463–471

Jasim B, Jimtha John C, Mathew J, Radhakrishnan EK (2013) Plant growth promoting potential of endophytic bacteria isolated from Piper nigrum. Plant Growth Regul.

Jensen JB, Egsgaard H, Van Onckelen H, Jochimsen BU (1995) Catabolism of Indole-3-acetic acid and 4- and 5-Chloroindole-3-acetic acid in Bradyrhizobium japonicum. J Bacteriol 177: 5762–5766

Jill E, Clarridge III (2004) Impact of 16S rRNA gene sequence analysis for identification of bacteria on clinical microbiology and infectious diseases. Clin Microbiol Rev 17:840–862

Karnwal A (2009) Production of indole acetic acid by fluorescent Pseudomonas In the presence of L-tryptophan and rice root exudates. J Plant Pathol 91(1):61–63

Lodewyckx C, Vangronsveld J, Porteous F, Moore ERB, Taghavi S (2002) Endophytic bacteria and their potential applications. CRC Crit Rev Plant Sci 21:583–606

Mano H, Morisaki H (2008) Endophytic bacteria in the rice plant. Microbes Environ 23:109–117

Matsuda F, Miyazawa H, Wakasa K, Miyagawa H (2005) Quantification of indole-3-acetic acid and amino acid conjugates in rice by liquid chromatography-electrospray ionization-tandem mass spectrometry. Biosci Biotechnol Biochem 69:778–783

Miethke M, Marahiel MA (2007) Siderophore-based iron acquisition and pathogen control. Microbiol Mol Biol Rev 71:413–451

Panchal H, Ingle S (2011) Isolation and characterization of endophytes from the root of medicinal plant Chlorophytum borivilianum (Safed musli). J Adv Dev Res 2:205–209

Patten CL, Glick BR (2002) Role of Pseudomonas putida indoleacetic acid in development of the host plant root system. Appl Environ Microbiol 68:3795–3801

Rahman A, Sitepu IR, Tang S-Y, Hashidoko Y (2010) Salkowski's reagent test as a primary screening index for functionalities of Rhizobacteria isolated from wild dipterocarp saplings growing naturally on medium-strongly acidic tropical peat soil. Biosci Biotechnol Biochem 74:2202–2208

Ramirez-Ahumada Mdel C, Timmermann BN, Gang DR (2006) Biosynthesis of curcuminoids and gingerols in turmeric (Curcuma longa) and ginger (Zingiber officinale): identification of curcuminoid synthase and hydroxycinnamoyl-CoA thioesterases. Phytochemistry 67:2017–2029

Rosenblueth M, Martínez-Romero E (2006) Bacterial endophytes and their interactions with hosts. Mol Plant Microbe Interact 19:827–837

Ryan RP, Monchy S, Cardinale M, Taghavi S, Crossman L, Avison MB, Berg G, van der Lelie D, Dow JM (2009) The versatility and adaptation of bacteria from the genus Stenotrophomonas. Nat Rev Microbiol 7:514–525

Schwyn B, Neilands JB (1987) Universal chemical assay for the detection and determination of siderophores. Anal Biochem 160(1):47–56

Sergeeva E, Danielle Hirkala LM, Louise NM (2007) Production of indole-3-acetic acid, aromatic amino acid aminotransferase activities and plant growth promotion by Pantoea agglomerans rhizosphere isolates. Plant Soil 297:1–13

Shi Y, Lou K, Li C (2009) Isolation, quantity distribution and characterization of endophytic microorganisms within sugar beet. Afr J Biotechnol 8:835–840

Spaepen S, Vanderleyden J, Remans R (2007) Indole-3-acetic acid in microbial and microorganism-plant Signaling. FEMS Microbiol Rev 31:1–24

Strobel GA (2003) Endophytes as sources of bioactive products. Microbes Infect 5:535–544

Taghavi S, Garafola C, Monchy S, Newman L, Hoffman A, Weyens N, Barac T, Vangronsveld J, van der Lelie D (2009) Genome survey and characterization of endophytic bacteria exhibiting a beneficial effect on growth and development of poplar trees. Appl Environ Microbiol 75:748–757

Tamura K, Peterson D, Peterson N, Stecher G, Nei M, Kumar S (2011) MEGA5: molecular evolutionary genetics analysis using maximum likelihood, evolutionary distance, and maximum parsimony methods. Mol Biol Evol 28:2731–2739

Thomas P, Soly TA (2009) Endophytic bacteria associated with growing shoot tips of banana (Musa sp.) cv. Grand Naine and the affinity of endophytes to the host. Microb Ecol 58:952–964

Tian F, Ding Y, Zhu H, Yao L, Du B (2009) Genetic diversity of siderophore-producing bacteria of tobacco rhizosphere. Brazilian J Microbiol 40:276–284

Wilson MK, Abergel RJ, Arceneaux JE, Raymond KN, Byers BR (2010) Temporal production of the two *Bacillus anthracis* siderophores, petrobactin and bacillibactin. Biometals 23:129–134

Zhang Z, Schwartz S, Wagner L, Miller W (2000) A greedy algorithm for aligning DNA sequences. J Comput Biol 7:203–214

Zinniel DK, Lambrecht P, Harris NB, Feng Z, Kuczmarski D, Higley P, Ishimaru CA, Arunakumari A, Barletta RG, Vidaver AK (2002) Isolation and characterization of endophytic colonizing bacteria from agronomic crops and prairie plants. Appl Environ Microbiol 68:2198–2208

19

Molecular and in situ characterization of cadmium-resistant diversified extremophilic strains of *Pseudomonas* for their bioremediation potential

Sourabh Jain · Arun Bhatt

Abstract Cadmium-resistant strains psychrotolerant *Pseudomonas putida* SB32 and alkalophilic *Pseudomonas monteilli* SB35 were originally isolated from the soil of Semera mines, Palamau, Jharkhand, India. Further, to unravel the mechanism involved in cadmium resistance, plasmid DNA was isolated from the strains and subjected to amplification of the *czc* gene, which is responsible for the efflux of three metal cations, viz. Co, Zn and Cd, from the cell. Furthermore, the amplicon was cloned into pDrive cloning vector and sequenced. When compared with the available database, the sequence homology of the cloned gene showed the presence of a partial *czcA* gene sequence, thereby indicating the presence of a plasmid-mediated efflux mechanism for resistance in both strains. These results were further confirmed by atomic absorption spectroscopy and transmission electron microscopy. Moreover, the strains were characterized functionally for their bioremediation potential in cadmium-contaminated soil by performing an in situ experiment using soybean plant. A marked increase in agronomical parameters was observed in presence of both strains. Further, the concentration of metal ions decreased in both plants and soil in the presence of these bioinoculants.

S. Jain
Uttarakhand Technical University, Dehradun, India

S. Jain
National Research Centre on Plant Biotechnology,
IARI Campus, Pusa, New Delhi, India

A. Bhatt (✉)
Department of Crop Improvement, Uttarakhand University
of Horticulture and Forestry, Uttarakhand, India
e-mail: bhatt.gbpec@gmail.com

Keywords AAS · Psychrotolerant · Alkaliphile · Bioremediation · Cadmium

Introduction

Naturally occurring bacteria that are capable of metal accumulation have been extensively studied since it is difficult to imagine that a single bacterium would remove all heavy metals from its polluted site (Clausen 2000). Therefore, diversified microorganisms have to be classified morphologically, physiologically and biochemically to understand their taxonomical variation and evolutionary distance from the parental linage.

Cadmium is a non-redox metal unable to produce active oxygen species (AOS) via Fenton and IQ Haber–Weiss reaction (Sanita Di Toppi and Gabbrielli 1999). However, several reports demonstrate that Cd can indirectly promote the generation of AOS (Sandalio et al. 2001). Cadmium-increased lipid peroxidation has been demonstrated in *Phaseolus vulgaris* roots and leaves (Chaoui et al. 1997). When accumulated in the plant tissue, it causes alteration in catalytic efficacy of enzymes (Somashekaraiah et al. 1992; Romero-Puertas et al. 1999; Piqueras et al. 1999), damage to the cellular membranes (Tu and Brovillett 1987) and inhibits the root growth. This metal enters the environment mainly from fertilizers and is transferred to animals and humans through industrial processes (Wagner 1993) leading to serious damage to human health (Alloway 1990). It affects cell proliferation, differentiation, apoptosis and increases oncogene activation to carcinogenesis (Naidu et al. 1997).

Microbial survival in polluted soil depends on intrinsic biochemical and structural properties, physiological and/or genetic adaptations including morphological changes in cells and environmental modifications of metal speciation (Wuertz and Mergeay 1997). Microbes apply various types of resistance mechanisms in response to heavy metals (Nies 2003). Microbial methods of environment purification and cleanup are known to be the most promising because of their safety, efficiency, and cost-effectiveness (He et al. 2011). Many microorganisms can absorb and concentrate heavy metals thus, providing resistance (Burke and Pfister 1986). On the other hand, studying metal ion resistances give us important insights into environmental processes and provide an understanding of basic living processes. Furthermore, genes for resistance to inorganic salts of soft metals are found both on plasmids and in chromosomes. The physiological role of plasmid-encoded determinants is generally to confer resistance; however, mostly chromosomally encoded systems may include metal ion homeostasis (Wuertz and Mergeay 1997).

Two well-studied genetic mechanisms of metal resistance in bacteria include heavy metal efflux systems (Nies and Silver 1995) and the presence of metal binding proteins (Robinson et al. 1990). Many operons of efflux system are known. In the Gram-positive bacteria, the plasmid-encoded Cd efflux system, called the CadA resistance system, utilizes the CadA protein, which is a P-type ATPase (Tsai and Linet 1993). However, Cd resistance in Gram-negative organisms is due to a multi-protein chemiosmotic antiport system (Silver 1996). Primarily, the *czc* system detoxifies the cell by cation efflux, the three metal cations, viz. cobalt, zinc and cadmium, which are taken up into the cell by fast and unspecific transport system for Mg^{2+} are actively extruded from cell by products of *czc* resistance determinants (Nies et al. 1989b). The protein complex is composed of three subunits, *czc*C, *czc*B and *czc*A (Nies et al. 1989a). *Czc*A is a RND protein (Tseng et al. 1999) and contains two hydrophilic domains, which are located in the periplasm (Goldberg et al. 1999), as the other two subunits of *czc* complex, *czc*B and *czc*C. There are at least six regulatory proteins (*Czc*D, *Czc*R, *Czc*S, *Czc*N, *Czc*I and an unknown sigma factor, "RpoX") involved in regulation of the three structural genes *czcCBA*. The proteins *czc*A and *czc*B alone are capable of pumping zinc ions out of the cell, even if *czc*C is not present (Nies et al. 1989b). Therefore, *czc*A is proposed to function as the actual efflux-transportation protein; however, *czc*A's specificity for exportation of these ions is apparently regulated by the presence of two other structural proteins (Nies and Silver 1995). The cell surfaces of all microorganisms are negatively charged due to the presence of various anionic structures which gives bacteria the ability to bind metal cations. So, isolation of heavy metal-resistant bacteria may be useful to improve the application of microorganism in environment protection.

In this study, an effort has been made to unravel the mechanism involved in Cd resistance in psychrotolerant *P.*

putida SB32 and alkalophilic *P. monteilli* SB35. This includes molecular characterization of the strains to identify the location of *czc* gene and to find out whether the resistance to Cd was due to efflux mechanism or due to accumulation of metal in the bacterial cell. Furthermore, in situ trials on soybean plant in presence of the two isolates were performed to analyze the bioremediation potential of strains.

Materials and methods

Isolation of Cd-resistant extremophilic isolates

Psychrotolerants *P. putida* SB32 and alkalophilic *P. monteilli* SB35 strains were isolated from the soil sample of Semera Mines, Palamau, Jharkhand, India. The strains were grown in nutrient broth (Himedia Laboratories Private Limited, Mumbai, India) containing 1 mM cadmium chloride ($CdCl_2$) and was kept overnight at 30 °C. Furthermore, the grown culture was also streaked on nutrient agar plates (Himedia Laboratories) and the plates were kept at 4 °C until further use. They were optimized for their respective parameters, i.e., temperature for psychrotolerant *P. putida* SB32 (optimum growth at 20 °C) and pH for alkalophilic *P. monteilli* SB35 (pH optima 9.0) and are found to be possessing high resistance to Cd.

Plasmid and genomic DNA isolation

The plasmid DNA was isolated using previously described protocol (Anderson and McKay 1983) and genomic DNA by Qiagen Bacterial genomic DNA isolation kit, Hilden, Germany. The isolated DNA was made to run on 0.8 % agarose gel and visualized subsequently.

Amplification of czc gene

The *czc* gene was amplified from both plasmid and genomic DNA of strains SB32 and SB35 using primers *czc*F (AAC CAG ATC TCG CGC GAG AAC) and *czc*R (CGG CAACAC CAG TAG GGT CAG). Polymerase chain reaction (50 μl) mixture contained 0.5 μM of each primer, 200 μM dNTPS, 1.0 U Taq DNA Polmerase, PCR buffer supplied with the enzyme and 1 μl (75 ng) of template DNA. The total volume of the reaction mixture was maintained with sterilized triple distilled water. PCR was performed in 'BioRad iQ^{TM}5 multicolor real-time PCR detection system' and was carried out as follows: a single denaturation step at 95 °C for 5 min followed by a 36-cycle program which included denaturation at 94 °C for 1 min, annealing at 61.6 °C for 30 s and extension 72 °C for 1 min and a final extension at 72 °C for 10 min.

Cloning and sequencing of PCR products

The purified products of *czc* gene were ligated with pDrive vector overnight at 4 °C using Qiagen cloning kit, Hilden, Germany and were transformed into *E. coli* DH5α. DNA sequencing of single pass of *czc* gene was done using primer T7. The sequence thus obtained was analyzed using BLASTn search (http://www.ncbi.nlm.nih.gov/Blast).

Atomic absorption spectroscopy (AAS)
and transmission electron microscopy (TEM)

To confirm efflux or intracellular accumulation of metal, AAS studies were done by combining the protocols given by Roane et al. 2001 and Vasudevan et al. 2001. The strains (SB32 and SB35) were grown both in the presence and absence (control) of 0.1 mM $CdCl_2$ and 25 ml of sample was drawn at different intervals of time, i.e., during the lag, log, stationary and death phase. The samples were centrifuged at 7,000 rpm for 10 min at 4 °C. The supernatant was collected and passed through 0.22 μm pore size filters with the help of 10 ml disposables syringe and stored at 4 °C for further metal analysis by flame atomic absorption spectrophotometer. The pellets were dried overnight at 90 °C in the oven and weight was noted. The dried pellets (4–200 mg) were then acid lysed overnight by 5 ml concentrated HNO_3 and incinerated on sand bath for 4–6 h at slow heating (45–50 °C). Further, 1 ml of concentrated HNO_3 and per-chloric acid (60 %) in the ratio 6:1 was added and again kept for incineration on sand bath for 3 h till white residue is formed. The white residue was dissolved in 50 ml of deionized water. The concentration of Cd was determined by flame atomic absorption spectrophotometer at 228 nm with lamp current 3 mA.

Consequently, samples of both control and treated cells were drawn during the late log phase for TEM analysis. The strains were grown in 25 ml nutrient broth in the presence and absence of 0.1 mM $CdCl_2$ and 5.0 ml sample from each experimental flask was drawn after 24 h. The samples were centrifuged at 10,000 rpm for 10 min at 4 °C. The supernatant was discarded and the pellet was washed with 5 ml PBS (pH 7.4, 0.1 M) for four times. Centrifugation was done at 8,000 rpm for 10 min at 4 °C. The pelleted cells were fixed overnight in 1 ml fixative [2.5 % glutaraldehyde and 2 % paraformaldehyde in (pH 7.4) sodium phosphate buffer] and rinsed twice with phosphate buffer for 1 h. Further, the samples were washed three times with 0.1 M phosphate buffer saline at pH 7.4 for 15 min and then fixed for 1 h at room temperature in 1 % osmium tetraoxide. The samples were dehydrated in three changes of 50 % alcohol for 15 min, four changes of absolute alcohol for 15 min and two changes of 100 % toluene for 30 min each, before being transferred to a mixture of equal parts of araldite and toluene overnight at room temperature. Impregnation was carried out in the fresh change of araldite and continued for 2 days. The samples were finally embedded in fresh araldite and polymerized for 3 days. The ultra thin sections (70–80 nm) were cut on a microtome and mounted on uncoated copper grids. The sections were stained in a saturated solution of uranyl acetate in 50 % alcohol for 15 min followed by lead citrate for 15 min and examined in 100 kV transmission electron microscope (JEOL, JEM 1011).

In situ characterization

Seeds were sown in the pot filled with alkaline soil (pH was approximately 8.6 ± 0.2). To detect the remediation ability of strains and subsequent effect to soybean, 124 μM $CdCl_2$ was added to the soil (whereas, the bacterial strains used in this study can tolerate much higher concentration, but at the concentration >124 μM, plants were unable to grow even in presence of bioinoculants). Each treatment was taken individually as indicated below:

1. Uninoculated soil (control)
2. Uninoculated soil with Cd
3. Inoculated soil with SB35
4. Inoculated with SB35 in Cd-contaminated soil
5. Inoculated soil with SB32
6. Inoculated with SB32 in Cd-contaminated soil

The inoculated pots of soybean were kept in net house at a temperature of 30 ± 5 °C. Pots were irrigated with tap water to maintain moisture content. Fifteen seeds were sown per pot, with three replicates per treatment. Plants were uprooted after 60 days of cultivation. Plant height and wet weight were measured before the plant biomass was oven dried. The agronomical parameters of plant were measured after harvesting.

Heavy metal analysis

Plant materials

Plants were harvested and roots were washed extensively in several changes of a solution containing 5 mM Tris HCl (pH 6.0) and 5 mM EDTA, and then distilled water to remove non-specifically bound metal ions. Shoots and roots were oven dried separately at 60 °C for 24 h followed by 70 °C for 3 h. Aliquots (1 g) of dried leaves were ground in a porcelain mortar while dried roots were extensively minced with a razor blade and then the samples were analyzed by modified methods of wet ashing (Burd et al. 2000). Wet ashing was performed by placing aliquots of dried material in 10 ml of concentrated nitric acid at 70 °C till the brown vapor subsides. Subsequently, 10 ml of diacid mixture (HNO_3 and

HClO₄, 3:1 ratio) was added and left till they reduced to 1 ml. Later 5 ml of 6 NHCl for 1 h was added and cooled. The digested samples were brought to 50 ml with deionized water and heavy metal analysis was done using a flame atomic absorption spectrophotometer. The instrument was zeroed with 1 % HNO₃ blanks. Samples in triplicate were taken.

Soil

Soil samples were collected periodically and analyzed for residual Cd concentration. To analyze heavy metals contents in soil, 10 g of dried soil was taken in 125 ml conical flasks. Then 20 ml of diethylene triamine penta acetic acid (DTPA) extracting solution [(1 l DTPA extracting solution: dissolve 13.1 ml reagent grade triethanolamine (TEA), 1.967 g DTPA (AR grade) and add 1.47 g of $CaCl_2 \cdot 2H_2O$ in 100 ml of triple distilled water, pH 7.3 ± 0.5) was shaken for 2 h at 120 cycles min^{-1} and filtrate was analyzed for Cd using flame atomic absorption spectrophotometer (Gupta 1993). Samples were taken in triplicate.

Results and discussion

The strains SB32 and SB35 chosen for this study were tolerant to 5 mM $CdCl_2$. Further, psychrotolerant strain SB32 showed optimum growth at 20 °C, while, alkalophilic strain SB35 exhibited optimum growth at pH 9.0. The partial sequencing of 16srDNA of the strains revealed the isolates to be *Pseudomonas putida* (SB32) and *Pseudomonas monteilli* (SB35) with accession numbers: HQ610451 and HQ864710, respectively.

Molecular characterization

Cadmium-resistant plasmid and chromosomal operons have also been reported by other workers (Lebrun et al. 1994; Horitsu et al. 1985; Lee et al. 2001; El-Deeb 2009; Mullapudi et al. 2010), therefore, to identify the location of the resistant gene; an attempt was made to amplify the *czc* gene from both plasmid and genomic DNA of the strains. The plasmid DNA profile revealed the presence of plasmid DNA of ∼5 kb (Fig. 1a). The plasmid DNA isolated from both strains, when subjected to PCR for amplification of the *czc* gene, produced an amplicon of approximate to 650 bp, whereas no amplicon was obtained in case of genomic DNA in both the strains (Fig. 1b). Also, *E. coli* DH5α which served as a negative control did not show any amplification. The presence of amplified product from the plasmid DNA of strains SB32 and SB35 indicated the presence of *czc* gene which is responsible for the efflux of three metal cations, viz. cobalt, zinc and cadmium thereby

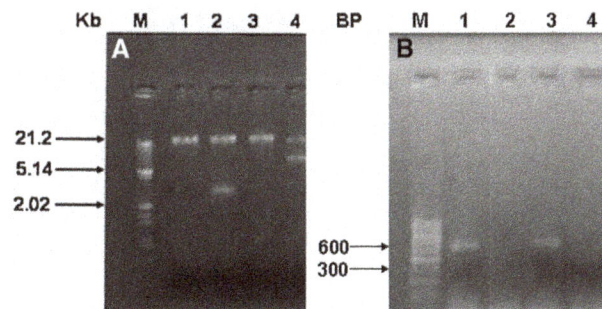

Fig. 1 a Plasmid DNA profile of cadmium-resistant strains. *Lane M* λ DNA/*Eco*R1/*Hin*dIII double digest; *lanes 1–4* cadmium-resistant strains grown in the absence and presence of $CdCl_2$; *1, 2* SB32; *3, 4* SB35. **b** Amplified czc gene from plasmid and genomic DNA of cadmium-resistant strains. *Lane M* 100 bp ladder; *lanes 1, 3* amplified *czc* gene from plasmid DNA of strains SB32 and SB35, respectively; *lanes 2, 4* amplified czc gene from genomic DNA of strains SB32 and SB35, respectively

giving a clear indication of efflux mechanism. The amplicons were sequenced and when compared with the available database using BLASTn search, sequence homology of the cloned genes showed that the amplicons contained a partial *czcA* gene sequence. The nucleotide sequence of the cloned gene of strains SB32 and SB35 showed 95 % similarity with the *czcA* gene of *Stenotrophomonas maltophilia* D457 (accession number HE798556) and 94 % similarity with putative heavy metal efflux pump *czc*A family of *S. maltophilia* JV3 (accession number CP002986). Further, alignment of the translated product showed similarity with the heavy metal efflux pump, *czcA* family of *S. maltophilia* R551, *Stenotrophomonas* sp. SKA14 and *Acidovorax delafieldii* 2AN, thereby indicating the presence of an plasmid-mediated efflux mechanism of resistance in the strains. The *czc* gene sequences of strains SB32 and SB35 were deposited in gene bank database under accession numbers: KC750207 and KC750208, respectively. The mechanism was further confirmed by AAS and TEM analysis.

Atomic absorption spectroscopy and transmission electron microscopy

AAS analysis of strain SB32 and SB35 showed that the concentration of Cd in strain SB32, increased in the bacterial cells with the start of log phase (12 h) and reached a maximum of 7.5 $\mu g\ mL^{-1}$ of dry weight of the cell. However, in the supernatant, the concentration of Cd followed was reversed, i.e., the Cd concentration decreased with the start of log phase (Fig. 2a).

Further, in strain SB35, the AAS results showed that the concentration of Cadmium increased with the start of log phase (12 h) and reached to a maximum of 8 $\mu g\ ml^{-1}$ of the dry weight and then decreased after 12 h, i.e., with the

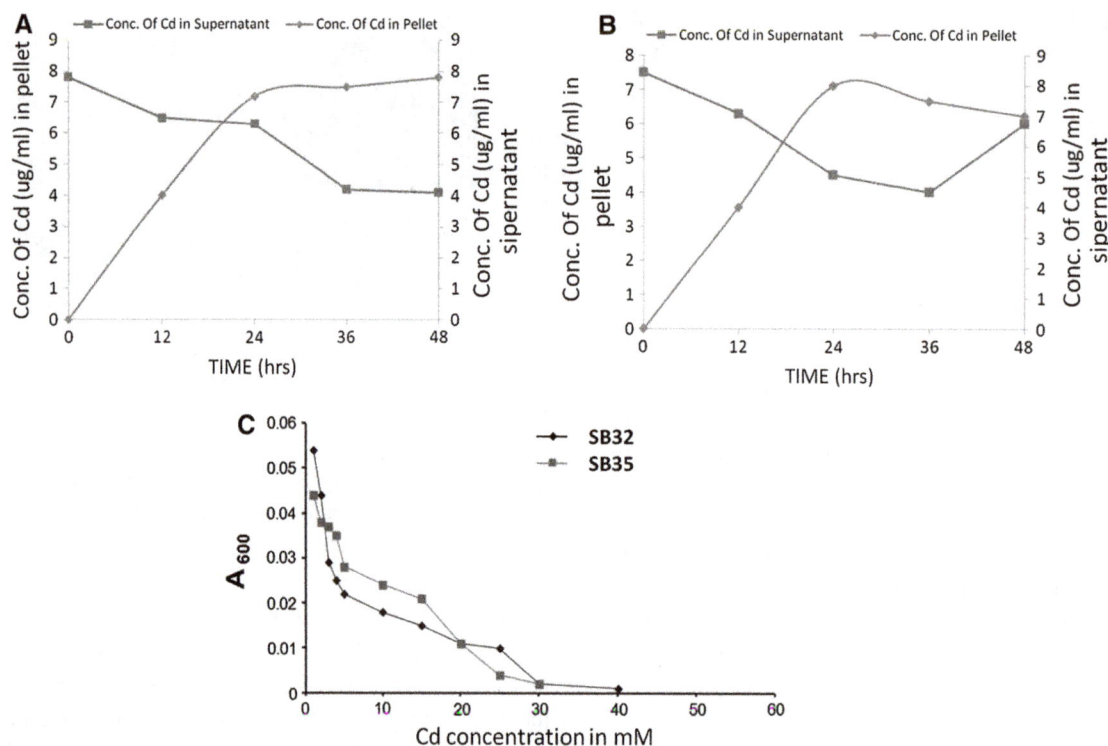

Fig. 2 Cadmium content in pellet and supernatant fraction of **a** *P. putida* strain SB32 and **b** *P. monteilli* strain SB35 at different time intervals of the growth profile. **c** Correlation between growth profile and concentrations of cadmium isolates SB32 and SB35. Cells were grown in nutrient broth for 18 h

end of log phase. However, in the supernatant the concentration of cadmium followed the reverse trend, i.e., the concentration decreased with the start of log phase followed by an increase in the late log phase (Fig. 2b). This clearly indicates the Cd uptake/bioremediation ability of both *Pseudomonas* strains SB32 and SB35. Furthermore, the AAS data were substantiated by TEM.

TEM analysis revealed an increase in cell size of both strains in the presence of Cd (Fig. 3). Cell length was found increased by 51.13 % in SB32, while, in SB35 the increase was of 25.2 %, although there was no change in cell width. It has been reported previously that SEM analysis of *Pseudomonas aeruginosa* strain MCCB 102 showed an increase in cell size due to Cd together with lead accumulation in the cell wall and along the external cell surfaces (Zolgharnein et al. 2010). However, in this study, the cell length of strain SB32 and SB35 increased due to the entry of metal, but no further deformations were observed. This may be due to the presence of the *czc* gene which is responsible for the efflux of metal ions from the cells. The metal, which is taken into the cells by fast and unspecific transport for Mg^{2+} ions, is actively extruded by products of *czc* resistance determinants (Nies 2000). The ability of a bioinoculant to colonize aggressively makes them the preferred choice for bioremediation studies. Considerable information is available with respect to their

use in natural environment but little is known about the effect of pH on bacterial survival, persistence and its subsequent effect on their bioremediation ability. So considering this point, in situ study was conducted with these diversified extremophilic Cd-resistant strains to establish a relationship between pH, temperature, their survival and subsequent effect on soybean growth.

Impact of bioinoculants on Cd toxicity in soybean

After 60 days of germination, the plants were harvested and comparative cadmium accumulation in roots and shoots was measured. Both of the strains were compared for their ability to inhibit cadmium accumulation in plants. Cd toxicity caused a significant reduction in shoot length, root length, fresh and dry weight of soybean, and the magnitude of reduction was 5.0, 30.2, 15.45 and 25.6 % relative to control, respectively (Table 1). Nevertheless, bioinoculation of strains (SB32 and SB35) in Cd-polluted soil increased the agronomical parameters in comparison with uninoculated soil. Such comparison revealed that *P. monteilli* SB35 strain was able to reduce cadmium accumulation more than *P. putida* SB32 strain (Table 1) and reduction was found to be 47.5 and 56.9 % by *P. monteilli* SB35, than 26.9 and 17.6 % by *P. putida* SB32 in root and shoot, respectively (Fig. 4a). While comparing the effects

Fig. 3 Transmission electron micrographs of **a** and **b** *P. monteilli* SB35 in absence and presence of Cd, respectively, at a magnification of ×50,000; **c**, and **d** *P. putida* SB32 in absence and presence of Cd, respectively at a magnification of ×40,000

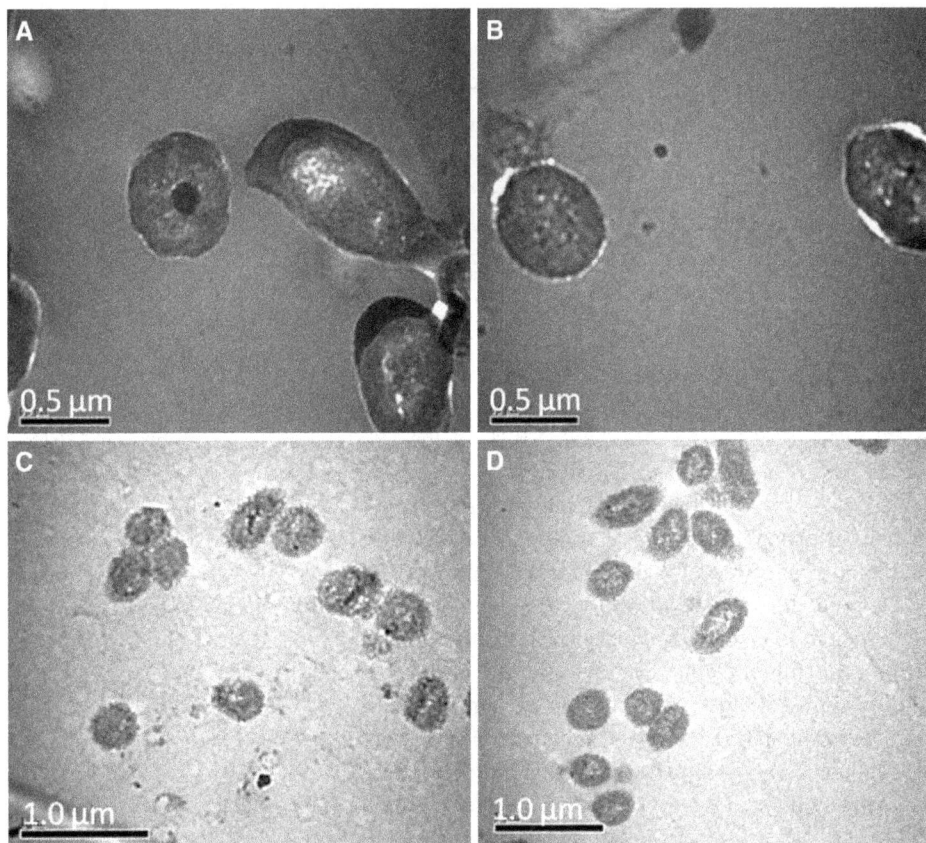

Table 1 Two way ANOVA depicting the effect of *P. putida* SB32 and *P. monteilli* SB35 on Soybean under greenhouse conditions in alkaline soil (30 ± 5 °C) after 60 days of germination, respectively

	Root length[a] (cm)	Shoot length[a] (cm)	Wet weight[a] (g)	Dry weight[a] (g)	Chlorophyll[b] (mg g^{-1})
In absence of cadmium					
Mean (control)	11.4 ± 0.11[d]	33.7 ± 0.18	4.16 ± 0.17	2.41 ± 0.63	3.1 ± 0.17
Mean (SB35 strain treated)	12.9 ± 0.85 (13.1 %)[e]	39.7 ± 0.16 (17.8 %)	4.83 ± 0.14 (16.1 %)	3.07 ± 0.11 (6.0 %)	3.3 ± 0.12 (3.5 %)
Mean (SB32 strain treated)	12.5 ± 0.61 (9.6 %)	37.2 ± 0.24 (10.3 %)	2.78 ± 0.10 (46.9 %)	2.96 ± 0.70 (45.4 %)	3.05 ± 0.18 (7.1 %)
In presence of cadmium					
Mean (control)	8.32 ± 0.45	28.1 ± 0.16	2.23 ± 0.18	1.34 ± 0.87	1.7 ± 0.15
Mean (SB35 strain treated)	10.5 ± 0.55 (26.2 %)	31.4 ± 0.17 (11.7 %)	3.5 ± 0.11 (56.9 %)	2.12 ± 0.95 (58.2 %)	2.5 ± 0.21 (47 %)
Mean (SB32 strain treated)	12.1 ± 0.27 (45.4 %)	29.7 ± 0.18 (5.6 %)	2.47 ± 0.88 (10.7 %)	1.66 ± 0.92 (23.8 %)	1.97 ± 0.33 (15.8 %)
Critical difference at 5 %					
	2.22	4.66	0.44	0.26	0.41

[a] Mean of ten replicates

[b] Mean of ten replicates

[c] Mean of four replicates

[d] The reported values are ±SEM

[e] Values indicate the % increase over the respective control

Fig. 4 a Comparative cadmium
accumulation in soybean in the
presence of *P. monteilli SB35*
and *P. putida* SB32,
respectively (*n* = 3,
mean ± SEM). **b** Comparative
DTPA extractable residual
cadmium accumulation in the
presence of *P. monteilli* SB35
and *P. putida* SB32,
respectively (*n* = 3,
mean ± SEM)

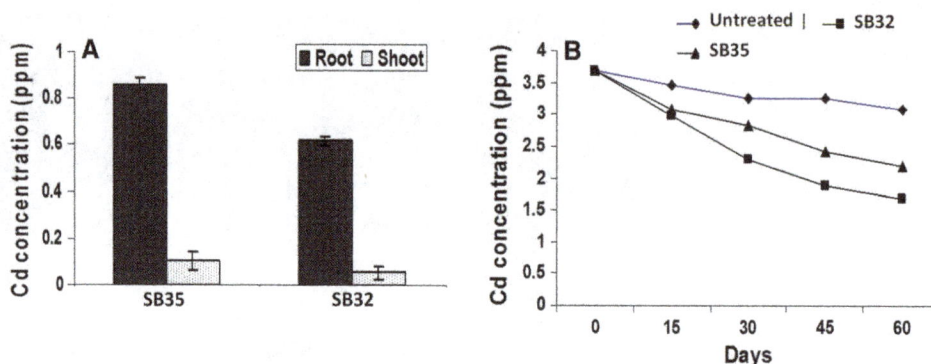

of the *P. putida* SB32 and *P. monteilli* SB35 on the Cd
content reduction in soil (Table 1), it appeared that the
strain SB35 was more effective in reducing Cd concentration than SB32 and it reduces 1.33 times more than latter
in alkaline soil type (Fig. 4b). As evident from the data
(Table 1), in alkaline soil, *P. monteilli* SB35 was found to
be more efficient in enhancing plant growth than *P. putida*
SB32 strain. Cadmium mainly occurs as the free metal ion
Cd^{2+} (Bingham and Page 1975) and ion exchange mechanisms have a dominating influence on metal concentration
in soil solution. Changes in pH exert both a biological and
chemical effect on metal ion toxicity (Campbell and Stokes
1985). Low pH favors greater metal ion solubility, and, in
the absence of complexing ions, reduced speciation of the
metal ion, which tends to increase toxicity compared to
higher pH. However, low pH also enhances competition
between H^+ and metal ion for cell surface binding sites,
which tends to decrease metal ion toxicity.

The increase in agronomical parameters in the presence
of cadmium in the soil documented the growth promotory
as well as bioremediation potential of bioinoculants. Similar observations were made earlier by Burd et al. (2000).
However, toxicity of these metals depends on the soil pH.
Further, cadmium content analysis revealed alkalophilic *P.
monteilli* SB35 was more effective than *P. putida* SB32
strain. Both of the strains were able to reduce the cadmium
accumulation in plants and soil significantly. This shows
the importance of abiotic factor such as pH for survival and
growth of bioinoculants for the establishment of a threshold population of viable inoculant which is an important
prerequisite for the successful bioremediation strategy.

Conclusion

It is clear from the above results that resistance to Cd in
both the diversified extremophilic strains SB32 and SB35
of *Pseudomonas* is due to the *czc* gene present on the
plasmid DNA and involves metal binding and/or an efflux
mechanism of resistance. Furthermore, the results of in situ

trial, AAS, TEM and EDAX analysis suggest that the
strains may have considerable potential as an agent for
bioremediation under natural conditions with reference to
Cd. Moreover, knowledge of the gene employed and the
mechanism involved in Cd resistance may be used to
design a tailor-made plant growth promotory bioinoculant.

Acknowledgments This study was supported by Junior research
fellowship to SJ by DBT-India during his tenure at GBPUAT, India.

Conflict of interest None.

References

Alloway BJ (1990) Cadmium. In: Alloway BJ (ed) Heavy metals in
    soils. Wiley, New York, pp 100–124
Anderson DG, Mckay LL (1983) Simple and rapid method for
    isolating large plasmid DNA from *Lactic streptococci*. Appl
    Environ Microbiol 46:549–552
Bingham FT, Page AL (1975) In: Hutchinson TC (ed) International
    conference on heavy metals in the environment, Toronto,
    pp 433–441
Burd GI, Dixon GD, Glick BR (2000) A plant growth promoting
    bacterium that decreases heavy metal toxicity in plants. Can J
    Microbiol 46:237–245
Burke BE, Pfister RM (1986) Cadmium transport by a Cd^{2+} sensitive
    and a Cd^{2+} resistant strain of *Bacillus subtilis*. Can J Microbiol
    32:539–542
Campbell PGC, Stokes PM (1985) Acidification and toxicity of
    metals to aquatic biota. Can J Fish Aquat Sci 42:2034–2049
Chaoui A, Mazhouri S, Ghorbal MH, Ferjani EE (1997) Cadmium
    and zinc induction of lipid peroxidation and effects of antiox-
    idant enzyme activities in bean (*Phaseolus vulgaris* L.). Plant Sci
    127:139–147
Clausen CA (2000) Isolating metal-tolerant bacteria capable of
    removing copper, chromium, and arsenic from treated wood.
    Waste Manag Res 18:264–268
El-Deeb B (2009) Plasmid mediated tolerance and removal of heavy
    metals by *Enterobacter* sp. Am J Biochem Biotechnol
    5(1):47–53

Goldberg M, Pribyl T, Jhunke S, Nies DH (1999) Energetics and topology of *CzcA*, a cation/proton antiporter of the RND protein family. J Biol Chem 274:26065–26070

Gupta VK (1993) Soil analysis for available micro-nutrients. In: Tandon HLS (ed) Methods of analysis of soils, plants, waters and fertilizers. Fertilize Development and Consultation Organization, New Delhi, India, pp 26–46

He M, Li X, Liu H, Miller SJ, Wang G, Rensing C (2011) Characterization and genomic analysis of a highly chromate resistant and reducing bacterial strain *Lysinibacillus fusiformis* ZC1. J Hazard Mater 185:682–688

Horitsu H, Yamamoto K, Wachi S, Kawai K, Fukuchi A (1985) Plasmid-determined cadmium resistance in *Pseudomonas putida* GAM-1 isolated from soil. J Bacteriol 165(1):334–335

Lebrun M, Audurier A, Cossart P (1994) Plasmid-borne cadmium resistance genes in *Listeria monocytogenes* are similar to CadA and CadC of *Staphylococcus aureus* and are induced by cadmium. J Bacteriol 176(10):3040–3048

Lee SW, Glickman E, Cooksey DA (2001) Chromosomal locus for cadmium resistance in *Pseudomonas putida* consisting of a cadmium-transporting ATPase and a MerR family response regulator. Appl Environ Microbiol 67(4):1437–1444

Mullapudi S, Siletzky RM, Kathariou S (2010) Diverse cadmium resistance determinants in *Listeria monocytogenes* isolates from the Turkey processing plant environment. Appl Environ Microbiol 76(2):627–630

Naidu R, Kookana RS, Sumner ME, Harter RD, Tiller KG (1997) Cadmium sorption and transport in variable charge soils: a review. J Environ Qual 26:602–617

Nies DH (2000) Heavy metal-resistant bacteria as extremphiles:molecular physiology and biotechnological use of *Ralstonia* sp. CH34. Extremophiles 4:77–82

Nies DH (2003) Eflux-mediated heavy metal resistance in prokaryotes. FEMES Microbiol Rev 27:313–339

Nies DH, Silver S (1995) Ion efflux systems involved in bacterial metal resistances. J Ind Microbiol 14:186–199

Nies DH, Nies A, Chu L, Silver S (1989a) Expression and nucleotide sequence of a plasmid determined divalent cation efflux system from *Alcaligenes eutrophus*. Proc Natl Acad Sci USA 86:7351–7355

Nies A, Nies DH, Silver S (1989b) Cloning and expression of plasmid genes encoding resistance to chromate and cobalt in *Alcaligenes eutrophus* CH34. J Bacteriol 171:5065–5070

Piqueras A, Olmos E, Martinez-solano JR, Hellin E (1999) Cd induced oxidative burst in tobacco BY2 cells: time course, subcellular location and antioxidant response. Free Radic Res 31:S33–S38

Roane TM, Josephon KL, Pepper IL (2001) Dual-bioaugumentation strategy to enhance remediation of co-contaminated soil. Appl Environ Microbiol 67:3208–3215

Robinson NJ, Gupta A, Fordham-Skelton AP, Croy RRD, Whitton BA, Huckle JW (1990) Prokaryotic metallothionein gene characterization and expression: chromosome crawling by ligation-mediated PCR. Proc R Soc Lond B 242:241–247

Romero-Puertas MC, McCarthy I, Sandalio LM, Palma IM, Corpas FI, Gomez M, Del Rio LA (1999) Cadmium toxicity and oxidative metabolism of pea leaf peroxisomes. Free Radic Res 31:525–531

Sandalio LM, Dalurzo HC, Gomez M, Romero-Puertas MC, Del Rio LA (2001) Cadmium-induced changes in the growth and oxidative metabolism of pea plants. J Exp Bot 52:2115–2126

Sanita Di Toppi L, Gabbrielli R (1999) Response to cadmium in higher plants. Environ Exp Bot 41:105–130

Silver S (1996) Bacterial resistances to toxic metal ions - a review. Gene 179:9–19

Somashekaraiah BV, Padmaja K, Prasad ARK (1992) Phytotoxicity of cadmium ions on germinating seedling of mungbean (*Phaseolus vulgaris*): involvement of lipid peroxides in chlorophyll degradation. Physiol Plant 85:85–89

Tsai KJ, Linet AL (1993) Formation of a phosphorylated enzyme intermediate by the cadA Cd^{2+}-ATPase. Arch Biochem Biophys 305:267–270

Tseng TT, Gratwick KS, Kollman J, Park D, Nies DH, Goffeau A, Saier MHJ (1999) The RND superfamily: an ancient, ubiquitous and diverse family that includes human disease and development proteins. J Mol Microbiol Biotechnol 1:107–125

Tu SI, Brovillett JN (1987) Metal ion inhibition of cotton root plasma membrane ATPase. Phytochemistry 26:65–69

Vasudevan P, Padmavarthy V, Tewari N, Dhingra SC (2001) Biosorption of heavy metal ions. J Sci Ind Res 60:112–120

Wagner GJ (1993) Accumulation of cadmium in crop plants and its consequence to human health. Adv Agron 51:173–212

Wuertz S, Mergeay M (1997) The impact of heavy metals on soil microbial communities and their activities. In: Van Elsas JD, Wenington EMH, Trevors JT (eds) Modern soil microbiology. Marcel Decker, NY, pp 1–20

Zolgharnein H, Karami K, Mazaheri Assadi M, Dadolahi Sohrab A (2010) Investigation of heavy metals biosorption on *Pseudomonas aeruginosa* strain MCCB 102 isolated from the Persian gulf. Asian J Biotechnol 2(2):99–109

20

Studies on heavy metal removal efficiency and antibacterial activity of chitosan prepared from shrimp shell waste

V. Mohanasrinivasan · Mudit Mishra ·
Jeny Singh Paliwal · Suneet Kr. Singh ·
E. Selvarajan · V. Suganthi · C. Subathra Devi

Abstract Chitosan, a natural biopolymer composed of a linear polysaccharide of α (1–4)-linked 2-amino 2-deoxy β-D glucopyranose was synthesized by deacetylation of chitin, which is one of the major structural elements, that forms the exoskeleton of crustacean shrimps. The present study was undertaken to prepare chitosan from shrimp shell waste. The physiochemical properties like degree of deacetylation (74.82 %), ash content (2.28 %), and yield (17 %) of prepared chitosan indicated that that shrimp shell waste is a good source of chitosan. Functional property like water-binding capacity (1,136 %) and fat-binding capacity (772 %) of prepared chitosan are in total concurrence with commercially available chitosan. Fourier Transform Infra Red spectrum shows characteristic peaks of amide at 1,629.85 cm^{-1} and hydroxyl at 3,450.65 cm^{-1}. X-ray diffraction pattern was employed to characterize the crystallinity of prepared chitosan and it indicated two characteristic peaks at 10° and 20° at (2θ). Scanning electron microscopy analysis was performed to determine the surface morphology. Heavy metal removal efficiency of prepared chitosan was determined using atomic absorption spectrophotometer. Chitosan was found to be effective in removing metal ions Cu(II), Zn(II), Fe(II) and Cr(IV) from industrial effluent. Antibacterial activity of the prepared chitosan was also determined against *Xanthomonas* sp. isolated from leaves affected with citrus canker.

Keywords Shrimp shells · Chitosan · Deacetylation · Metal removal efficiency · Antibacterial activity

Abbreviations
DD   Degree of deacetylation
WBC Water-binding capacity
FBC Fat-binding capacity
FTIR Fourier transform infrared
XRD X-ray diffraction
SEM Scanning electron microscopy
AAS Atomic absorption spectroscopy
OD   Optical density

Introduction

Chitosan is one of the most important derivatives of chitin, which is the second most abundant natural biopolymer found on earth after cellulose (No and Meyers 1989) and is a major component of the shells of crustaceans such as crabs and shrimps. Chitosan can be obtained by N-deacetylation of chitin and it is a co-polymer of glucosamine and N-acetylglucosamine units linked by 1–4 glucosidic bonds (Fig. 1). Chitosan is a fiber-like cellulose only but unlike plant fibers, it possesses some unique properties including the ability to form films, optical structural characteristics, and much more. Chitosan have the ability to chemically bind with negatively charged fats, lipids, and bile acids and this ability is because of the presence of a positive ionic charge (Sandford 1992). In acidic conditions (pH < 6), chitosan becomes water soluble that enables the formation of biocompatible and very often biodegradable polymers with optimized properties in homogenous

V. Mohanasrinivasan (✉) · M. Mishra · J. S. Paliwal ·
S. Kr. Singh · E. Selvarajan · V. Suganthi · C. Subathra Devi
School of Biosciences and Technology, VIT University,
Vellore 14, Tamil Nadu, India
e-mail: v.mohan@vit.ac.in

Fig. 1 Showing the molecular structure of chitosan (Jayakumar et al. 2011)

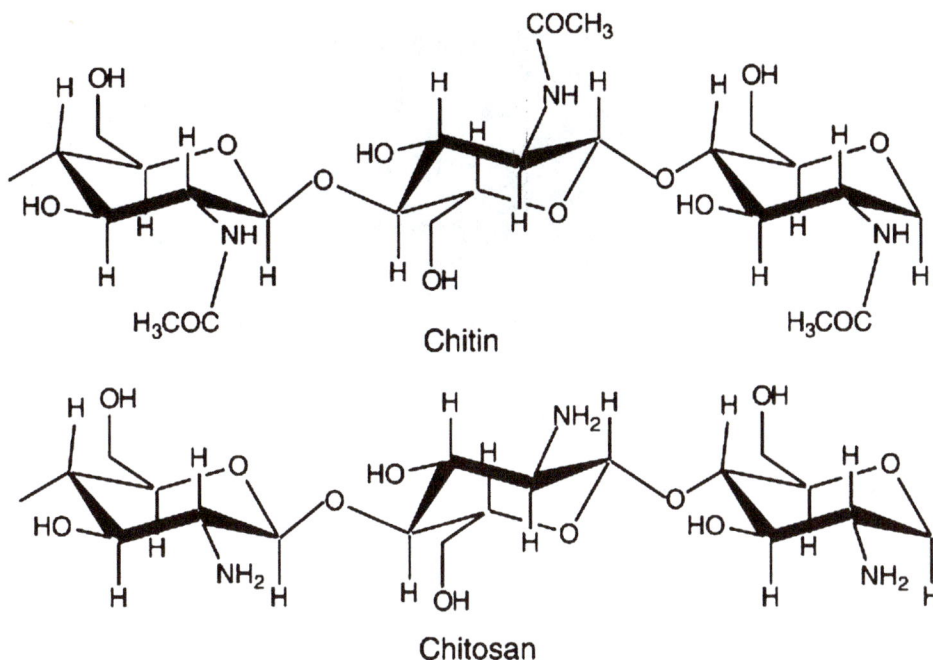

Chitin

Chitosan

solutions. Chitosan being a non-toxic, biodegradable, and biocompatible polysaccharide polymer have received enormous worldwide attention as one of the promising renewable polymeric materials for their extensive applications in industrial and biomedical areas such as paper production, textile finishes, photographic products, cements, heavy metal chelation, waste water treatment fiber, and film formations (Rathke and Hudson 1994). It can also be used in biomedical industries for enzyme immobilization and purification, in chemical plants for wastewater treatment, and in food industries for food formulations as binding, gelling, thickening, and stabilizing agent (Knorr 1984).

As chitosan can be readily converted into fibers, films, coatings, and beads as well as powders and solutions, further enhance its applications. The functional properties of chitosan are dependent on its molecular weight and its viscosity (No and Lee 1995). The presence of both free hydroxyl and amine groups enables chitosan to be modified readily to prepare different chitosan derivatives (Kurita 2001) that give some sophisticated functional polymers with exquisite properties quite different from those of the synthetic polymers. With its positive charge, chitosan can be used for coagulation and recovery of proteinaceous materials present in food processing operations (Knorr 1991). Chitosan has largely been employed as a non-toxic flocculent in the treatment of organic polluted wastewater and as a chelating agent for the removal of toxic (heavy and reactive) metals from industrial wastewater (An et al. 2001). There are some metals which exist in aqueous

solutions as anions like chromium (Rhazi et al. 2002). Chitin can be effectively extracted from prawn shells following deprotienization using 5 % NaOH and demineralization using 1 % HCl. Low molecular mass chitosan samples with degree of deacetylation (DD) >64 % and Mw of the major component <104 can be obtained by treating the chitin with 50 % NaOH at 100 °C for up to 10 h (Mohammed et al. 2012). At pH close to neutral the amine groups of chitosan binds to metal cations. At lower pH it is able to bind more of anions by electrostatic attraction as chitosan gets more protonated (Guibal 2004). Chitosan can be readily used as a biosorbent as it is cheaply available cationic biopolymer. Chitosan has antimicrobial activity, haemostatic activity, anti-tumor activity, accelerates wound healing, can be used tissue-engineering scaffolds and also for drug delivery (Burkatovskaya et al. 2006). The antimicrobial activity and antifungal activity of chitosan is largely because of its polycationic nature (Ziani et al. 2009 and Choi et al. 2001). It displays broad spectrum of antibacterial activity against both gram positive and gram negative bacteria and also antifungal activity against *Aspergillus niger, Alternaria alternata, Rhizopus oryzae, Phomopsis asparagi,* and *Rhizopus stolonifer* (Guerra-Sánchez et al. 2009; Zhong et al. 2007; Ziani et al. 2009). Chitin exhibited a bacteriostatic effect on gram-negative bacteria, *Escherichia coli* ATCC 25922, *Vibrio cholerae, Shigella dysenteriae,* and *Bacteroides fragilis* (Benhabiles et al. 2012).

The current research is to prepare chitosan from shrimp shell waste and to estimate the prepared chitosan by both

qualitatively and quantitatively. Other applications such as antibacterial activity against *Xanthomonas* sp. and metal removal efficiency have also been studied for the prepared chitosan.

Materials and methods

Materials

Shrimps shell waste material was collected from local market of Vellore. The chemicals such as Hydrochloric acid and Sodium hydroxide pellets were procured from Hi-Media Laboratory, Mumbai. Distilled water was used throughout the process.

Preparation of chitosan

The shrimp shells obtained from the local market of Vellore shown in Fig. 2a were first suspended in 4 % HCl at room temperature in the ratio of 1:14 (w/v) for 36 h. This causes the Demineralization of shells after which they were washed with water to remove acid and calcium chloride. Deproteinization of shells was done by treating the demineralized shells with 5 % NaOH at 90 °C for 24 h with a solvent to solid ratio of 12:1 (v/w). After the incubation time the shells were washed to neutrality in running tap water and sun dried. The product obtained was chitin. Chitosan preparation involves the deacetlyation of the obtained chitin (Dutta et al. 2004). Deacetylation of chitin involves the removal of acetyl groups from chitin and that was done by employing 70 % NaOH solution with a solid to solvent ratio of 1:14 (w/v) and incubated at room temperature for 72 h. Stirring is mandatory to obtain a homogenous reaction as shown in Fig. 3a–c. The residue obtained after 72 h was washed with running tap water to neutrality and rinsed with deionized water. It was then filtered, sun dried and finely grinded shown in Fig. 4a. The resultant whitish flakes obtained after grinding are chitosan as shown in Fig. 4b.

Determination of chitosan yield

Yield was determined by taking the dry weight of shrimp shells before treatment and the dry weight of prepared chitosan.

Determination of ash content

The ash content of the prepared chitosan was determined by placing 0.5706 g of chitosan into previously ignited, cooled, and tarred crucible. The samples were heated in a muffle furnace preheated to 600 °C for 6 h. The crucibles were allowed to cool in the furnace to <200 °C and then placed into desiccators with a vented top. Crucible and ash was weighed (AOAC 1990).

$$\% \, \text{Ash} = \frac{\text{Weight of residue (g)} \times 100}{\text{Sample weight (g)}}$$

Determination of moisture content

Moisture content was determined by employing the gravimetric method (Black 1965). The water mass was determined by drying the sample to constant weight and measuring the sample after and before drying. The water mass (or weight) was the difference between the weights of the wet and oven dry samples. Then moisture content was calculated using the following relationship:

$$\% \, \text{Moisture content} = \frac{\text{Wet weight (g)} - \text{Dry weight (g)} \times 100}{\text{Wet weight (g)}}$$

Determination of solubility

The solubility of prepared chitosan was determined by taking 200 mg of chitosan and then adding 200 ml of water and the same method was followed with 1 % acetic acid solution.

FTIR analysis

The samples of prepared chitosan were characterized in KBr pellets by using an infrared spectrophotometer in the range of 400–4,000 cm^{-1}.

Fig. 3 Flasks showing
shrimp shells during
a demineralization,
b deprotienization, and
c deacetylation steps

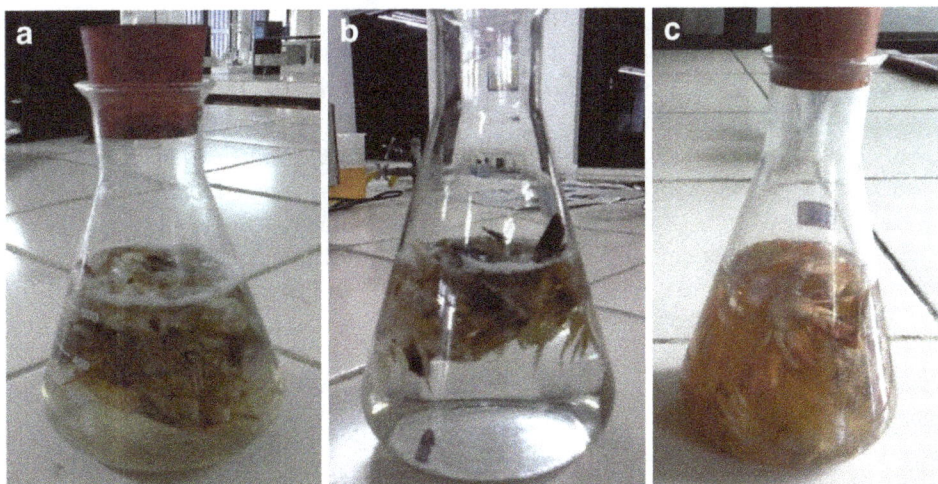

Fig. 3 Flasks showing shrimp shells during a demineralization, b deprotienization, and c deacetylation steps

Fig. 4 Prepared chitosan a before grinding and b chitosan flakes after grinding

Degree of deacetylation

The DD of chitosan was determined using a Fourier Transform Infra Red (FTIR) instrument with frequency of 4,000–400 cm^{-1}. The following equation (Struszczyk 1987) was used, where the absorbance at A1629.85 and A3450.65 cm^{-1} are the absolute heights of absorption bands of amide and hydroxyl groups, respectively.

$$DD = 100 - \frac{(A1629.85\,cm^{-1} - A3450.65\,cm^{-1}) \times 100}{1.33}$$

The factor '1.33' denoted the value of the ratio of A1629.85/A3450.65 for fully N-acetylated chitosan.

Determination of WBC

This property of chitosan was determined by using the modified method of (Wang and Kinsella 1976). For water-binding capacity (WBC) 0.5 g of chitosan sample was taken in a centrifuge tube of 50 ml which was weighed initially and then adding 10 ml of water, and mixing on a vortex mixer for 1 min to disperse the sample. The contents were later left at ambient temperature for 30 min with intermittent shaking for 5 s every 10 min and then centrifuged at 3,200 rpm for 25 min. After centrifugation the

supernatant was decanted and the tube was weighed again and WBC was calculated using the following relationship.

$$WBC\,(\%) = \frac{water\ bound\ (g)}{initial\ sample\ weight\ (g)} \times 100$$

Determination of FBC

Fat-binding capacity of prepared chitosan was calculated using the modified equation of (Wang and Kinsella 1976). For FBC 0.5 g of chitosan sample was taken in a 50 ml centrifuge tube which was weighed initially and 10 ml of gingelly oil was taken followed by mixing on a vortex mixer for 1 min to disperse the sample. The contents were later left at ambient temperature for 30 min with intermittent shaking for 5 s every 10 min and then centrifuged at 3,200 rpm for 25 min. After the centrifugation the supernatant was decanted and the tube was weighed again and FBC was calculated using the following relationship.

$$FBC\,(\%) = \frac{fat\ bound\ (g)}{initial\ sample\ weight\ (g)} \times 100$$

XRD analysis

The prepared chitosan was characterized by X-ray diffraction (XRD) technique using an X-ray diffractometer

(Bruker Germany, D8 Advance, 2.2 KW Cu Anode, Ceramic X-ray) with CuKα radiation ($\lambda = 1.5406$ Å). The measurement was in the scanning range of 5–70 at a scanning speed of 50 s^{-1}.

SEM analysis

Chitosan prepared from shrimp shell waste was examined by scanning electron microscopy (SEM) having a magnification range of 5,000 and accelerating voltage 20 kV.

Determination of heavy metals removal efficiency in industrial effluents by AAS

0.1 g of chitosan was mixed with 40 ml of industrial effluent (obtained from leather industry located at Ranipet) and its pH was measured, followed by an incubation of 3 h at 22 °C. The original effluent was used as control. The industrial effluent mixed with chitosan was kept for incubation for 3 h at 22 °C. The contents were then centrifuged at 7,000 rpm (revolutions per minute) for 5 min, and supernatant was filtered using Whatman filter paper no 2. The metal ions namely Cr(IV), Fe(II), Zn(II) and Cu(II) were analyzed for their residual metal concentration using atomic absorption spectrophotometer (AAS) (VARIAN, AA240). The standards of these metals were prepared (Gamage and Shahidi 2007).

Determination of inhibitory activity of chitosan against Xanthomonas sp.

Leaves showing symptoms of cankerous growth were plucked from lemon tree (*Citrus limon*) and were surface sterilized with sodium sulfite for consecutively seven times followed by distilled water. Leaves were crushed in mortar and pestle with 1 ml deionized water and the extract obtained was spread on nutrient agar plates supplemented with 5 % sucrose.

Method employed for evaluating the antimicrobial activity was growth inhibition in liquid medium. The antimicrobial effect of prepared chitosan was studied in liquid nutrient medium. Nutrient broth supplemented with 5 % sucrose was used. The flasks were marked as blank that contained only the media, control (standard) that had the bacterial culture only and the test which contained the prepared chitosan and *Xanthomonas* sp. cultures. The freshly grown inoculums were allowed to incubate in the presence of 0.2 g of chitosan to observe the bacterial growth pattern at 310 K (37 °C) and 150 rpm. In liquid medium, growth of *Xanthomonas* sp. was indexed by measuring the optical density (OD). Optical density measurements were carried out at $\lambda_{max} = 600$ nm after every 1 h interval up to 24 h. Graph was plotted to interpret the results.

Results and discussion

Yield

The prepared chitosan had a percentage yield of 17 % as shown in Table 1, which was at par when compared to the percentage yield obtained by (Brzeski 1982) who reported 14 % yield of chitosan from krill and showed no significant difference to the percentage yield of 18.6 % from prawn waste (Alimuniar and Zainuddin 1992).

Ash content

The prepared chitosan had an ash content of 2.28 % as shown in Table 1, which when compared to commercial chitosan which had an ash content of 2 % shows that the chitosan prepared had a standard percentage of ash content which can be used for commercial applications, as the ash content in chitosan is an important parameter that affects its solubility, viscosity and also other important characteristics.

Moisture content

The moisture content of chitosan obtained from shrimp shells was measured to be 1.25 % as shown in Table 1, which is in agreement with (Islam et al. 2011 and Hossein et al. 2008) who reported moisture content in the range of 1–1.30 obtained from brine shrimp shells. Although Li et al. (1992) reported that commercial chitosan products may contain <10 % moisture content.

Solubility

The prepared chitosan from shrimp shells waste was found to be soluble in 1 % acetic acid solution and partially soluble in water as shown in Table 1.

Degree of deacetylation

In the present study DD of the prepared chitosan was found to be 74.82 % (Table 1). It is an important parameter that

Table 1 Physiochemical and functional properties of chitosan

Properties	Percentage
Yield	17
Moisture content	1.25
Ash	2.28
DD	74.82
WBC	1,136
FBC	772
Solubility	1 % CH$_3$COOH

influences other properties like solubility, chemical reactivity and biodegradability. DD of the commercially available chitosan has values that range between 75 and 85 %. The value of DD depends on various factors such as the source and procedure of preparation and the values ranges from 30 to 95 % (Martino et al. 2005) and also on the type of analytical methods employed, sample preparation, type of instruments used, and other conditions may also influence the analysis of DD (Khan et al. 2002).

WBC and FBC

Water-binding capacity and FBC are functional properties that vary with the method of preparation. Chitosan prepared from shrimp shells in the present study has WBC and FBC of 1,136 and 772 % (Table 1) and these are in agreement with studies reported by (Rout 2001).

XRD analysis

The XRD pattern of chitosan prepared from shrimp shells waste illustrates two characteristic broad diffraction peaks at $(2\theta) = 10°$ and $20°$ that are typical fingerprints of semi-crystalline chitosan as shown in Fig. 5 (Bangyekan et al. 2006). The XRD pattern of standard chitosan procured from Sigma Aldrich shows similar peaks as shown in Fig. 6. The peaks around $2\theta = 10°$ and $2\theta = 20°$ are related to crystal I and crystal II in chitosan structure (Ebru et al. 2007; Marguerite 2006) and both these peaks attributes a high degree of crystallinity to the prepared chitosan (Julkapli and Akil 2007) as shown in Fig. 5.

FTIR analysis

The structure of the prepared chitosan was confirmed by FTIR analysis. The spectra of chitosan shows a broad absorption band in the region of 3,450.65 cm^{-1} that corresponds to OH stretching vibrations of water and hydroxyls and NH stretching vibrations of free amino groups as shown in Fig. 7. The band observed at 2,924.09 and 2,852.72 corresponds to asymmetric stretching of CH_3 and CH_2 in the prepared chitosan (Guo et al. 2005). The intensive peak around 1,629.85 cm^{-1} corresponds to bending vibration of NH_2 which is a characteristic feature of chitosan polysaccharide and also indicates the occurrence of deacetylation (Zhang et al. 2011 and Radhakumary et al. 2003).

SEM analysis

The SEM micrograph illustrates the morphology of the prepared chitosan from shrimp shells. The micrographs showed non-homogenous and non-smooth surface as shown in Fig. 8.

Heavy metals removal efficiency in industrial effluents by AAS

Industrial effluent obtained from Ranipet region contained traces of heavy metals namely Cu(II), Zn(II), Fe(II) and Cr(IV) that was confirmed by AAS. The results indicated that the prepared chitosan has the ability to adsorb the metal ions that were present in the industrial effluent as shown in Table 2. Out of all the metal ions Cu(II) was best absorbed

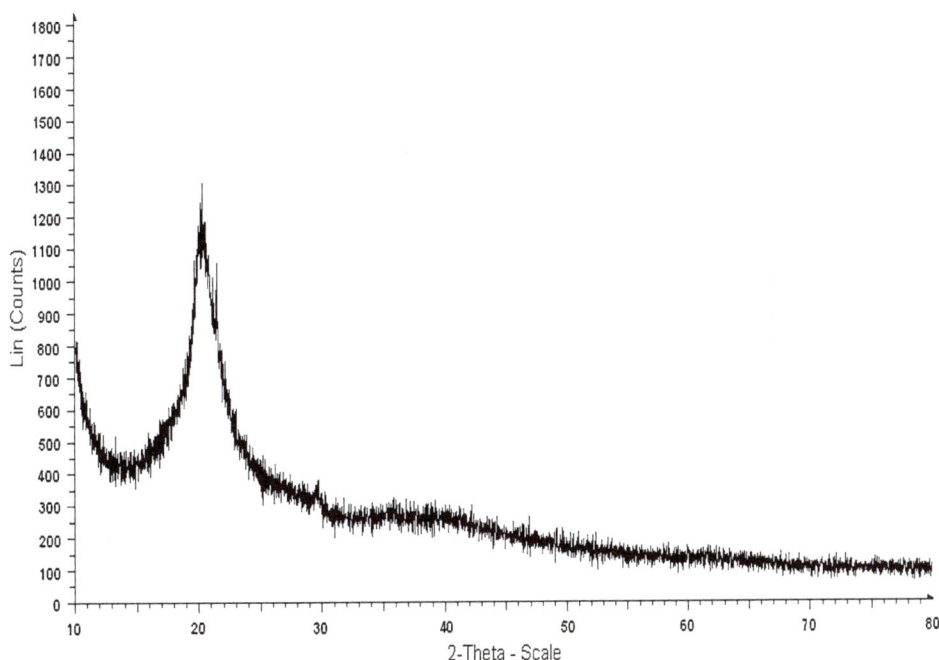

Fig. 5 XRD of prepared chitosan

Fig. 6 XRD of standard chitosan

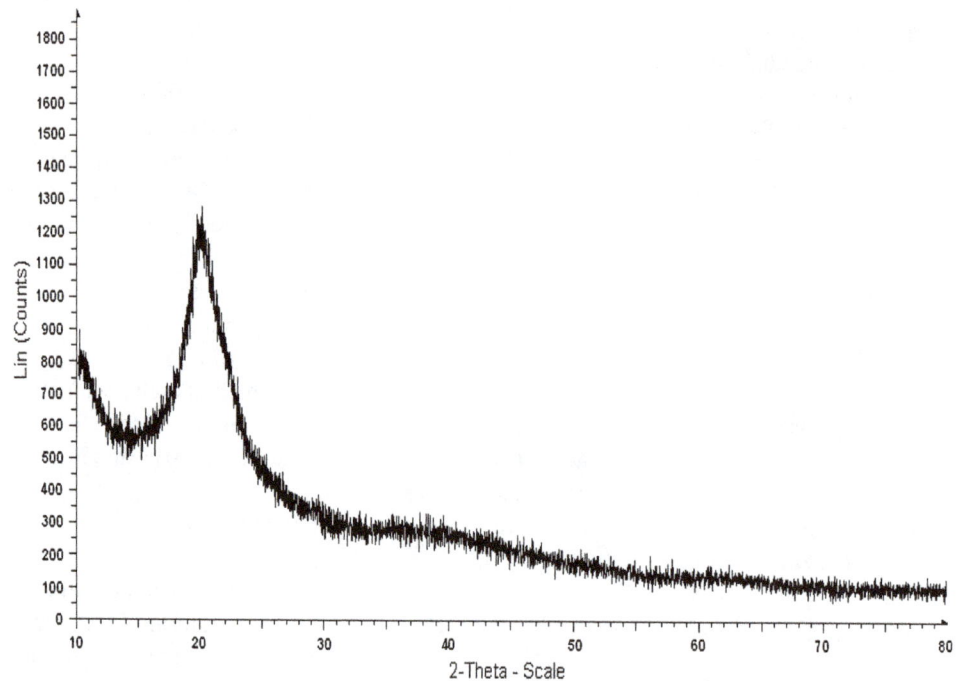

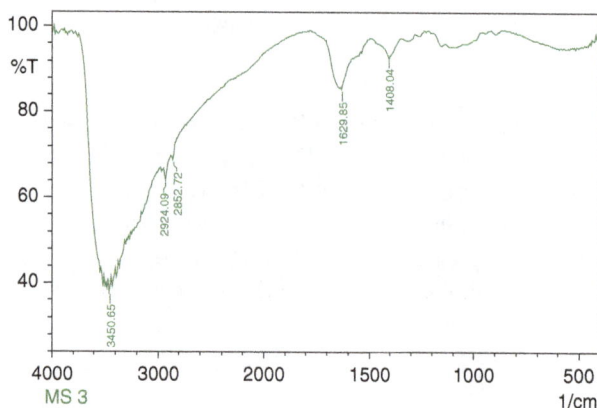

Fig. 7 FTIR of prepared chitosan

Fig. 8 SEM image of prepared chitosan

showing removal of 98.97 %. Zinc and its salts have high acute and chronic toxicity to aquatic life in polluted water. Zinc toxicity causes problems like nausea, vomiting, diarrhea and also sometimes abdominal cramps (Elinder and Piscator 1979). Zinc present in the industrial effluent was successfully absorbed by the prepared chitosan to around 86.15 % AAS results strongly indicates the removal of metal ions where the sample 1 used is the untreated one and sample 2 is the treated one. Fe(II) which is responsible for the unpleasant organoleptic properties in drinking water (Muzzzarelli et al. 1989) has also been absorbed by the prepared chitosan to less percentage of around 65.2 %. The efficacy of chitosan treated and untreated effulents were shown in Fig. 9. Therefore it can be concluded that the prepared chitosan has the potential to be used as a adsorbent in the treatment of industrial wastewater.

Table 2 Heavy metal removal percentage

Metal	Removal (%)
Cu(II)	98.97
Cr(IV)	37.51
Fe(II)	65.2
Zn(II)	86.15

Cu copper, *Cr* chromium, *Fe* Iron, *Zn* zinc

Inhibitory activity of chitosan on *Xanthomonas* sp.

Optical density (OD) measurements were indexed and turbidity was observed to evaluate the inhibitory activity of

Fig. 9 The untreated and treated effluents showing change in color before and after treatment with chitosan

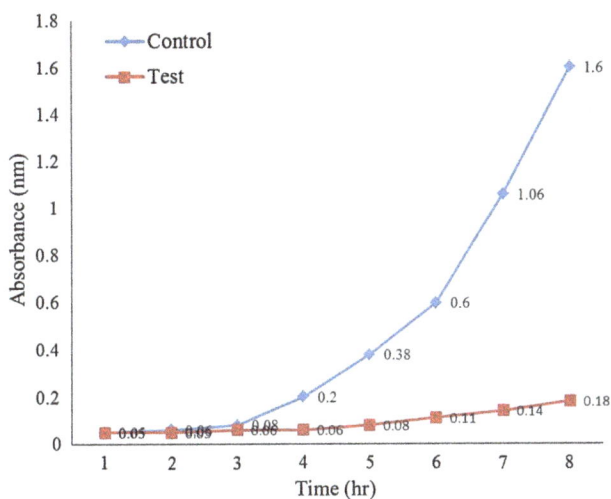

Fig. 10 Graph showing inhibitory activity of chitosan on *Xanthomonas* sp.

chitosan. Growth of *Xanthomonas* sp. was inhibited by the chitosan in the liquid medium. Very less turbidity was there in the test flask which contained both the chitosan and the organism. Whereas, the standard (control) that had only the organism was turbid and by the increase in OD readings showed that there was growth. Graph was plotted to determine the difference in the growth pattern of test ant control as shown in Fig. 10. No activity was observed in the blank.

Conclusion

The present study observations indicate that chitosan has been successfully prepared from shrimp shell waste. The functional, physiochemical properties, XRD and FTIR of the prepared chitosan showed that it can be used commercially and can be supplemented in food and drug preparation. The prepared chitosan was found effective in removing metals from industrial effluent and the result clearly indicated that the metal ion percentage was reduced to mere negligible level. Inhibition in growth of *Xanthomonas* sp. was observed in presence of chitosan prepared from shrimp shells. Since chitosan has the potential to be used as an antibacterial agent to control plant diseases. Method employed in this study prepares chitosan in a very economical way and it can also be a way to control pollution as shrimp shell waste is being used which is otherwise discarded.

Acknowledgments The authors are grateful to the management of VIT University, Vellore, India for providing us the facilities to conduct the research work.

Conflict of interest The authors declare that they have no conflict of interest.

References

Alimuniar A, Zainuddin R (1992) An economical technique for producing chitosan. In: Brine CJ, Sanford PA, Zikakis JP (eds) Advances in chitin and chitosan. Elsevier Applied Science, London and New York, pp 627–632

An HK, Park BY, Kim DS (2001) Crab shell for the removal of heavy metals from aqueous solutions. Water Res 35:3551–3556

AOAC (1990) Official methods of analysis, 15th edn, Association of Official Analytical Chemists, Washington DC, 1990

Bangyekan C, Aht-Ong D, Srikulkit K (2006) Preparation and properties evaluation of chitosan-coated cassava starch films. Carbohyd Polym 63(1):61–71

Benhabiles OMS, Salah R, Lounici H, Drouiche N, Goosen MFA, Mameri N (2012) Antibacterial activity of chitin, chitosan and its oligomers prepared from shrimp shell waste. Food Hydrocolloid 29(1):48–56

Black CA (1965) methods of soil analysis: part I physical and mineralogical properties. American Society of agronomy. Madison, Wisconsin, pp 671–698

Brzeski MM (1982) Concept of chitin/chitosan isolation from Antarctic Krill (*Euphausia superba*) shells on a technique scale. In: Hirano S, Tokura S (eds) Proceedings of the 2nd international conference on chitin and chitosan. The Japan Society of Chitin and Chitosan, Sapporo, Japan, pp 15–29

Burkatovskaya M, Tegos GP, Swietlik E, Demidova TN, Castano AP, Hamblin MR (2006) Use of chitosan bandage to prevent fatal infections developing from highly contaminated wounds in mice. Biomaterials 27:4157–4164

Choi BK, Kim KY, Yoo YJ, Oh SJ, Choi JH, Kim CY (2001) In vitro antimicrobial activity of a chito oligosaccharide mixture against *Actinobacillus, actinomycete mcomitans* and *Streptococcus mutans*. Int J Antimicrob Agents 18:553–557

Dutta PK, Dutta J, Tripathi VS (2004) Chitin and chitosan: chemistry, property and application. J sci Ind Res 63:20–31

Ebru G, Dilay P, Cuney H, Unlu OA, Nurfer G (2007) Synthesis and characterization of chtiosan—MMT bio composite systems. Carbohyd Polym 67:358

Elinder CG, Piscator M (1979) Zinc. In: Friberg L, Norberg GEF, Vouk B (eds) Handbook on the toxicology of metals. Elsevier, North Holland Biomedical Press, Amsterdam, pp 675–680

Gamage A, Shahidi F (2007) Use of chitosan for the removal of metal ion contaminants and proteins from water. J food chemistry 104:989–996

Guerra-Sánchez MG, Vega-Pérez J, Velázquez-del Valle MG, Hernández-Lauzardo AN (2009) Antifungal activity and release of compounds on Rhizopusstolonifer (Ehrenb.:Fr.) Vuill. by effect of chitosan with different molecular weights. Pestic Biochem Phys 93(1):18–22

Guibal E (2004) Interactions of metal ions with chitosan-based sorbents: a review. Sep Purif Technol 38:43–74

Guo M, Diao P, Cai S (2005) Hydrothermal growth of well aligned ZnO nanorodarrays: dependence of morphology and alignment ordering upon preparing conditions. J Solid State Chem 178(6): 1864–1873

Hossein T, Mehran M, Seyed MRR, Amir ME, Farnood SSJ (2008) Preparation of chitosan from brine shrimp (Artemiaurmiana) cyst shells and effects of different chemical processing sequences on the physicochemical and functional properties of the product. Molecules 13:1263–1274

Islam MM, Masum SM, Rahman MM, Ashraful M, Molla I, Shaikh AA, Roy SK (2011) Preparation of chitosan from shrimp shell and investigation of its properties. Int J Basic Appl Sci 11(1): 116–130

Jayakumar R, Prabaharan M, Sudheesh Kumar PT, Nair SV, Tamura H (2011) Biomaterials based on chitin and chitosan in wound dressing applications. Biotechnol Adv 29:322–337

Julkapli MN, Akil MH (2007) X-ray powder diffraction (XRD) studies on kenaf dust filled chitosan bio-composites. AIP Conf Proc 989:11

Khan T, Peh K, Ch'ng HS (2002) Reporting degree of deacetylation values of chitosan: the influence of analytical methods. J Pharm Pharmaceut Sci 5(3):205–212

Knorr D (1984) Use of chitinous polymers in food—a challenge for food research and development. Food Technol 38(1):85–97

Knorr D (1991) Recovery and utilization of chitin and chitosan in food processing waste management. Food Technol 45:114–122

Kurita K (2001) Controlled functionalization of the polysaccharide chitin. Prog Polym Sci 269:1921–1971

Li Q, Dunn ET, Grandmaison EW, Goosen MFA (1992) Applications and properties of chitosan. J Bioactive Compatible Polym 7:370–397

Marguerite R (2006) Chitin and chitosan: properties and applications. Prog Polym Sci 31(7):603–632

Martino AD, Sittinger M, Risbud MV (2005) Chitosan: a versatile biopolymer for orthopedic tissue engineering. Biomaterials 26:5983–5990

Mohammed MH, Williams PA, Tverezovskaya O (2012) Extraction of chitin from prawn shells and conversion to low molecular mass chitosan. Food Hydrocolloid 31:166–171

Muzzzarelli RAA, Weckx M, Filippini O (1989) Removal of trace metal ions from industrial waters, nuclear effluents and drinking water, with the aid of cross-linked N-carboxymethyl chitosan. Carbohyd Polym 11:293–306

No HK, Lee MY (1995) Isolation of chitin from crab shell waste. J Korean Soc Food Nutr 24:105–113

No HK, Meyers SP (1989) Crawfish chitosan as a coagulant in recovery of organic compounds from seafood processing streams. J Agric Food Chem 37(3):580–583

Radhakumary C, Divya G, Nair PD, Mathew S, Reghunadhan Nair CP (2003) Graft copolymerization of 2-hydroxy ethyl methacrylate onto chitosan with cerium(iv) ion i synthesis and characterization. J Macromol Sci A 40:715–730

Rathke TD, Hudson SM (1994) Review of chitin and chitosan as fiber and film formers. Macromolecular Science Rev Macromol Chem Phys 34:375–437

Rhazi M, Desbrières J, Tolaimate A, Rinaudo M, Vottero P, Alagui A (2002) Contribution to the study of the complexation of copper by chitosan and oligomers. Polymer 43:1267–1276

Rout SK (2001) Physicochemical, functional, and spectroscopic analysis of crawfish chitin and chitosan as affected by process modification. Dissertation, Louisiana State University, Baton Rouge, LA

Sandford PA (1992) High purity chitosan and alginate: preparation, analysis, and applications. Front Carbohydr Res 2:250–269

Struszczyk H (1987) Preparation and properties of microcrystalline chitosan. J Appl Polym Sci 33:177–189

Wang JC, Kinsella JE (1976) Functional properties of novel proteins: alfalfa leaf protein. J Food Sci 41:286–292

Zhang AJ, Qin QL, Zhang H, Wang HT, LiXuan ML, Wu YJ (2011) Preparation and characterization of food grade chitosan from housefly larvae. Czech J Food Sci 29(6):616–623

Zhong Z, Chen R, Xing R (2007) Synthesis and antifungal properties of sulfanilamide derivatives of chitosan. Carbohydr Res 342:2390–2395

Ziani K, Fern'andez-Pan I, Royo M, Mate JI (2009) Antifungal activity of films and solutions based on chitosan against typical seed fungi. Food Hydrocoll 23:2309–2314

21

Using small molecules as a new challenge to redirect metabolic pathway

Dina Morshedi · Farhang Aliakbari ·
Hamid Reza Nouri · Majid Lotfinia ·
Jafar Fallahi

Abstract The presence of acetate in the bacterial medium leads to a reduction in the growth rate of cells and recombinant protein production. In this study, three compounds including propionic acid, lithium chloride and butyric acid were added to the medium which decreased acetate levels and enhanced recombinant protein production (alpha-synuclein). In fact, propionic acid and lithium chloride are both known as acetate kinase inhibitors. The results obtained in the case of butyric acid were similar to those of the two other compounds indicating that butyric acid may act through a mechanism similar to propionic acid and lithium chloride. Consequently, it was shown that the presence of each of these supplements (5–200 µM) increased recombinant alpha-synuclein production and cell density by approximately 10–15 %. HPLC analysis showed that the levels of acetate in the media containing the supplements were considerably less than those of the control. Furthermore, pH values remained almost constant in the supplemented cultures. Growing the bacteria at lower temperatures (25 °C) indicated that the positive effects of these supplements were not as effective as at higher temperatures (37 °C), presumably due to the adequate balance between oxygen and carbon consumption. This study can confirm the viewpoint regarding the harmful effects of acetate on the recombinant protein production and cell density. Besides, such methods represent easy and complementary ways to increase target recombinant protein production without negatively affecting host cell density, and requiring complex genetic manipulation.

Keywords Acetate · *Escherichia coli* · Butyric acid · Lithium chloride · Propionic acid · Recombinant protein

D. Morshedi (✉) · F. Aliakbari · H. R. Nouri · J. Fallahi
Department of Industrial and Environmental Biotechnology,
National Institute of Genetic Engineering and Biotechnology,
Shahrak-e Pajoohesh, km 15, Tehran-Karaj Highway,
P. O. Box: 14965/161, Tehran, Iran
e-mail: morshedi@nigeb.ac.ir; morshedidina@yahoo.com

F. Aliakbari
Department of Biotechnology, Semnan University of Medical
Sciences, Semnan, Iran

M. Lotfinia
Department of Stem Cells and Developmental Biology, Cell
Science Research Center, Royan Institute for Stem Cell Biology
and Technology, ACECR, Tehran, Iran

M. Lotfinia
Department of Biochemistry, Pasteur Institute of Iran, Tehran,
Iran

Introduction

A number of bacteria, particularly *Escherichia coli*, are common hosts for the expression of a wide variety of recombinant proteins associated with therapeutic, diagnostic and industrial applications. However, in the case of *E. coli*, one of the problems during its growth is the reduction and sometimes elimination of recombinant protein production. One of the main reasons for these unfavorable outcomes is the generation of acetic acid as a harmful by-product (Eiteman and Altman 2006; Pflug et al. 2007). During glucose consumption, bacteria release acetate into the medium. It is believed that acetate can have different undesirable effects on bacterial growth and productivity. A rising amount of acetate in the medium causes inhibition of cell growth (Jin et al. 2012; Luli and Strohl

1990; Roe 2002). In fact, the presence of acetate has negative effects on recombinant protein production (Jensen and Carlsen 1990), and by making the environment more acidic might influence bacterial growth (Desvaux 2006; Richmond et al. 2012). In addition, it has been shown that acetate causes the plasmid copy number in the host to drop off considerably (Cunningham et al. 2009; Pan et al. 2010). There has been much interest in how acetate accumulation can have broad negative effects, such as reduction in the pH of the culture medium. Another effect is the deficiency of certain essential amino acids like methionine (Roe 2002) or nucleic acid sources (Cunningham et al. 2009; Pan et al. 2010).

There are currently many efforts being attempted to block the formation of acetate and its release into the media. Industrial strategies tend toward reducing acetate production by the modification of external or internal parameters connected to acetate production. Some of these methods have achieved reduction in acetate production and subsequent increase in recombinant protein production (Aristidou and San 1995; De Anda et al. 2006; Vemuri et al. 2006). The external parameters known as "process controlling" include medium modification, limitation of glucose consumption as well as aeration. For example, controlling glucose consumption rate by complex glucose feeding schemes has successfully reduced acetate accumulation in *E. coli* cultures (Akesson et al. 2001; Phue et al. 2005; Shiloach et al. 1996).

The internal genotype of the host cell can also be altered (De Mey et al. 2007; Papagianni 2012). Some of these approaches include engineering strains to modify the glucose uptake rate (glucose phosphotransferase system *ptsG*) (De Anda et al. 2006; Knabben et al. 2011), redirecting the carbon flux toward less inhibitory by-products (e.g., acetoin by acetolactate synthase *als*) (Aristidou and San 1995), ethanol production through the pet operon (Diaz-Ricci et al. 1992; Ingram and Conway 1988), storage of excess carbon as glycogen (Dedhia et al. 1994) and elimination of the major acetate formation pathway (acetate kinase *ackA*, phosphotransacetylase *pta*) (De Mey et al. 2007; Phue et al. 2010).

The objective of this study was to modify the medium content in a simple and moderate way to improve recombinant protein (alpha-synuclein) production without negative effects on cell density. Accordingly, propionic acid, butyric acid and lithium chloride were added to the culture media to observe their impact on recombinant protein production and acetate reduction. In fact, propionic acid and lithium chloride are known to act as acetate kinase inhibitors (Fox and Roseman 1986). And, on the other hand, it was assumed that butyric acid might act as an acetic acid analogue and thus affect acetate production. We just focused on recombinant protein production and acetate

reduction by these compounds. Consequently, the effects of the above mentioned compounds on the amount of acetate released into the medium, cell density, pH of the culture medium and also recombinant protein production were examined. Alpha-synuclein was expressed as a recombinant protein to examine the quality of the expression. This protein is known to be associated with neurodegenerative diseases. It has been produced in vitro for several years to study the mechanism of amyloid formation, and screen fibril inhibitors, or as a biomaterial in nanobiotechnology. Alpha-synuclein can enter into the preplasmic space of *E. coli* without the host specific signal peptides (Huang et al. 2005). We explored the negative effects of acetate on this preplasmic protein in the presence or absence of the mentioned compounds.

Materials and methods

Biochemicals and reagents

Most of the biochemicals and reagents used in this study were obtained from Merck (Germany) and were of analytical grade or higher. Isopropyl β-D-thiogalactopyranoside (IPTG), kanamycin and chloramphenicol were purchased from Sigma-Aldrich (USA).

Plasmid and strain

The expression plasmid pNIC28-Bsa4 (7,284 bp) containing the human α-synuclein cDNA and a kanamycin resistance gene carrying the T7-lacO promoter were transformed into *E. coli* BL21 (DE3)-pLysS.

Expression of recombinant α-synuclein

The transformed cells were screened on Luria–Bertani Agar (LB) medium containing kanamycin (50 μg/mL) and chloramphenicol (34 μg/mL). An overnight culture derived from a single colony of the transformed *E. coli* was prepared in LB broth containing the same concentrations of the antibiotics. Terrific broth (TB) was used for the expression of the protein, and was based upon tryptone (12 g/L), yeast extract (24 g/L), KH_2PO_4 (2.3 g/L), K_2HPO_4 (12.5 g/L), and glucose (10 g/L) instead of glycerol. Subcultures at a dilution of 1/50 from an overnight starter were prepared and incubated at temperatures of 37 and 25 °C with shaking at 200 rpm. When the optical density (OD_{600}) approached 0.6–0.7, protein expression was induced using isopropyl β-D-thiogalactopyranoside (IPTG) at the final concentration of 500 μM, and the cells were then grown overnight (approximately 15–16 h), under the same conditions mentioned above. Sampling was carried out 7 h after induction and overnight cultivation. All samples were centrifuged at 5,000g for 10 min. The pellets were then used for the analyses of recombinant protein production.

Addition of supplements

Propionic acid, at different concentrations of 5, 10, 20 and 200 μM, was added to the individual cultures at the two stages of inoculation and induction. Lithium chloride and butyric acid were also added to individual cultures at the same mentioned stages. TB medium with no supplementation of propionic acid, lithium chloride and butyric acid was used as control.

Cell density measurement

To measure cell density, the optical density of culture samples was recorded at 600 nm (OD$_{600}$) using a Beckman DU 500, UV–visible spectrophotometer (USA). Samples were diluted with TB medium to enable photometric measurement in the linear range between 0.1 and 0.5 OD. One unit of OD$_{600}$ corresponds to a dry cell weight of 0.34 g/L.

Estimation of protein concentrations

Protein expression was examined by sodium dodecyl sulfate polyacrylamide gel electrophoresis (SDS-PAGE). Protein samples were loaded onto a 12 % SDS–polyacrylamide gel and the percentage of recombinant alpha-synuclein production was analyzed by the AlphaEase FC image analysis software, version 6.0.0. Western blotting was carried out with the alpha-synuclein monoclonal antibody (Amersham Biosciences, UK).

Analysis of acetate using high-performance liquid chromatography

Acetate accumulation in the culture media was determined by high-performance liquid chromatography (HPLC) using a Cecil 4200 HPLC system (CE 4200 England) equipped with an ODS-3 C18 column (MZ, Germany). The mobile phase consisted of acetonitrile—ddH$_2$O (9/1) plus 0.1 % H$_3$PO$_4$, and a flow rate of 0.5 mL/min was applied without any gradient application. An injection volume of 20 μL was applied and detection of acetate was carried out at a wavelength of 215 nm using a UV–visible detector. A standard curve was prepared using different concentrations of acetate (0–20 μg in 20 μL) that were diluted in the same mobile phase as shown in Supplementary Fig. 1A. The retention times of propionic acid and butyric acid were less than that of acetic acid, as shown in Supplementary Fig. 1B.

Assessment of pH

The pH values of samples from overnight cultivations in the presence and absence of propionic acid were analyzed by the Beckman Phi 72 pH meter.

Statistical analysis

All experiments were repeated three times. The data obtained were analyzed for significant differences between the control and experimental groups using the SPSS software version 16.0, involving the unpaired Student's t test. $P_{value} <0.05$ was considered as significant.

Results

Effect of propionic acid on recombinant protein production and cell density

As mentioned previously, cell density and protein production in *E. coli* could be amplified by reducing the accumulation of acetate in the bacterial culture medium (Fox and Roseman 1986). Accordingly, this study investigated whether adding the aforementioned compounds to the culture medium could reduce acetate accumulation and subsequently enhance recombinant protein production. Initially, the effects of propionic acid as an acetic acid analogue and acetate kinase competitive inhibitor were investigated (Fox and Roseman 1986). The effect of propionic acid on recombinant protein production was verified using shake flask cultivations. Figure 1 shows the polyacrylamide gel electrophoresis patterns of protein in the presence of different concentrations of propionic acid (5, 10, 20 and 200 μM), after 7 h of growth (a) and overnight (b) cultivation. Cultures were supplemented with propionic acid in two steps (inoculation and induction times). Western blotting also confirmed that the protein was alpha-synuclein, as indicated in Supplementary Fig. 2A. Supplementary Fig. 3 which is derived from Fig. 1 indicates the percentage of recombinant protein production as estimated by the AlphaEase FC image analysis software. The production of alpha-synuclein in the presence of propionic acid increased by up to 9 % after 7 h of growth and 12 % following overnight cultivation, when compared to that of the control at 37 °C. Cell density also rose in the presence of propionic acid. The growth rate and the biomass of the cultivation were determined using optical density (OD) measurements at 600 nm. As shown in Fig. 2a, an increase in OD was observed for the samples treated with 200 μM of propionic acid, when compared to that of the control, especially regarding the overnight cultivation (approximately 15–16 h). Figure 2b shows the effects of different

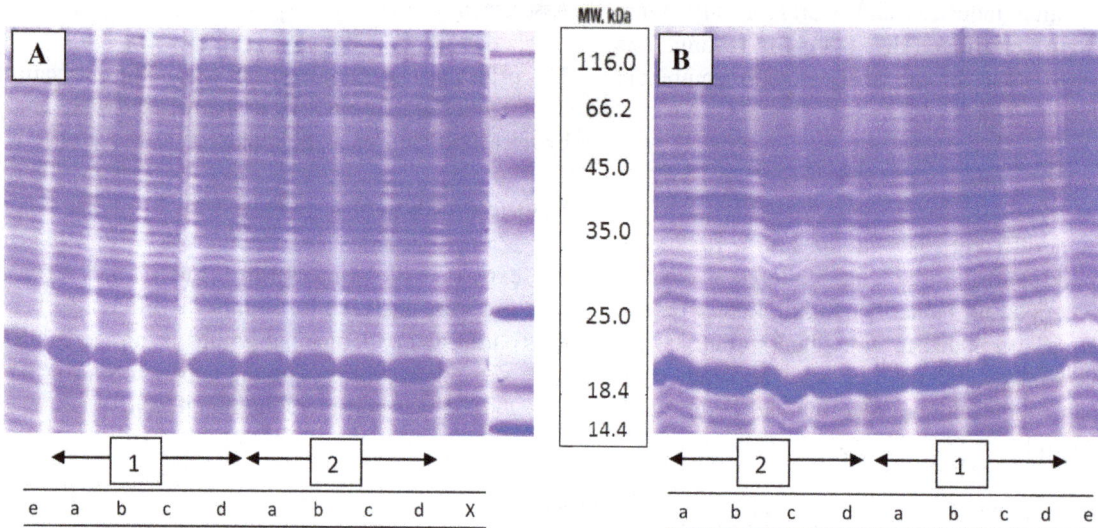

Fig. 1 SDS-PAGE pattern of total proteins. SDS-PAGE pattern in the presence of propionic acid at different concentrations (*a* 5 μM, *b* 10 μM, *c* 20 μM and *d* 200 μM), as compared to the control

(*e*) which were added at inoculation (*1*) and (*2*) induction times, 7 h of incubation (**a**) and overnight cultivations (**b**). *x* before induction

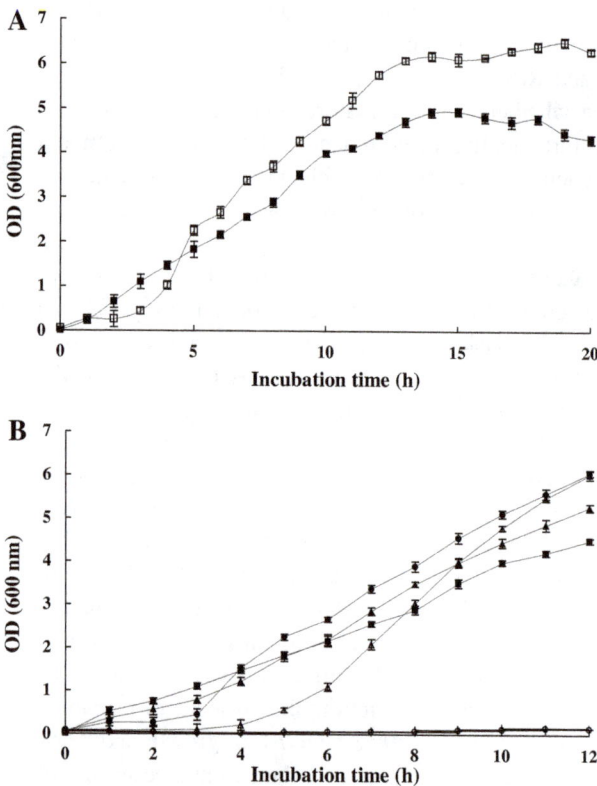

Fig. 2 (**a**) Growth rate in the absence (*filled square*) and the presence of 200 mM propionic acid (*unfilled square*). (**b**) Growth rate in the presence of different concentrations of propionic acid that include: 0 μM (*filled square*), 20 μM (*filled triangle*), 200 μM (*filled circle*), 500 μM (*unfilled triangle*), and 4 mM (*unfilled circle*). Propionic acid was added at the time of inoculation. The samples were grown at 37 °C in shaking flasks. The stated errors are the SDs of three repeats

concentrations of propionic acid on the growth curve. It seems that the presence of propionic acid at the different concentrations has significant effects on both the cell density and protein production. In fact, by increasing the amount of propionic acid, a longer lag phase is observed. Furthermore, in the presence of high levels of propionic acid, cell growth is completely suppressed.

Effects of propionic acid on acetate production

Factors, such as changing the balance between glucose metabolism and oxygen consumption, cause an increase in acetate accumulation during cultivation. Considering that propionic acid enhanced cell density and protein production, it was thus assumed that this carboxylic acid may have an impact on acetate accumulation. Therefore, using HPLC, as mentioned briefly in the materials and methods, acetate accumulation was monitored during cultivation in the presence of propionic acid at the concentration of 200 μM. Figure 3 demonstrates the kinetics of acetate production in the presence of 200 μM propionic acid. In culture media supplemented with 200 μM propionic acid, acetate production decreased by 27.5 and 49.5 % following 7 h of growth and overnight (approximately 15 h) cultivation, respectively. According to the kinetics of acetate production in the presence of 200 μM propionic acid, a similar effect was also expected in other concentrations, after 7 and 15 h of cultivation (Supplementary Table 1) Fig. 3 and Supplementary Table 1 show that acetate production increased in the control medium throughout

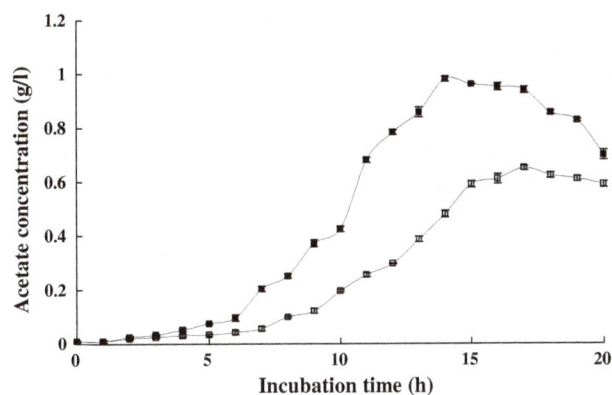

Fig. 3 Acetate concentration in a 1 L medium in the absence (*filled square*) and the presence (*unfilled square*) of 200 μM propionic acid. Propionic acid was added at the time of inoculation. The stated errors are the SDs of three repeats

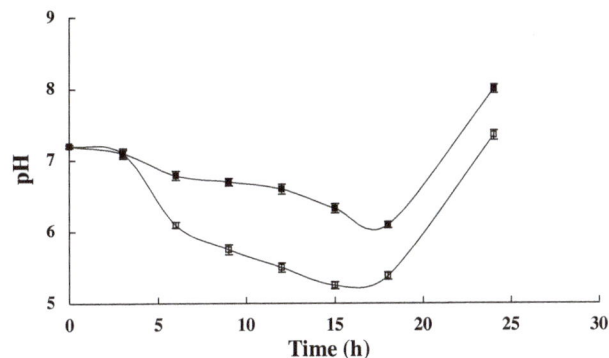

Fig. 4 Fluctuations in pH in the presence (*dark filled*) and the absence of propionic acid (*white filled*). 200 μM propionic acid was added at the time of inoculation. The stated errors are the SDs of three repeats

cultivation, whereas the addition of propionic acid to the medium had negative effects on acetate accumulation.

Effects of propionic acid on medium acidification

Release of acetate into the medium causes a change in the pH of the medium, therefore, in this study, the pH values of the culture media in the presence and absence of propionic acid were analyzed throughout growth. Results showed that pH of the culture medium supplemented with this compound remained almost constant throughout growth when compared to that of the control. As expected, the pH values in the control medium decreased significantly (Fig. 4), thus confirming the release and accumulation of acetate in the medium.

Effects of butyric acid and lithium chloride on recombinant protein production, cell density, acetate production and medium acidification

Other compounds that may have similar effects to that of propionic acid were also investigated. One of the candidates was butyric acid which was considered as an analogue of acetic acid and assumed to interrupt acetate release, thus paving the way for high production of recombinant protein. Despite previous studies that demonstrated butyric acid could not function as an analogue during acetate metabolism in *E. coli* (Fox and Roseman 1986), this investigation revealed that it can have the same effect as propionic acid on the acetate pathway in *E. coli*. Another candidate was lithium chloride which has been known to influence biological systems in different ways, including inhibitory effects on the enzyme acetate kinase (Fox and Roseman 1986). This research showed that both lithium chloride and butyric acid could also have positive effects, similar to that of propionic acid, regarding

recombinant protein production. The effects of two concentrations of butyric acid and lithium chloride on protein production are demonstrated in Supplementary Figure 4. Similar to propionic acid, these two compounds were also added to the culture medium in two steps during growth. The percentage of recombinant protein production was enhanced by up to 8–12 % after 7 h of growth and 8.6–12.9 % following overnight cultivation (Supplementary Fig. 4). Supplementary Figures S2B and C show the results of the Western blotting procedure which confirmed that the protein was alpha-synuclein. Figure 5 demonstrates changes in OD (600 nm) and acetate concentration of the cultures in the presence and absence of lithium chloride and butyric acid (20 μM at inoculation time). Results show that cell density had almost increased in the presence of lithium chloride and butyric acid and acetate production had decreased.

As shown in Fig. 5a, an increase in the OD of samples treated with lithium chloride and butyric acid was observed relative to the control; especially regarding the cultivation of cultures involving approximately 13–16 h of incubation. In fact, cell density was increased by up to 21.16 and 23.8 %, in the presence of lithium chloride and butyric acid, respectively. Furthermore, acetate accumulation was measured during growth in the presence of the above-mentioned supplements. Figure 5b shows that acetate production was reduced following the addition of lithium chloride and butyric acid, when compared to that of the control. Acetate accumulation was decreased up to 40.98 and 48.36 %, in the presence of lithium chloride and butyric acid, respectively. HPLC analysis showed that in the presence of these supplements (either added at the time of inoculation or induction) there was a decrease in acetate accumulation. In the presence of lithium chloride and butyric acid, the amount of acetate released into the medium decreased to 46.45 and 32.59 %, respectively, in approximately 7–13 h. Furthermore, the addition of these

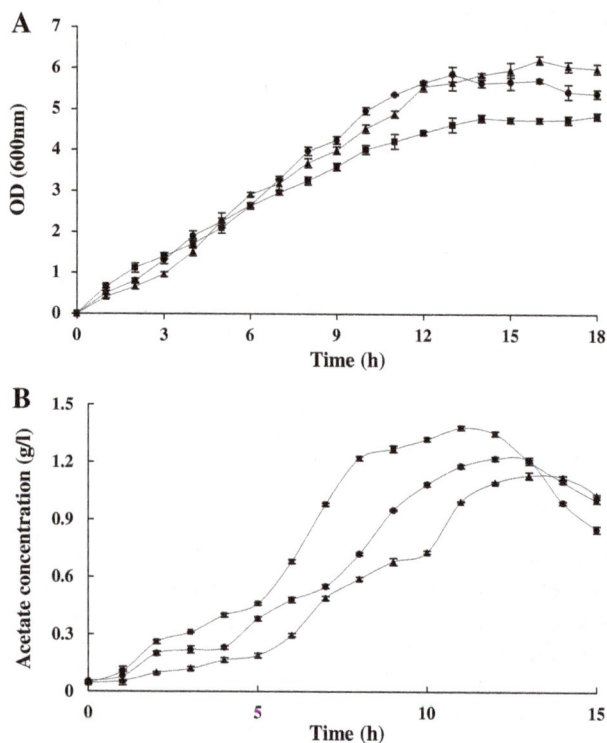

Fig. 5 Changes in OD_{600} (**a**) and acetate concentration (**b**) of the cultures in the absence (*filled square*) and the presence of lithium chloride (*filled circle*) and butyric acid (*filled triangle*). 20 µM of lithium chloride and butyric acid were added at the time of inoculation. The stated errors are the SDs of three repeats

two compounds to the culture media did not change the pH values of the media as much as those observed in the control medium during cultivation (data not shown).

In fact, propionic acid and lithium chloride are known to act as acetate kinase inhibitors (Fox and Roseman 1986). And, on the other hand, it was assumed that butyric acid might act as an acetic acid analogue and thus affect acetate production.

As stated above, propionic acid and lithium chloride are identified to act as acetate kinase inhibitors (Fox and Roseman 1986) and their effects on decrease of acetate levels and enhancement of protein production almost certainly due to their functions as acetate kinase inhibitors, and butyric acid might have the similar role on the acetate accumulation. In this regard, more investigation was carried out concerning the effects of butyric acid on the cell density and recombinant protein production including the addition of more concentrations of such compound to the culture medium and provided more evidences to support the hypothesis. Supplementary Figure 5 showed the SDS-PAGE pattern of the total proteins in presence of different concentrations of butyric acid (10, 50, 200, 400 µM). Supplementary Figure 6 depicts the growth rate of bacteria in the absence and the presence of different concentrations of butyric acid (200, 400, 4,000, 10,000 µM) compared to

that of control. As elucidated from this Figure, in spite of propionic acid, the higher concentrations of butyric acid (4 and 10 mM) did not have negative effects on growth rate, however, it inspired the lag phase to become longer. In this concern, it is worth mentioning that butyric acid smells terrible and it is hard to work with it. In addition it is not necessary to use the higher amount of such compound. Furthermore, lithium chloride is involved with several pathways in bacteria, so we concluded that propionic acid was a better compound and effortless to work. Therefore, the focus of this study is more on the propionic acid and the results for two other compounds support the idea that such compounds can decrease acetate accumulation and enhance recombinant protein production.

Effect of incubation conditions

Changing the incubation condition, such as temperature or speed of agitation during cultivation can highly affect bacterial metabolism and subsequently the growth rate and recombinant protein production. For this reason, the effects of the considered additives on acetate accumulation, recombinant protein production, pH and cell density were examined at a different temperature; 25 °C (Table 1; Supplementary Fig. 7) and agitation speed; 230 rpm after 7 h and overnight cultivations. It should be noted that the effect of temperature was investigated at a supplement concentration of 20 µM, which was added at the time of inoculation. Data obtained when using the other concentrations were found to be approximately the same (data not shown). The results indicated that cultivation of the supplemented cultures at 25 °C did not have any considerable positive effects on cell density and recombinant protein production, when compared to those obtained at 37 °C after 7 h of growth. However, following overnight cultivation at 25 °C, increases in recombinant protein production were observed. Moreover, the amount of acetate accumulation in the control medium was also found to be higher than in the supplemented samples at 25 °C (Table 1).

Changing the agitation speed can have an impact on growth rate by influencing oxygen transfer through the medium (Henzler and Schedel 1991). Figure 6 depicts an increase in the OD of the cultures at higher agitation speeds and in this case the propionic acid effect is obviously highlighted. In the sample treated with 200 µM propionic acid, acetate production was decreased by 70 % and growth rate was increased by up to 27.9 %, relative to the control.

Discussion

Propionic acid and lithium chloride have previously been identified as the inhibitors of enzymes involved in the

Table 1 Effects of propionic acid, butyric acid and lithium chloride on recombinant protein production, cell density, medium acidification and acetate production at 25 °C cultivations

	Protein production (%)		OD$_{600}$		pH variation (%)		Relative acetate reduction (%)	
	7 h	ON	7 h	ON	7 h	ON	7 h	ON
Control	22.1 ± 0.25	21.6 ± 0.4	4.35 ± 0.04	4.73 ± 0.06	6.22 ± 0.03	5.78 ± 0.02		
Prop 20 μM	23.7 ± 0.1	25.2 ± 0.2	4.21 ± 0.02	4.53 ± 0.05	6.45 ± 0.04	6.17 ± 0.06	8.025 ± 0.16	11.12 ± 0.15
Bt 20 μM	23.6 ± 0.21	26.7 ± 0.26	4.09 ± 0.09	4.54 ± 0.05	6.41 ± 0.01	6.08 ± 0.07	8.24 ± 0.18	13.12 ± 0.12
Li 20 μM	22.6 ± 0.17	24.4 ± 0.23	4.41 ± 0.04	4.68 ± 0.09	6.38 ± 0.02	5.96 ± 0.05	8.02 ± 0.17	12.6 ± 0.2

All results were significantly different from the control ($P < 0.05$)

h hours, *ON* overnight, *Prop* propionic acid, *Bt* butyric acid; *Li* lithium chloride

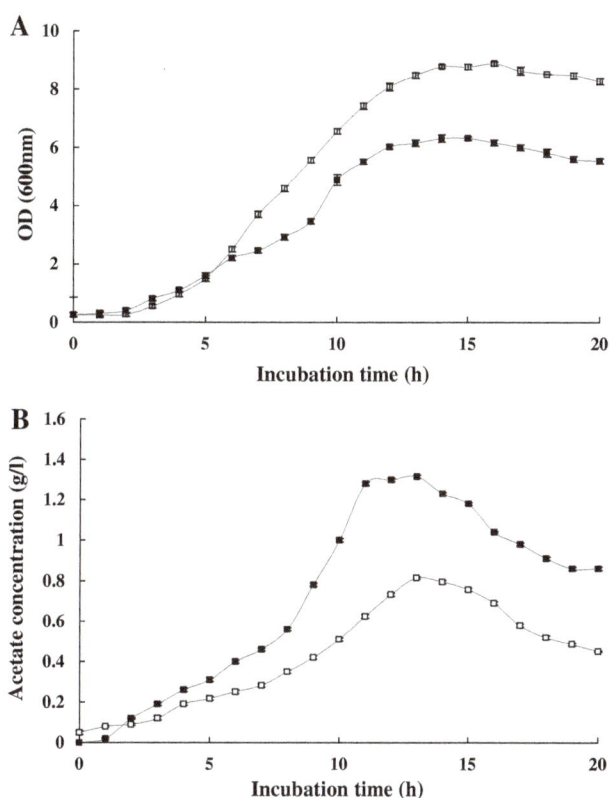

Fig. 6 Evaluating the influence of higher cell density on the effects of propionic: (**a**) OD and (**b**) the acetate concentration of the media in the absence (*filled square*) and the presence (*unfilled square*) of 200 μM propionic acid. Propionic acid was added at the time of inoculation. The agitation speed was 230 rpm. The stated errors are the SDs of three repeats

conversion of pyruvate to acetate (Supplementary Fig. 8) (Yang et al. 1999). There are contradictions regarding butyric acid as inhibitor of such enzymes (Fox and Roseman 1986). It seems that they would be able to decrease acetate accumulation. The results of this study have also confirmed this observation.

The addition of propionic acid to the medium at concentrations ranging from 5 to 200 μM had positive effects on bacterial cell density and recombinant protein production, leading to increases of 10 and 8 %, respectively, in an overnight cultivation (approximately 13–16 h). However, this elevation in recombinant protein production is rather less than the levels obtained when using other metabolic engineering methods (Pan et al. 2010). As the findings of this study showed, the addition of propionic acid either at the time of inoculation or before induction led to similar effects and so its efficiency has been stable. The induction time was nearly 1 h after inoculation time and the results for both times were comparable. As the results of both times were equivalent, it could be possible to use such supplements during the preparation of the medium. Alteration in the culture medium allowed an increase in cell density and recombinant protein production after 7 h of incubation, and even more significant results were obtained following overnight cultivation. In the control samples, however, the percentage of heterologous proteins against total proteins reduced dramatically after overnight incubation. There is an assumption that long-time incubation may lead to destabilization of plasmids, because of lower levels of carbon being available for nucleic acid and amino acid precursors (Cunningham et al. 2009). However, by inhibiting consumption of the carbon supply in unwanted products like acetate, the stability of plasmid becomes presumably enhanced (Cunningham et al. 2009). In this study, the production of the recombinant protein in the control samples was dramatically decreased. Nevertheless, protein production in supplemented cultures remained considerably stable during long-term cultivations. Study of the kinetics of acetate production revealed that there is a significant reduction in acetate accumulation in the media of the treated cultures, and this was highlighted approximately during 11–18 h of cultivation.

Furthermore, butyric acid was found to have effects similar to those of propionic acid and lithium chloride, with regard to cell density and recombinant protein production. The present results suggest that these compounds may act in a similar way on acetate accumulation and consequently recombinant protein production. The HPLC data showed that in their presence, the amount of acetate decreased remarkably, confirming this hypothesis.

Reductions in the pH of the culture medium as well as in the cytoplasmic space have been considered as some of the factors responsible for the negative effects of acetate (Desvaux 2006; Richmond et al. 2012). Protons can interact with membranes or diffuse across membrane bilayers and induce anion accumulation (Booth 1985; Stratford and Anslow 1998). The presence of propionic acid, lithium chloride or butyric acid showed that addition of the supplements protected culture media from extraordinary fluctuations in pH value.

Acetate formation is the result of a disturbed balance between oxygen consumption and aerobic glucose metabolism, so the excess carbon flux from glucose or other compounds leads to the repression of the TCA cycle enzymes, and the promotion of uncoupled metabolisms (El-Mansi 1989; Majewski and Domach 1990). The production of acetate represents a loss of carbon flux to cell growth as well as a loss of recombinant protein production. It has been demonstrated that the metabolic engineering methods which cause the inhibition of acetate production, have extremely positive effects on plasmid stability and recombinant protein production (Cunningham et al. 2009; Pan et al. 2010).

On the other hand, some results from the metabolic analysis of carbon redistribution have shown that the ackA–pta mutation or antisense RNA systems reduce acetate levels at the expense of cell density (Kim and Cha 2003; Yang et al. 1999). However, in the experiments of this study, cell densities did not reduce and also fairy improved. In addition, in the ackA–pta deficient strain, a much higher rate of lactate formation, and simultaneously, lower rates of formate and ethanol synthesis have been observed (Kim and Cha 2003; Knabben et al. 2011). In other studies, researchers have succeeded in increasing cell biomass dramatically by inducing glycogen formation from excess carbon via using certain other mutations or transforming cells with specific plasmids (Dedhia et al. 1994).

When E. coli cells are growing under anaerobic conditions, sugars are fermented to a variety of products (Aristidou and San 1995; Clark and Cronan 1980; Ni et al. 2011). These pathways are dominant in the bacteria which use carbohydrates as sources of energy (Desvaux 2006; Zhang et al. 2012). One important pathway involves the conversion of acetyl-CoA to acetaldehyde by means of acetyl-CoA dehydrogenase. The acetaldehyde is subsequently converted to ethanol by alcohol dehydrogenase. Adding propionic acid, butyric acid and lithium chloride perhaps reduces the carbon flux to acetate and to a smaller extent ethanol. Due to the aerobic process, the pathway shifts to convert the excess pyruvate into much less toxic compounds such as acetoin, lactate or TCA cycle-related

compounds (Aristidou and San 1995; Wolfe 2005; Yang et al. 1999).

Considering that these compounds might interfere with other enzymes or pathways in E. coli, further investigations are required to clarify their functions.

As mentioned previously, when carbon metabolism changes due to a sudden burst of cell growth, the conditions of growth change, leading to a situation similar to anaerobiosis and induction of anaerobic pathways, consequently lead to increased production of acetate (Hasona et al. 2004). The high rate of glucose consumption increases the uncoupled pathways instead of the TCA cycle because of a limited amount of intermediate compounds, such as NADH and oxaloacetate (Britten 1954; Majewski and Domach 1990). The relationship between carbohydrate consumption and acetate production is common in the bioreactor-used microorganisms such as Clostridium phytofermentans, and acetate production can influence the manufacturing performance (Jin et al. 2012). Certain strategies have been applied to limit glucose consumption and control the balance between oxygen consumption and glucose catabolism, for example, using continuous or fed-batch fermentation (Kayser et al. 2005; Weber et al. 2005). Co-culturing with specific strains which can consume acetate as a source of energy has also shown positive effects regarding the yield of product (Zhang et al. 2013). Another way is to slow down the first phase of the bacterial growth rate by incubating cells at a lower temperature (25 °C); the growth rate would then be slower than at higher temperatures (37 °C), which can then help to balance the oxygen consumption and carbon catabolism in the aerobic pathways and the TCA cycle. Accordingly, from the results obtained for cultures grown at 37 and 25 °C in this study, it can be assumed that by slowing down the growth rate at the lower temperature, lower levels of acetate will thus accumulate in the medium. In fact, results showed that under such a condition, the addition of compounds which can inhibit acetate production did not have significant effects on bacterial growth and protein production, especially in the case of short incubation times. Conversely, in another part of this study, by increasing the agitation, propionic acid was found to be more effective regarding growth rate and protein production.

Conclusion

In conclusion, it seems that the use of propionic acid, lithium chloride and butyric acid, which are capable of diverting metabolic pathways to decrease acetate production, is a good strategy for optimizing bacterial media and growth conditions. Thus, using such compounds decrease the carbon flux toward acetate synthesis, resulting in enhancement of recombinant protein production (alpha-synuclein) without negatively affecting host cell density. It should be noted that this effect is significant when production of acetate is high. Therefore, this strategy would be useful especially in higher cell density cultures.

Acknowledgments This work was supported by Grant NIGEB-455 from the National Institute of Genetic Engineering and Biotechnology. The authors would like to acknowledge Dr. Parvin Shariati for kindly editing the English language.

References

Akesson M, Hagander P, Axelsson JP (2001) Avoiding acetate accumulation in *Escherichia coli* cultures using feedback control of glucose feeding. Biotechnol Bioeng 73:223–230

Aristidou A, San K (1995) Metabolic engineering of *Escherichia coli* to enhance recombinant protein production through acetate reduction. Biotech Prog 11:475–478

Booth IR (1985) Regulation of cytoplasmic pH in bacteria. Microbiol Rev 49:359–378

Britten A (1954) Extracellular metabolic products of *Escherichia coli* during rapid growth. Science 119:578

Clark DP, Cronan JE Jr (1980) Acetaldehyde coenzyme a dehydrogenase of *Escherichia coli*. J Bacteriol 144:179–184

Cunningham DS, Koepsel RR, Ataai MM, Domach MM (2009) Factors affecting plasmid production in *Escherichia coli* from a resource allocation standpoint. Microb Cell Fact 8:27

De Anda R, Lara AR, Hernandez V, Hernandez-Montalvo V, Gosset G, Bolivar F, Ramirez OT (2006) Replacement of the glucose phosphotransferase transport system by galactose permease reduces acetate accumulation and improves process performance of *Escherichia coli* for recombinant protein production without impairment of growth rate. Metab Eng 8:281–290

De Mey M, Lequeux GJ, Beauprez JJ, Maertens J, Van Horen E, Soetaert WK, Vanrolleghem PA, Vandamme EJ (2007) Comparison of different strategies to reduce acetate formation in *Escherichia coli*. Biotechnol Prog 23:1053–1063

Dedhia NN, Hottiger T, Bailey JE (1994) Overproduction of glycogen in *Escherichia coli* blocked in the acetate pathway improves cell growth. Biotechnol Bioeng 44:132–139

Desvaux M (2006) Unravelling carbon metabolism in anaerobic cellulolytic bacteria. Biotechnol Prog 22:1229–1238

Diaz-Ricci JC, Tsu M, Bailey JE (1992) Influence of expression of the pet operon on intracellular metabolic fluxes of *Escherichia coli*. Biotechnol Bioeng 39:59–65

Eiteman MA, Altman E (2006) Overcoming acetate in *Escherichia coli* recombinant protein fermentations. Trends Biotechnol 24:530–536

El-Mansi EM (1989) Control of carbon flux to acetate excretion during growth of *Escherichia coli* in batch and continuous cultures. J Gen Microbiol 135:2875–2883

Fox DK, Roseman S (1986) Isolation and characterization of homogeneous acetate kinase from *Salmonella typhimurium* and *Escherichia coli*. J Biol Chem 261:13487–13497

Hasona A, Kim Y, Healy FG, Ingram LO, Shanmugam KT (2004) Pyruvate formate lyase and acetate kinase are essential for anaerobic growth of *Escherichia coli* on xylose. J Bacteriol 186:7593–7600

Henzler H, Schedel M (1991) Suitability of the shaking flask for oxygen supply to microbiological cultures. Bioprocess Eng 7:123–131

Huang C, Ren G, Zhou H, Wang CC (2005) A new method for purification of recombinant human alpha-synuclein in *Escherichia coli*. Protein Expr Purif 42:173–177

Ingram LO, Conway T (1988) Expression of different levels of ethanologenic enzymes from *Zymomonas mobilis* in recombinant strains of *Escherichia coli*. Appl Environ Microbiol 54:397–404

Jensen EB, Carlsen S (1990) Production of recombinant human growth hormone in *Escherichia coli*: expression of different precursors and physiological effects of glucose, acetate, and salts. Biotechnol Bioeng 36:1–11

Jin M, Gunawan C, Balan V, Dale BE (2012) Consolidated bioprocessing (CBP) of AFEX-pretreated corn stover for ethanol production using *Clostridium phytofermentans* at a high solids loading. Biotechnol Bioeng 109:1929–1936

Kayser A, Weber J, Hecht V, Rinas U (2005) Metabolic flux analysis of *Escherichia coli* in glucose-limited continuous culture. I. Growth-rate-dependent metabolic efficiency at steady state. Microbiology 151:693–706

Kim JY, Cha HJ (2003) Down-regulation of acetate pathway through antisense strategy in *Escherichia coli*: improved foreign protein production. Biotechnol Bioeng 83:841–853

Knabben I, Regestein L, Schauf J, Steinbusch S, Buchs J (2011) Linear correlation between online capacitance and offline biomass measurement up to high cell densities in *Escherichia coli* fermentations in a pilot-scale pressurized bioreactor. J Microbiol Biotechnol 21:204–211

Luli GW, Strohl WR (1990) Comparison of growth, acetate production, and acetate inhibition of *Escherichia coli* strains in batch and fed-batch fermentations. Appl Environ Microbiol 56:1004–1011

Majewski RA, Domach MM (1990) Simple constrained-optimization view of acetate overflow in *E. coli*. Biotechnol Bioeng 35:732–738

Ni BJ, Liu H, Nie YQ, Zeng RJ, Du GC, Chen J, Yu HQ (2011) Coupling glucose fermentation and homoacetogenesis for elevated acetate production: experimental and mathematical approaches. Biotechnol Bioeng 108:345–353

Pan Z, Cunningham DS, Zhu T, Ye K, Koepsel RR, Domach MM, Ataai MM (2010) Enhanced recombinant protein production in pyruvate kinase mutant of *Bacillus subtilis*. Appl Microbiol Biotechnol 85:1769–1778

Papagianni M (2012) Recent advances in engineering the central carbon metabolism of industrially important bacteria. Microb Cell Fact 11:50

Pflug S, Richter SM, Urlacher VB (2007) Development of a fed-batch process for the production of the cytochrome P450 monooxygenase CYP102A1 from *Bacillus megaterium* in *E. coli*. J Biotechnol 129:481–488

Phue JN, Noronha SB, Hattacharyya R, Wolfe AJ, Shiloach J (2005) Glucose metabolism at high density growth of *E. coli* B and *E. coli* K: differences in metabolic pathways are responsible for efficient glucose utilization in *E. coli* B as determined by microarrays and northern blot analyses. Biotechnol Bioeng 90:805–820

Phue JN, Lee SJ, Kaufman JB, Negrete A, Shiloach J (2010) Acetate accumulation through alternative metabolic pathways in ackA (−) pta (−) poxB (−) triple mutant in *E. coli* B (BL21). Biotechnol Lett 32:1897–1903

Richmond C, Ujo V, Ezeji CT (2012) Impact of syringaldehyde on the growth of *Clostridium beijerinckii* NCIMB 8052 and butanol production. 3 Biotech 2:159–167

Roe AJ (2002) Inhibition of *Escherichia coli* growth by acetic acid: a problem with methionine biosynthesis and homocysteine toxicity. Microbiology 148:2215–2222

Shiloach J, Kaufman J, Guillard AS, Fass R (1996) Effect of glucose supply strategy on acetate accumulation, growth, and recombinant protein production by *Escherichia coli* BL21 (lambdaDE3) and *Escherichia coli* JM109. Biotechnol Bioeng 49:421–428

Stratford M, Anslow PA (1998) Evidence that sorbic acid does not inhibit yeast as a classic 'weak acid preservative'. Lett Appl Microbiol 27:203–206

Vemuri GN, Altman E, Sangurdekar DP, Khodursky AB, Eiteman MA (2006) Overflow metabolism in *Escherichia coli* during steady-state growth: transcriptional regulation and effect of the redox ratio. Appl Environ Microbiol 72:3653–3661

Weber J, Kayser A, Rinas U (2005) Metabolic flux analysis of *Escherichia coli* in glucose-limited continuous culture. II. Dynamic response to famine and feast, activation of the methylglyoxal pathway and oscillatory behaviour. Microbiology 151:707–716

Wolfe AJ (2005) The acetate switch. Microbiol Mol Biol Rev 69:12–50

Yang YT, Aristidou AA, San KY, Bennett GN (1999) Metabolic flux analysis of *Escherichia coli* deficient in the acetate production pathway and expressing the *Bacillus subtilis* acetolactate synthase. Metab Eng 1:26–34

Zhang Y, Yu M, Yang ST (2012) Effects of ptb knockout on butyric acid fermentation by *Clostridium tyrobutyricum*. Biotechnol Prog 28:52–59

Zhang X, Ye X, Finneran KT, Zilles JL, Morgenroth E (2013) Interactions between *Clostridium beijerinckii* and *Geobacter metallireducens* in co-culture fermentation with anthrahydroquinone-2, 6-disulfonate (AH2QDS) for enhanced biohydrogen production from xylose. Biotechnol Bioeng 110:164–172

22

Antimicrobial resistance (AMR) and plant-derived antimicrobials (PDA$_m$S) as an alternative drug line to control infections

Jatin Srivastava · Harish Chandra ·
Anant R. Nautiyal · Swinder J. S. Kalra

Abstract Infectious diseases caused by antimicrobial-resistant microbes (ARMs) and the treatment are the serious problems in the field of medical science today world over. The development of alternative drug line to treat such infectious diseases is urgently required. Researches on ARMs revealed the presence of membrane proteins responsible for effusing the antibiotics from the bacterial cells. Such proteins have successfully been treated by plant-derived antimicrobials (PDA$_m$S) synergistically along with the commercially available antibiotics. Such synergistic action usually inhibits the efflux pump. The enhanced activity of plant-derived antimicrobials is being researched and is considered as the future treatment strategy to cure the incurable infections. The present paper reviews the advancement made in the researches on antimicrobial resistance along with the discovery and the development of more active PDA$_m$S.

Keywords Antimicrobial-resistant microbes · Efflux pumps · Antimicrobial resistance · Plant antimicrobial compounds

J. Srivastava (✉)
Department of Applied Sciences, Faculty of Environmental Science, Himalayan Institute of Technology and Management, BKT, NH 24, Lucknow 227005, UP, India
e-mail: jks_345@rediffmail.com

H. Chandra · A. R. Nautiyal
Department of Medicinal and Aromatic Plants, School of Agriculture and Allied Sciences, High Altitude Plant Physiology Research Center, H.N.B. Garhwal University, Srinagar, Uttrakhand, India

S. J. S. Kalra
Department of Chemistry, Dayanand Anglo Vedic College, Civil Lines, Kanpur, UP, India

Abbreviations

MDR	Multidrug-resistant
XDR	Extensively drug-resistant
AMR	Antimicrobial resistance
PDA$_m$	Plant-derived antimicrobials
ARM	Antimicrobial-resistant microbes
EPI	Efflux pump inhibitor
AMP	Antimicrobial peptides

Introduction

Increasing antimicrobial resistance (AMR) among microbes caused the emergence of new resistant phenotypes and further caused the development of new antimicrobial compounds (Goossens 2013). Infectious diseases caused by antimicrobial-resistant microbes (ARM) have been frequently reported since last few years (Vila and Pal 2010). About 440,000 new cases of multidrug-resistant tuberculosis (MDR-TB) are recorded annually, causing approximately 150,000 deaths all over the world. Recently, a joint meeting of medical societies, the first ever in India was held to tackle the challenges of antimicrobial resistance in developing world (Ghafur 2013). As a result of this conference "Chennai declaration" came into existence, initiating efforts through a national policy to control the rising trend of AMR in India and abroad (Ghafur et al. 2012).

Multidrug-resistant (MDR) microbes are resistant to three or more antibiotics (Styers et al. 2006), however; strains of *Mycobacterium tuberculosis*, resistant to virtually all classes of antimicrobials have also been identified in the Kwa Zulu Natal Province of South Africa (Gandhi et al. 2006), a typical example of Extremely Drug-Resistant

Tuberculosis (XDR TB) reported in 64 countries to date (World Health Organization 2011). The global emergence of MDRs is increasingly limiting the effectivity of the existing antibiotic drugs (Hancock 2005) for e.g. methicillin-resistant *Staphylococcus aureus* (MRSA) and vancomycin-resistant *Enterococci spp.* (Norrby et al. 2005). The development of resistance among the microbes is the result of continuous selection pressure of antibiotics and their surroundings causing genetic alterations (Bush 2004) which, are transferred to the next generation and reach out to the wider range of other geographical regions through the transfer of genetic information exchange between microbes (Amábile-Cuevas 2003) (Table 1 presents the examples of some of the common MDRs). In this review, attempt has been made to understand specific issues such as factors causing resistance, the role of developing world with a quick overview of plant-derived antimicrobials (PDA_m) and synergistic compounds as an alternative drug line.

Factors causing AMR

Microbes comprise 50 % of total living biomass and are well-survived life forms on earth. There exists a sharp distinction between microbes as pathogenic and non-pathogenic although; one-way exchange of genetic elements (Amábile-Cuevas 2003) may confer the pathogenic characters to the non-pathogenic microbe. Pathogenic microbes cause infectious diseases in humans and animals and are treated with antibiotics. Antibiotics also known as antimicrobials are chemical substances, toxic for most of the life forms. Irrational and deliberate use of antibiotics, migration of infected individuals to other communities (Memish et al. 2003), prolonged use of medical health care systems in hospitals, hunger and malnutrition are some of the main causes of the development of resistance against antibiotics in the microbes (Byarugaba 2004; Vila and Pal 2010). Antimicrobial use in veterinary practices especially as food additives is one of the causes of development of AMRs in zoonotics that may spread to humans (Memish et al. 2003) through the food chain. In this connection, reports of Schlegelova et al. (2008) suggest, least chances of spreading of a resistant strain through the dairy products, however; improperly processed raw meat is strongly discouraged for human consumption in developed nations (Threlfall 2002).

Molecular understanding of AMR

Microbes attain resistance very rapidly against most of the currently available antibiotics because of the adaptability feature conferred by plasmids. Table 1 presents the examples of such plasmids carrying integron and gene cassettes in most common MDRs which on transfer, widespread the resistance (Kumarasamy et al. 2010). Gram-negative (Kumarasamy et al. 2010) and Gram-positive bacteria (Grohman et al. 2003) both exhibit conjugative transfer of plasmids, a natural way of horizontal gene transfer for e.g. the horizontal transfer of plasmid in between *Vibrio fluvialis* and *Vibrio cholerae* conferring resistance to *V. fluvialis* (Rajpara et al. 2009). Recent cases of AMR development include *Pseudomonas aeruginosa* and *Acinetobacter baumannii* resistant to nearly all

Table 1 Examples of plasmids carrying integron integrase carrying gene cassettes imparting resistance against antimicrobials

Plasmid gene cassette	Resistance against	Microbes (isolation)	Conjugative transfer	References
pVN84	MDR	*Vibrio* spp.	✔	Rajpara et al. (2009)
MLS_B [*erm*(B) & *erm*(C)]	Erythromycin	*Staphylococcus* spp.	✗	Schlegelova et al. (2008)
grlA or *gyr A*	Ciprofloxacin	*Staphylococcus* spp.	✗	Campion et al. (2004)
pbp2X	β-Lactam	*Staphylococcus* spp.	✗	Coffey et al. (1991)
CTX-M				
aac(6′)-Ib	Aminoglycoside	*Klebsiella pneumoniae*	✔	Soge et al. (2006)
emr(B)	Macrolide-lincosamide-streptogramin B	*K. pneumoniae*	✔	
pla TEM-1	Ampicillin	*K. pneumoniae*	✔	
dfr	Trimethoprim	*K. pneumoniae*	✔	
p3iANG				
*dfr*A15	Trimathoprim	*Vibrio cholerae*	✔	Ceccarelli et al. (2006)
bla PI	β-Lactam	*V. cholerae*	✔	
qacH	Quaternary ammonia-compounds	*V. cholerae*	✔	
aadA8	aminoglycosides	*V. cholerae*	✔	
mecA	Methicillin (MDR)	*S. aureus*	✗	Hiramatsu et al. (2002)
qnr (carried on class 1 integron)	Ciprofloxacin	*V. Cholerae*	✗	Fonseca et al. (2008)
bla_{MDL-1}	Carbapenem	*Enterobacteriaceae*	✔	Kumarasamy et al. (2010)

antibiotics including the carbanems (Huang and Hsueh 2008). Antibiotic inactivation (degradation of antibiotics by the microbial enzymes e.g. transferase and β-lactamase) causes resistance in microbes (Wright 2005; Jacoby and Munoz-Price 2005), more than 1,000 such β-lactamases are identified till date (Bush and Fisher 2011). Different antibiotics have different mode of actions, therefore, their use is largely dependent on variety of traits other than resistance (Amábile-Cuevas 2010) which either undergo rapid enzymatic degradation or actively effused by the resistant bacteria. Efflux pump in MDRs was first described by Roberts (1996) for tetracycline and macrolide antibiotics. In general, efflux pumps act through membrane proteins of substrate specificity, effuse the antibiotics from the bacterial cell, resulting in a low intracellular ineffective concentration of the drug (Gibbons 2004; Thorrold et al. 2007) altering the permeability of membrane. In a study, staphylococcal accessory regulator (*sarA*) was reported to contribute promising role, imparting resistance in *S. aureus* (Riordan et al. 2006). In addition, Kuete et al. (2011) reported two efflux pumps viz., AcerAB-TolC (Enterobacteriaceae) and MexAB-OprM (*Pseudomonas aeruginosa*) imparting resistance in Gram-negative bacteria against natural products. AMR is a genetically-modified manifestation, linked to the point mutation in bacterial non-chromosomal DNA. As in case of MRSA, the resistance to methicillin is associated with acquisition of a mobile genetic element, *SCCmec*, which contains *mecA*-resistant gene (Okuma et al. 2002). Analytical procedure followed on *Escherichia coli* showed reversible function of class 1 integron integrase gene machinery under selective pressure (Díaz-Mejía et al. 2008). Similar results were also observed by Hsu et al. (2006) whereby *E. coli* MDR was found associated with the class 1 integron gene. Detailed mechanism of development of AMR among microbes has been extensively reviewed by Byarugaba (2010).

Developing world: the factory of MDRs

Developing world especially the countries of South East Asia, Western and Central Africa, India and Pakistan are the most vulnerable for various infectious pandemic diseases. Byarugaba (2004) comprehensively reviewed and reported the AMR in developing countries. Several factors are associated with the AMR development including nosocomial infections, unsafe disposal of biomedical waste, inappropriately used antibiotics, self drug abuse, shortfall of antibiotic course and lack of mass awareness of infectious diseases and personal hygiene (Okeke et al. 2005a, b). In addition to these, lack of surveillance data, providing information of microbial infections common to a geographic location and the invasive microbial species have

been suggested as the major causes of MDRs development in developing countries (Okeke et al. 2005a, b; Giske and Cornaglia 2010; Kartikeyan et al. 2010; Lalitha et al. 2013). Giske and Cornaglia (2010) emphasized on the surveillance practices especially the monitoring and sampling techniques of invasive microbial isolates. Surveillance of resistance in many developing countries is suboptimal (Okeke et al. 2005b) and unable to present the real picture of infectious diseases and the medication. Recent reports of Lalitha et al. (2013) showed the feasibility of proper surveillance of resistance by carrying experimental surveillance study on the school children in different geographic locations of Indian subcontinent. In India for *Salmonella typhi*, MDR has become a norm in strains. This widespread resistant bacterium is associated with contaminated water supply in developing countries and through food products such as contaminated meat in developed countries (Threlfall 2002). Remarkable report of Kumarasamy et al. (2010) provides sufficient evidences in support of the positive role of developing world in the development of ARMs. Resistance to carbapenem conferred by plasmid encoded New Delhi metallo-β-lactamase-1 (bla_{NDM-1}) is a worldwide health problem, especially in UK, (Kumarasamy et al. 2010) having the roots in India and Pakistan. The selective pressure on the bacterial cells is associated with the adaptations causing resistance among microbes for multiple antimicrobials for e.g. genes encoding NDM-1, OXA-23 and OXA-51 enzymes (hydrolyzing specific antibiotics) were observed in three different isolates of *Acinetobacter baumannii* in India (Kartikeyan et al. 2010). Alterations in gene structure were reported in *A. baumannii* as a result of selection pressure of antibiotics (Kartikeyan et al. 2010). The literature suggest, substandard surveillance of resistance, non-prescribed antibiotic usage causes huge selection pressure resulting in the development of AMR in developing countries and their suburbs (Byarugaba 2004; Okeke et al. 2005b; Kumarasamy et al. 2010). Figure 1 shows a schematic diagram showing the development of MDR microbe in community.

Plants derived antimicrobial (PDA$_m$): a ray of hope

Antimicrobial resistance is rapidly increasing along with the development of classical antibiotics consequently, there is an urgent need to develop a different drug line to treat and control MDR bacterial infections. Medicinal values of plants were known to earlier traditional medical practitioners (Emeka et al. 2012). PDA$_m$ substances are plant-originated secondary metabolites and have great concern because of their antibiotic activity without conferring resistance (Baris et al. 2006; Palaniappan and Holley 2010).

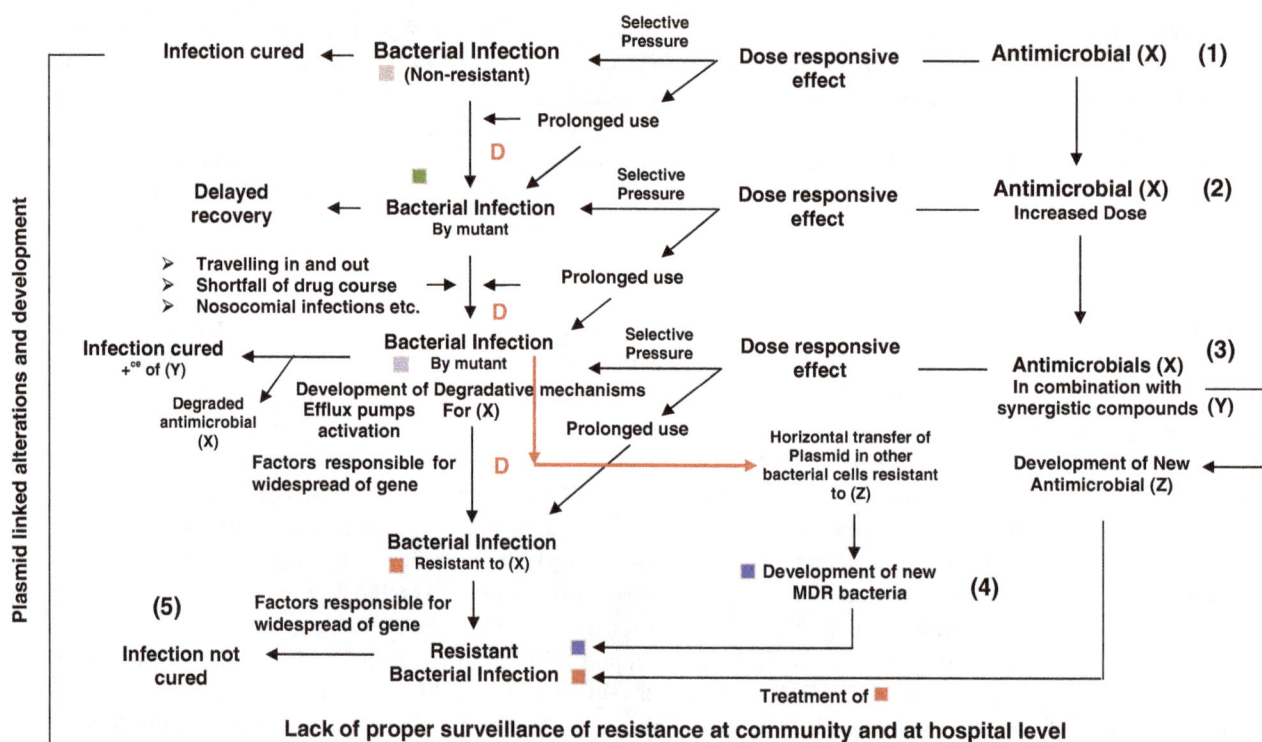

Fig. 1 Illustrative sketch of the development of MDR microbes. The sketch is divided into various segments: (*1*) Bacterial infection was treated with calculated amount of antimicrobial drug (X) followed by complete cure, in the same time prolonged use of drug (X) put selective pressure causing point mutation (D). (*2*) Second infection (in a community only) was treated with same drug (X) with a higher dose, a delayed response was displayed because of mutant bacterial strain, (*3*) Third time infection (in a community only) trigger the resistance, in particular microbe for a particular drug (X); therefore, synergistic compounds (Y) were administered along with (X) may be for clinical trials, the successful treatment, leading to the production of new

antimicrobial drug (Z), (*4*) Since the earlier bacteria attained resistance in due course of time for the drug (X) transferred the resistant gene into another strain of same species of bacteria resistant to the drug (Z) which was introduced in this community from the other one, gene cassettes got recombined on the plasmid to confer multi-drug resistant status to the new introduced bacteria. Infection caused by both these bacteria might be having same symptoms which would be treated with the newly developed drug (Z) keeping the resistance against (X) in consideration. (*5*) Infection could not be cured because the drug was applied to cure the (X) drug-resistant bacteria however; another bacteria having resistance against (Z) remained as such

PDA$_m$s are classified as antimicrobial on the basis of dose ranging from 100 to 1,000 μg ml^{-1} for the minimum inhibitory concentration (MIC) susceptibility test performed on bacteria (Tegos et al. 2002). Table 2 presents few of the examples of plants and their active antimicrobial compounds. Plants have unlimited ability to produce wide variety of secondary metabolites most of which are aromatic compounds including alkaloids, glycosides, terpenoids, saponins, steroids, flavonoids, tannins, quinones and coumarins (Das et al. 2010) forming the basis of PDA$_m$ compounds (Table 3). Target specific plant's secondary metabolites having potential to treat and control the infections are being screened out globally for e.g. Coumarins having specificity on *Staphylococcus aureus* and ineffective on Gram-negative bacteria (Lewis and Ausubel 2006). The literature such as Cowan (1999); Lewis and Ausubel (2006) and González-Lomothe et al. (2009) provides comprehensive information on the major secondary metabolites of plant origin. Precise mechanistic approach of PDA$_m$ and their activity on microbes has been discussed by Lewis and

Ausubel (2006). In general, PDA$_m$s (mostly secondary metabolites) are phenol derivatives, sufficiently able to control microbes by reducing pH, increasing membrane permeability, altering efflux pumping. Examples mentioned in Table 2 followed by recent studies of (Machado et al. 2003; Ram et al. 2004; McGaw et al. 2008; Renisheya et al. 2011; Ahmed et al. 2012; Emeka et al. 2012; Upadhyaya 2013) and the references there in, suggest the antimicrobial potential of various local and exotic plant species, although very few reports have suggested the mechanism of their actions. The affectivity of PDA$_m$s largely depends upon the extraction methods (Das et al. 2010). In a study carried out by our group, methanolic, ethanolic and water extracts of several plants species viz., *Argemone maxicana, Callistomon lanceolatus, Allium sativum, Swietenia mahogany, Citrulus colocynthis, Salvadora persica, Madhuca Indica, Acacia nilotica* and *Pongamia pinnata* were assayed for their antimicrobial activity on most of the common MDRs viz., *Staphylococcus aureus, Bacillus cereus, B. pumilus, Klebsiella pneumonia, Salmonella typhi, E. coli* exhibiting

Table 2 Plant derivatives as antimicrobial for the treatment of microbial infections

Plants	Plant derivatives	Effective against	References
Medicago sativa	Saponins, canavanine	*Enterococcus faecium Staphylococcus aureus*	Aliahmadi et al. (2012)
Onobrychis sativa	AMPs (antimicrobial peptides)	*E. faecium, S. aureus*	Aliahmadi et al. (2012)
Allium sativum	Organosulfur compounds (phenolic compounds)	*Campylobacter jejuni*	Lu et al. (2011)
Raphanus sativum	RsAFP2 (Antifungal peptide)	*Candida albicans*	Aerts et al. (2009)
Vetiveria zizanioides L. Nash	Vetivone (vetiver oil)	*Enterobacter* spp.	Srivastava et al. (2007)
Chelidonium majus	Glycoprotein	*B. cereus, Staphylococcus* spp.	Janovska et al. (2003)
Sanguisorba officinalis	Alkaloids, antimicrobial peptides	*Ps. aeruginosa, E. coli*	Janovska et al. (2003)
Cinnamomum osmophloeum	Cinnamaldehyde (in essential oil)	*Legionella pneumophila*	Chang et al. (2008)
Ocimum basilicum	Essential oil	*Salmonella typhi*	Wan et al. (1998)
Micromeria nervosa	Ethanolic extract	*Proteus vulgaris*	Ali-Shtayeh et al. (1997)
Rabdosia trichocarpa	Trichorabdal A	*Helicobacter pylori*	Kadota et al. (1997)
Melaleuca alternifolia and *Eucalyptus* sp.	Essential oil	*Staphylococcus* spp. and *Streptococcus* spp.	Warnke et al. (2009)
Anthrocephalous cadamba and *Pterocarpus santalinus*	Ethanolic extract	MDRs[M]	Dubey et al. (2012)
Lantana camara L.	Leaf extract in dichloromethane & methanol	MDRsG + ve and MDRsG−ve	Dubey and Padhy (2013)
Butea monosperma Lam.	Ethanolic and hot water extract of leaf	MDRs[M]	Sahu and Padhy (2013)
Jatropha curcas (Linn.)	Ethanolic and methanolic extract	MDRsG + ve + *Micrococcus* sp. & MDRsG−ve + *Shigella* sp. + *Bacillus* sp.	Igbinosa et al. (2009)
Ficus exasperate and *Nauclea latifolia*	Methanolic extract of leaf and stem	*E. coli, Shigella dysenteriae, S. typhi, C. albicans, P. aeruginosa*	Tekwu et al. (2012)
Rhus coriaria	Ethanolic extract	MDR *P. aeruginosa*	Adwan et al. (2010)

MDRsM = *Staphylococcus aureus* + *Acinetobacter* sp. + *Citrobacter freundii* + *Chromobacterium violaceum* + *Escherichia coli* + *Klebsiella* sp. + *Proteus* sp. + *Pseudomonas aeruginosa* + *Salmonella typhi* + *Vibrio cholera*; MDRsG + ve = *S. aureus* (MRSA) + *Streptococcus pyogenes* + *Enterococcus faecalis* (VRE); MDRsG−ve = *Acinetobacter baumannii* + *Citrobacter freundii* + *Proteus mirabilis* + *Proteus vulgaris* + *Pseudomonas aeruginosa*

activity of all the extracts, however; the target specificity of plant extracts could not be established because of uncertain mechanism of plant-derived antimicrobial compounds. A generalized mechanism of PDA_ms on microbes suggests the effects of efflux pumping on MDRs: increasing permeability and reduce selection pressure (Lewis and Ausubel 2006). Antimicrobial peptides (AMPs) are also produced by plants against the infections also called as defensins. Plant defensins are small basic peptides, having characteristic 3D folding pattern, stabilized by eight disulfide linked cysteines (Thomma et al. 2002). AMPs have antimicrobial properties too (Li et al. 2012) and have been suggested as an alternative approach to improve treatment outcome (Brouwer et al. 2011), for e.g. IbAMP1, a plant originated disulfide linked β-sheet antimicrobial peptide (Wang et al. 2009).

Synergistic actions of PDA_ms

The AMR is conferred by several factors which have already been reviewed in previous sections. Plasmid encoded resistance facilitate bacterial cells to develop resistance of various degrees. For instance, unlike Gram-positive, MDR Gram-negative bacterial species have developed a sophisticated permeability barrier as outer membrane comprised of hydrophilic lipopolysaccharide restricting the entry of hydrophobic (quinones and alkaloids) and amphipathic antibiotic compounds (Lewis and Ausubel 2006). The biased effect of PDA_ms on Gram-positive and -negative species has been a key to the discovery of the synergistic compounds of plant origin (Lewis 2001). Plant antimicrobials act well in combinations with other amphipathic compounds. In addition to this, resistance in MDRs conferred by efflux pumping can be treated with the synergistic combinations of antimicrobial with an efflux pump inhibitor (EPI) and altering outer membrane permeability of MDR bacteria providing an effective drug (Savage 2001; Gibbons 2004; Baskaran et al. 2009). Studies of Chusri et al. (2009) reported another example of synergistic effect of plant-derived phenolics such as Ellagic acid (a derivative of Gallic acid) a non-antimicrobial, administered as EPI in combination with classical antibiotic to control *Acinetobacter baumannii*. Another example belongs to the well-studied plant *Berberis fremontii* and its

Table 3 Examples of plant derivatives and their antimicrobial activities

Plant-derived antimicrobial groups	Structure	Chemical properties	Effective on microbes	References
Quinones		Conjugated cyclic-dione structure with molecular formula $C_6H_4O_2$ e.g. Anthraquinone from *Cassia italica*	*Pseudomonas pseudomallei, Bacillus anthracis, Corynebacterium pseudodiphthericum, Pseudomonas aeruginosa*	Kazmi et al. (1994)
		6-(4,7 Dihydroxy-heptyl)quinone	*Staphylococcus aureus, Bacillus subtilis, Proteus vulgaris*	Ignacimuthu et al. (2009)
Alkaloids		Naturally occurring amines having nitrogen in heterocyclic ring of compounds and are the derivative amino acids e.g. glabradine from tubers of *Stephania glabra*	*S. aureus, S. mutans, Microsporum gypseum, M. canis, Trichophyton rubrum*	Semwal and Rawat (2009)
		L-Proline derived Monophyllidin from *Zanthoxylum monophyllum*	*Enterococcus faecalis*	Patino and Cuca (2011)
Lectins and polypeptides	–	Lectins are carbohydrate binding proteins (phytoaglutinin) with MW around 17,000–400,000	*E. coli, P. aeruginosa, Enterococcus hirae, Candida albicans* (fungi)	(Zhang and Lewis (1997)
Flavones/ flavonoids/ flavonols		Are ubiquitous in plant's parts, fruits, seeds, flowers and even honey. Flavones are hydroxylated phenolics containing one carbonyl group	MDR *Klebsiella pneumoniae, P. aeruginosa, E. coli*	Özçelik et al. (2008); Edziri et al. (2012)
Coumarins		Coumarins are phenolic substances made of fused benzene and alpha pyrone ring forming toxic compounds found in plants such as *Dipteryx odorata, Anthoxanthum odoratum* etc	*S. mutans, S. viridans, S. aureus*	Widelski et al. (2009); Lewis and Ausubel (2006)
Terpenoids and essential oils		Isoprene derivatives having a general formula $C_{10}H_{16}$ therefore also called as Isoprenoids. Well-known examples include menthol	*S. viridans, S. aureus, E. coli, B. subtilis, Shigella sonnei* (highly active) *P. aeruginosa, E. coli, S. aureus, T. mentagrophytes* (low activity)	Banso (2009); Ragasa et al. (2008)
Tannins		Large polyphenolic compound containing sufficient hydroxyls and other suitable groups	*S. aureus, S. typhimurium,*	Moneim et al. (2007)

Chemical structure given in front of corresponding group of antimicrobials is not to be considered as generalized one, the references are in correspondence with bacteria

amphipathic cation berberine inhibits the NorA MDR pump of *Staphylococcus aureus* when applied in combination with 5'-MHC (5'-methoxyhydnocarpin, an amphipathic weak acid) a real inhibitor of the pump enhancing the activity of berberine (Stermitz et al. 2000). Similar non-antimicrobial compounds known to enhance effectivity of antimicrobials have been discussed by Lewis (2001). Detailed mechanism of PDA$_m$s on MDR *S. aureus* has been discussed in the review by Gibbons (2004). Wang et al. (2009) defined that the role of AMP plant defensin Ib-AMP1 isolated from plant *Impatiens balsamina* have a

prime target, intercellular components, forming small channels that permit the transit of ions or protons across the bacterial membrane, the same activity was also observed in the linear analogs of this peptide.

Future studies

Researches on the AMR and alternating drug system are endless and a lot of scope is there in the field of ethno-pharmacology. Scientists are working on the development of safe and effective antimicrobials all over the world. Future studies may involve the development of new plant-derived synergistic compounds capable of enhancing the activity of PDA_ms. A lot of research potential is also there to answer the questions for e.g. mechanism of resistance in different bacterial species, development of XDRs and their control.

Conclusion

AMR is a worldwide problem. Research literatures suggest that the substandard living in major parts of developing world is one of the major causes of the development of resistance among bacteria. The developed world is also vulnerable of getting widespread infections for e.g. USA is surrounded by the developing countries having high rates of resistance development. Nosocomial, water borne, health care systems and food products especially meats are some of the most common means of widespread of resistant gene globally. Thanks to the modern molecular approaches for making better understanding of the pathways of resistance development and its remedy. Pharmacologists are developing new antibiotic drugs to treat and control various infections, however; the chances of the development of resistance are equal to the emergence of new drugs. In addition, research suggest that the combinations of PDA_ms and the synergistic compounds work efficiently on resistant strains ensuring no further resistance development. Moreover; concerted efforts have been solicited by the world community because poor countries are worst affected by the antimicrobial resistance and the developed countries are no longer safe (Diáz-Granados et al. 2008). In this regard, PDA_ms in combination with plant-derived synergistic compounds may be the cost-effective approach to deal with global antimicrobial resistance.

Acknowledgments The Corresponding and the one of the main contributors of this review article Dr. Jatin K Srivastava is acknowledged to the chairman of Global Group of Institutions Lucknow for providing the necessary facilities during the compilation of this review paper.

Conflict of interest Authors have no conflict of interest with any of the organization, funding agencies or any person.

References

Adwan G, Abu-Shanab B, Adwan K (2010) Antibacterial activities of some plant extracts alone and in combination with different antimicrobials against multidrug resistant *Pseudomonas aeruginosa* strains. Asian Pac J Trop Med 3(4):266–269

Aerts AM, Carmona-Gutierrez D, Lefevre S, Govaert G, François IE, Madeo F, Santos R, Cammue BP, Thevissen K (2009) The antifungal plant defensin RsAFP2 from radish induces apoptosis in a metacaspase independent way in *Candida albicans*. FEBS Lett 583(15):2513–2516

Ahmed AS, Elgorashi EE, Moodley N, McGaw LJ, Naidoo V, Eloff JN (2012) The antimicrobial, antioxidative, anti-inflammatory activity and cytotoxicity of different fractions of four South African *Bauhinia* species used traditionally to treat diarrhea. J Ethnopharmacol 143(3):826–839

Aliahmadi A, Roghanian R, Emtiazi G, Mirzajani F, Ghassempour A (2012) Identification and primary characterization of a plant antimicrobial peptide with remarkable inhibitory effects against antibiotic resistant bacteria. Afr J Biotechnol 11(40):9672–9676

Ali-Shtayeh MS, Al-Nuri MA, Yaghmour RMR, Faidi YR (1997) Antimicrobial activity of *Micromeria nervosa* from the Palestinian area. J Ethnopharmacol 58:143–147

Amábile-Cuevas CF (2003) Gathering of resistance genes in Gram-negative bacteria: an overview. In: Amábile-Cuevas CF (ed) Multidrug resistant bacteria. Horizon Scientific Press, Wymondham, pp 9–31

Amábile-Cuevas CF (2010) Global perspective of antibiotic resistance. In: de-J-Soso A et al (eds) Antimicrobial resistance in developing countries. Springer, New York, pp 3–14

Banso A (2009) Phytochemical and antibacterial investigation of bark extracts of *Acacia nilotica*. J Med Plants Res 3(2):82–85

Baris O, Gulluce M, Sahin F, Ozer H, Kilic HH, Ozkan H, Sokmen M, Ozbek T (2006) Biological activities of the essential oil and methanolic extract of *Achillea biebersteinii* Afan. (Asteraceae). Turk J Biol 30:65–73

Baskaran SA, Kazmer GW, Hinckley L, Andrew JM, Venkitanarayanan K (2009) Antimicrobial effect of plant derived antimicrobials on major bacterial mastitis pathogens in vitro. J Dairy Sci 92(4):1423–1429

Brouwer CPJM, Rahman M, Welling MM (2011) Discovery and development of a synthetic peptide derived from lactoferrin for clinical use. Peptide 32(9):1953–1963

Bush K (2004) Antibacterial drug discovery in the 21st century. Clin Microbiol Infect 10(S4):10–17

Bush K, Fisher JF (2011) Epidemiological expansion, structural studies, and clinical challenges of new β-lactamases from Gram-negative bacteria. Annu Rev Microbiol 65:455–478

Byarugaba DK (2004) Antimicrobial resistance in developing countries and responsible risk factors. Int J Antimicrob Agents 24(2):105–110

Byarugaba DK (2010) Mechanism of antimicrobial resistance. In: de-J-Soso A et al (eds) Antimicrobial resistance in developing countries. Springer, New York, pp 15–26

Campion JJ, McNamara PJ, Evans ME (2004) Evolution of ciprofloxacin-resistant *Staphylococcus aureus* in in vitro pharmacokinetic environments. Antimicrob Agents Chemother 48(12):4733–4744

Ceccarelli D, Salvia AM, Sami J, Cappuccinelli P, Colombo MM (2006) New cluster of plasmid-located class 1 integrons in *Vibrio cholerae* O1 and a *dfrA15* cassette containing integron in *Vibrio parahaemolyticus* isolated in Angola. Antimicrob Agents Chemother 50:2493–2499

Chang C, Chang W, Chang S, Cheng S (2008) Antibacterial activities of plant essential oils against *Legionella pneumophila*. Water Res 42:278–286

Chusri S, Villanueva I, Voravuthikunchai SP, Davies J (2009) Enhancing antibiotic activity: a strategy to control *Acinetobacter* infections. J Antimicrob Chemother 16:1203–1211

Coffey TJ, Dowson CG, Daniels M, Zhou J, Martin C, Spratt BG, Musser JM (1991) Horizontal transfer of multiple penicillin-binding protein genes and capsular biosynthetic genes in natural populations of *Streptococcus pneumoniae*. Mol Microbiol 5(9): 2255–2260

Cowan MM (1999) Plant products as antimicrobial agents. Clin Microbiol Rev 12(4):564–582

Das K, Tiwari RKS, Shrivastava DK (2010) Techniques for evaluation of medicinal plant products as antimicrobial agent: current methods and future trends. J Med Plant Res 4(2):104–111

Díaz-Granados CA, Cardo DM, McGowan-Jr JE (2008) Antimicrobial resistance: international control strategies with a focus on limited resource settings. Int J Antimicrob Agents 32(1):1–9

Díaz-Mejía JJ, Amábile-Cuevas CF, Rosas I, Souza V (2008) An analysis of the evolutionary relationships of integron integrases with emphasis on the prevalence of class1 integron in *Escherichia coli* isolates from clinical and environmental origins. Microbiol 154:94–102

Dubey D, Padhy RN (2013) Antibacterial activity of *Lantana camara* L. against multidrug resistant pathogens from ICU patients of a teaching hospital. JHerb Med (In press).

Dubey D, Sahu MC, Rath S, Paty BP, Debata NK, Padhy RN (2012) Antimicrobial activity of medicinal plants used by aborigines of Kalahandi, Orissa, India against multidrug resistant bacteria. Asian Pac J Trop Biomed 2(2):S846–S854

Edziri H, Mastouri M, Mahjoub MA, Mighri Z, Mahjoub A, Verschaeve L (2012) Antibacterial, antifungal and cytotoxic activities of two flavonoids from *Retama raetam* flowers. Molecules 17:7284–7293

Emeka PM, Badger-Emeka LI, Fateru F (2012) In-vitro antimicrobial activities of *Acalypha ornata* leaf extracts on bacterial and fungal clinical isolates. J Herb Med 2(4):136–142

Fonseca EL, dos Santos Freitas FF, Vieira VV, Vicente ACP (2008) New *qnr* gene cassettes associated with superintegron repeats in *Vibrio cholerae* O1. Emerg Infect Dis 14:1129–1131

Gandhi NR, Moll A, Sturm AW, Pawinski R, Govender T, Lalloo U, Zeller K, Andrews J, Friedland G (2006) Extensively drug-resistant tuberculosis as a cause of death in patients co-infected with tuberculosis and HIV in a rural area of South Africa. Lancet Infect Dis 368:1575–1580

Ghafur A (2013) The Chennai declaration: an Indian perspective on the antimicrobial resistance challenge. J Global Antimicrob Resist 1(1):5–6

Ghafur A, Mathai D, Muruganathan A, Jayalal JA, Kant R, Chaudhary D, Prabhash K, Abraham OC, Gopalakrishnan R, Ramasubramanian V, Shah SN, Pardeshi R, Huilgol A, Kapil A, Gill JPS, Singh S, RIssam HS, Todi S, Hegde BM, Parikh P (2012) The Chennai declaration: recommendations of "A roadmap to tackle the challenge of antimicrobial resistance"— a joint meeting of medical societies of India. Indian J Cancer 49(4):84–94

Gibbons S (2004) Anti-staphylococcal plant natural products. Nat Prod Rep 21:263–277

Giske CG, Cornaglia G (2010) Supranational surveillance of antimicrobial resistance: the legacy of the last decade and proposals for the future. Drug Resist Updat 13(4–5):93–98

González-Lomothe R, Mitchell G, Gattuso M, Diarra MS, Malouim F, Bouarab K (2009) Plant antimicrobial agents and their effects on plant and human pathogens. Int J Mol Sci 10:3400–3419

Goossens H (2013) The Chennai declaration on antimicrobial resistance in India. Lancet Infect Dis 13(2):105–106

Grohman E, Muth G, Espinosa M (2003) Conjugative plasmid transfer in Gram-positive bacteria. Microbial Mol Biol Rev 67(2):277–301

Hancock EW (2005) Mechanisms of action of newer antibiotics for Gram-positive pathogens. Lancet Infect Dis 5(4):209–218

Hiramatsu K, Katayama Y, Yuzawa H, Ito T (2002) Molecular genetics of methicillin-resistant *Staphylococcus aureus*. Int J Med Microbiol 292:67–74

Hsu S, Chiu T, Pang J, Hsuan-Yuan C, Chang G, Tsen H (2006) Characterization of antimicrobial resistance patterns and class 1 integrons among *Escherichia coli* and *Salmonella enterica* serovar Choleraesuis strains isolated from humans and swine in Taiwan. Int J Antimicrob Agents 27(5):383–391

Huang Y, Hsueh P (2008) Antimicrobial drug resistance in Taiwan. Int J Antimicrob Agents 32(3):S174–S178

Igbinosa OO, Igbinosa EO, Aiyegoro OA (2009) Antimicrobial activity and phytochemical screening of stem bark extracts from *Jatropha curcas* Linn. Afr J Pharma Phramacol 3(2):58–62

Ignacimuthu S, Pavunraj M, Duraipandiyan V, Raja N, Muthu C (2009) Antibacterial activity of a novel quinone from the leaves of *Pergularia daemia* (Forsk.), a traditional medicinal plant. Asian J Trad Med 4(1):36–40

Jacoby GA, Munoz-Price LS (2005) The new B-lactamases. N Engl J Med 352:380–391

Janovska D, Kubikova K, Kokoska L (2003) Screening for antimicrobial activity of some medicinal plants species of traditional Chinese medicine. Czech J Food Sci 21(2):107–110

Kadota S, Basnet P, Ishii E, Tamura T, Namba T (1997) Antibacterial activity of trichorabdal A from *Rabdosia trichocarpa* against *Helicobacter pylori*. Zentralbl Bakteriol 286(1):63–67

Kartikeyan K, Thirunarayan MA, Krishnan P (2010) Coexistence of $bla_{OXA\text{-}23}$ with bla_{NDM1} and *armA* in clinical isolates of *Acinetobacter baumannii* from India. J Antimicrob Chemother 65:2253–2254

Kazmi MH, Malik A, Hameed S, Akhtar N, Noor AS (1994) An anthraquinone derivative from *Cassia italica*. Photochemistry 36:761–763

Kuete V, Alibert-Franco S, Eyong KO, Ngameni B, Folefoc GN, Nguemeving JR, Tangmovo JG, Fatso GW, Komguem J, Ouahouo BMW, Bolla JM, Chevalier J, Ngadjui BT, Nkengfack AE, Pages JM (2011) Antibacterial activity of some natural products against bacteria expressing a multi-drug-resistant phenotype. Int J Antimicrob Agents 37(2):156–161

Kumarasamy KK, Toleman MA, Walsh TR, Bagaria J, Butt F, Balakrishnan P, Chaudhary U, Doumith M, Giske CG, Irfan S, Krishna P, Kumar AV, Maharajan S, Mushtaq S, Noorie T, Paterson DL, Pearson A, Perry C, Pike C, Rao B, Ray U, Sarma JB, Sharma M, Sheridan E, Thirunarayan MA, Turton J, Upadhyay S, Warner M, Welfare W, Livemore DM, Woodford N (2010) Emergence of a new antibiotic resistance mechanism in India, Pakistan and UK: a molecular, biological and epidemiological study. Lancet Infect Dis 10(9):597–602

Lalitha MK, David T, Thomas K (2013) Nasopharyngeal swabs of school children, useful in rapid assessment of community antimicrobial resistance patterns in *Streptococcus pneumoniae* and *Haemophilus influenza*. J Clin Epidemiol 66(1):44–51

Lewis K (2001) In search of natural substrates and inhibitors of MDR pumps. J Mol Microbiol 3:247–254

Lewis K, Ausubel FM (2006) Prospects for plant–derived antimicrobials. Nat Biotechnol 24:1504–1507

Li Y, Xiang Q, Zhang Q, Huang Y, Su Z (2012) Overview on the recent study of antimicrobial peptides: origin, functions, relative mechanisms and application. Peptides 37(2):207–215

Lu X, Rasco BA, Jabal JM, Aston DE, Lin M, Konkel ME (2011) Investigating antibacterial effects of garlic (*Allium sativum*) concentrate and garlic-derived organosulfur compounds on *Campylobacter jejuni* by using fourier transform infrared spectroscopy, Raman spectroscopy, and electron microscopy. Appl Environ Microbiol 77(15):5257–5269

Machado TB, Pinto AV, Pinto MCFR, Leal ICR, Silva MG, Amaral ACF, Kuster RM, Netto-dossantos KR (2003) In-vitro activity of Brazilian medicinal plants, naturally occurring naphthoquinone and their analogues, against methicillin resistant *Staphylococcus aureus*. Int J Antimicrob Agents 21(3):279–284

McGaw LJ, Lall N, Meyer JJM, Eloff JN (2008) The potential of South African plants against *Mycobacterium* infections. J Ethnopharmacol 119(3):482–500

Memish ZA, Venkatesh S, Shibi AM (2003) Impact of travel on international spread of antimicrobial resistance. Int J Antimicrob Agents 21(2):135–142

Moneim A, Suleman E, Issa FM, Elkhalifa EA (2007) Quantitative determination of tannin content in some sorghum cultivars and evaluation of its antimicrobial activity. Res J Microbiol 2(3):284–288

Norrby RS, Nord CE, Finch R (2005) Lack of development of new antimicrobial drugs: a potential serious threat to public health. Lancet Infect Dis 5(2):115–119

Okeke IN, Laxminarayan R, Bhutta ZA, Duse AG, Jenkins P, O'Brien TF, Pablos-Mendez A, Klugman KP (2005a) Antimicrobial resistance in developing countries. Part I: recent trends and current status. Lancet Infect Dis 5(8):481–493

Okeke IN, Klugman KP, Bhutta ZA, Duse AG, Jenkins P, O'Brien TF, Pablos-Mendez A, Laxminarayan R (2005b) Antimicrobial resistance in developing countries. Part II: strategies for containment. Lancet Infect Dis 5(9):568–580

Okuma K, Iwakawa K, Turnidge JD, Grubb WB, Bell JM, O'Brien FG, Coombs GW, Pearman JW, Tenover FC, Kapi M, Tiensasitorn C, Ito T, Hiramatsu K (2002) Dissemination of new methicillin resistant *Staphylococcus aureus* clones in the community. J Clin Microbiol 40(11):4289–4294

Özçelik B, Deliorman OD, Özgen S, Ergun F (2008) Antimicrobial activity of flavonoids against Extended Spectrum β-Lactamase (ESBL) producing *Klebsiella pneumoniae*. Trop J Pharma Res 7(4):1151–1157

Palaniappan K, Holley RA (2010) Use of natural antimicrobials to increase antibiotic susceptibility of drug resistant bacteria. Int J Food Microbiol 140(2–3):164–168

Patino OJ, Cuca LE (2011) Monophyllidin, a new alkaloid L-Proline derivative from *Zanthoxylum monophyllum*. Phytochem Let 4:22–25

Ragasa CY, Ha HKP, Hasika M, Maridable J, Gaspillo P, Rideout J (2008) Antimicrobial and cytotoxic terpenoids from *Cymbopogon citratus* Stapf. Phillipin Sci 45(1):111–122

Rajpara N, Patel A, Tiwari N, Bahuguna J, Antony A, Choudhury I, Ghosh A, Jain R, Bhardwaj AK (2009) Mechanism of drug resistance in a clinical isolate of *Vibrio fluvialis*: involvement of multiple plasmids and integrons. Int J Antimicrob Agents 34:220–225

Ram AJM, Bhakshu L, Raju RRV (2004) In vitro antimicrobial activity of certain medicinal plants from Eastern Ghats, India, used for skin diseases. J Ethnopharmacol 90(2–3):353–357

Renisheya JJMT, Johnson M, Mary MU, Arthy A (2011) Antimicrobial activity of ethanolic extracts of selected medicinal plants against human pathogens. Asian Pac J Trop Biomed 1(1):S76–S78

Riordan JT, O'Leary JO, Gustafson JE (2006) Contribution of SigB and SarA to distinct multiple antimicrobial resistance mechanisms of *Staphylococcus aureus*. Int J Antimicrob Agents 28(1):54–61

Roberts MC (1996) Tetracycline resistance determinants: mechanisms of action, regulation of expression, genetic mobility, and distribution. FEMS Microbiol Rev 19:1–24

Sahu MC, Padhy RN (2013) In vitro antimicrobial potency of *Butea monosperma* Lam. against 12 clinically isolated multidrug resistant bacteria. Asia Pac J Trop Disease 3(3):217–226

Savage PB (2001) Multidrug resistant bacteria: overcoming antibiotic permeability barriers of Gram-negative bacteria. Ann Med 33:167–171

Schlegelova J, Vlkova H, Babak V, Holasova M, Jaglic Z (2008) Resistance to erythromycin of *Staphylococcus* spp. isolates from the food chain. Veterinarni Med 53(6):307–314

Semwal DK, Rawat U (2009) Antimicrobial hasubanalactam alkaloid from *Stephania glabra*. Planta Med 75(4):378–380

Soge OO, Adeniyi BA, Roberts MC (2006) New antibiotic resistance genes associated with CTX-M plasmids from uropathogenic Nigerian *Klebsiella pneumoniae*. J Antimicrob Chemother 58:1048–1053

Srivastava J, Chandra H, Singh N (2007) Allelopathic response of *Vetiveria zizanioides* (L.) Nash on members of the family Enterobacteriaceae and *Pseudomonas* spp. Environmentalist 27:253–260

Stermitz FR, Lorenz P, Tawara JN, Zenewicz LA, Lewis K (2000) Synergy in a medicinal plant: antimicrobial action of berberine potentiated by 5′-methoxyhydnocarpin, a multidrug pump inhibitor. Appl Biol Sci 97(4):1433–1437

Styers D, Sheehan DJ, Hogan P, Sahm DF (2006) Laboratory-based surveillance of current antimicrobial resistance patterns and trends among *Staphylococcus aureus*: 2005 status in the United States. Ann Clin Microb Antimicrob 5:2

Tegos G, Stremitz FR, Lomovskaya O, Lewis K (2002) Multidrug pump inhibitors uncover remarkable activity of plant antimicrobials. Antimicrob Agents Chemother 46(10):3133–3141

Tekwu EM, Pieme AC, Beng VP (2012) Investigations of antimicrobial activity of some Cameroonian medicinal plant extracts against bacteria and yeast with gastrointestinal relevance. J Ethnopharmacol 142(1):265–273

Thomma BPHT, Cammue BPA, Thevissen K (2002) Plant defensins. Planta 216:193–202

Thorrold CA, Letsoalo ME, Duse AG, Marais E (2007) Efflux pump activity in fluoroquinolone and tetracycline resistant *Salmonella* and *E. coli* implicated in reduced susceptibility to household antimicrobial cleaning agents. Int J Food Microbiol 113(3):315–320

Threlfall EJ (2002) Antimicrobial drug resistance in Salmonella: problems and perspective in food and water-borne infections. FEMS Microbiol Rev 26(2):141–148

Upadhyaya S (2013) Screening of phytochemicals, nutritional status, antioxidant and antimicrobial activity of *Paderia foetida* Linn. From different localities of Assam, India. J Pharm Res 7(1):139–141

Vila J, Pal T (2010) Update on antimicrobial resistance in low income countries: factors favouring the emergence of resistance. Open Infect Dis J 4:38–54

Wan J, Wilcock A, Coventry MJ (1998) The effect of essential oil of basil on the growth of *Aeromonas hydrophila* and *Pseudomonas fluorescens*. J Appl Microbiol 84:152–158

Wang P, Bang JK, Kim HJ, Kim JK, Kim Y, Shin SY (2009) Antimicrobial specificity and mechanism of action of disulfide— removed linear analogs of the plant derived cys-rich antimicrobial peptides Ib-AMP1. Peptides 30(12):2144–2149

Warnke PH, Becker ST, Podschun R, Sivanathan S, Springer IN, Russo PAI, Wiltfang J, Fickenscher H, Sherry E (2009) The battle against multi resistant strains: renaissance of antimicrobial essential oils as a promising force to fight hospital acquired infections. J Cranio Maxillofacial Surg 37(7): 392–397

WHO (World Health Organization) (2011) Combat antimicrobial resistance. http://www.who.int/world-health-day/2011

Widelski J, Popova M, Graikou K, Glowniak K, Chinou I (2009) Coumarins from *Angelica lucida* L.—antibacterial activities. Molecule 14:2729–2734

Wright GD (2005) Bacterial resistance to antibiotics: enzymatic degradation and modification. Adv Drug Deliv Rev 57(10):1451–1470

Zhang Y, Lewis K (1997) Febatins: new antimicrobial plant peptides. FEMS Microbiol Lett 149:59–64

23

Oxidative stress induced by the chemotherapeutic agent arsenic trioxide

Mathews V. Varghese · Alex Manju ·
M. Abhilash · M. V. Sauganth Paul ·
S. Abhilash · R. Harikumaran Nair

Abstract Arsenic compounds have been used for medicinal purposes throughout history. Arsenic trioxide (As_2O_3) achieved dramatic remissions in patients with acute promyelocytic leukaemia. Unfortunately, the clinical usefulness of As_2O_3 has been limited by its toxicity. The present study was designed to investigate the toxic effects of As_2O_3 at its clinical concentrations. Experimental rats were administered with As_2O_3 2, 4 and 8 mg/kg body weight for a period of 45 days and the serum glucose, creatine kinase, lactate dehydrogenase, lipid peroxidation and antioxidant status were measured. As_2O_3-treated rats showed elevated serum glucose, creatine kinase and lactate dehydrogenase concentrations. Lipid peroxidation product malondialdehyde was found to be produced more in arsenic-treated rats. Reduced glutathione and glutathione-dependant antioxidant enzymes, glutathione-*S*-transferase and glutathione peroxidase, and the antiperoxidative enzymes, superoxide dismutase and catalase, concentrations were reduced with the As_2O_3 treatment. All these toxic effects were found increased with the increase in concentration of As_2O_3. The results of the study indicate that As_2O_3 produced dose-dependant toxic side effects at its clinical concentrations.

Keywords Arsenic trioxide · Antioxidants · Lipid peroxidation · Glutathione · Oxidative stress

Introduction

Arsenic is widely distributed in air, water and soil in the form of either metalloids or chemical compound (Basu et al. 2001). The most toxicologically potent arsenic compounds are in the trivalent oxidation state (Hughes 2002). Exposure to arsenic occurs via ingestion, inhalation, dermal contact, and the parenteral route (Tchounwou et al. 1999). The health effects that are associated with arsenic exposure include cardiovascular and peripheral vascular disease, developmental anomalies, neurologic and neurobehavioral disorders, diabetes, portal fibrosis, and multiple cancers (Tseng et al. 2003). Haematological abnormalities associated with arsenic intoxication are haemoglobinuria, intravascular coagulation, bone marrow depression, severe pancytopenia, and normocytic normochromic anaemia and basophilic stippling (Ratnaike 2003).

The treatment with low doses of arsenic trioxide (As_2O_3) caused high rates of complete remission in patients suffering from acute promyelocytic leukaemia (APL) (Wang and Chen 2008). The adverse effects associated with As_2O_3 treatment include hyperleukocytosis, APL differentiation syndrome, electrocardiographic abnormalities (QTc-interval prolongation), peripheral neuropathy, skin rash, gastrointestinal reactions and hyperglycaemia (Rust and Soignet 2001). Arsenic compounds also showed genotoxic effects; induced gene amplification, inhibit DNA repair, and induce expression of the oxidative stress protein heme oxygenase in mammalian cells (Vega et al. 1995; Keyse et al. 1990).

Due to the toxic side effects of arsenic, it carries significant risks in their therapeutic regiment. The mechanisms of arsenic-induced toxic effects during clinical trials were not fully elucidated. So the present study was

M. V. Varghese · A. Manju · M. Abhilash ·
M. V. S. Paul · S. Abhilash · R. H. Nair (✉)
School of Biosciences, Mahatma Gandhi University,
P.D Hills P.O, Kottayam 686560, Kerala, India
e-mail: harinair@mgu.ac.in

performed to understand the toxicological association between biochemical parameters and the blood antioxidant status at different clinically relevant concentrations of As_2O_3.

Materials and methods

Chemicals

Arsenic trioxide, sodium pyruvate, thiobarbituric acid and triton X-100, phenazine methosulphate, nitroblue tetrazolium were obtained from Sigma-Aldrich (Bangalore, India). L-aspartate, α-oxoglutarate, 2,4-dinitro phenyl hydrazine, nicotinamide adenine dinucleotide (reduced), 1-chloro-2,4-dinitrobenzene (CDNB), 5,5′-dithiobis-(2-nitrobenzoic acid), nicotinamide adenine dinucleotide phosphate and reduced glutathione were purchased from Merck Specialties Pvt Ltd (Mumbai, India). All other chemicals were purchased from Sisco Research Laboratories (Mumbai, Maharashtra, India).

Experimental animals

Male albino rats of Wistar strain (200–220 g) were purchased from Small Animal Breeding Station of Government Veterinary College, Mannuthy, Thrissur, Kerala, India. The animals were housed in polypropylene cages kept in the animal house of School of Biosciences, Mahatma Gandhi University, Kottayam. All the animals were maintained under standard laboratory conditions of temperature (22 ± 3 °C) and 12-h light and dark cycles throughout the experimental period. Rats were provided with laboratory chow (Hindustan Unilever Ltd., Mumbai, India) and water ad libitum. Experiments were conducted as per the guidelines of Institutional Animal Ethical Committee, School of Biosciences, Mahatma Gandhi University (Reg. No. B1662009/2). After 2 weeks of acclimation, animals were randomly divided into four groups with six animals in each group.

(a) Group 1: Normal Control
(b) Group 2: As_2O_3 2 mg/kg b.wt
(c) Group 3: As_2O_3 4 mg/kg b.wt
(d) Group 4: As_2O_3 8 mg/kg b.wt

The duration of the study was 45 days and the route of administration was daily by oral intubation. At the end of the experimental period, blood was drawn from the orbital sinus of the rat's eye. Anticoagulated blood was used for the analysis of blood antioxidant status and lipid peroxidation. The blood samples collected in another set of test tubes without anticoagulant were centrifuged at 3,000 rpm at 4 °C for 20 min; the clear serum obtained was used for the biochemical assays.

Analysis of serum glucose, creatine kinase and lactate dehydrogenase

Serum Glucose, Creatine Kinase (CK) and Lactate dehydrogenase (LDH) were detected (Agappe Diagnostic Ltd., Ernakulam, Kerala, India) using semi-auto analyzer (RMS, India).

Analysis of blood antioxidant status and lipid peroxidation

Haemoglobin concentration in blood was determined according to the method of Drabkin and Austin (1932). Superoxide dismutase (SOD) activity was determined by the method of Paoletti et al. (1986). Catalase (CAT) activity was measured in the sample according to the method of Aebi (1984) by measuring the decrease in absorbance of hydrogen peroxide (H_2O_2) at 240 nm. Glutathione-S-Transferase (GST) activity was estimated by determining the rate of formation of glutathione and CDNB conjugates (Beutler et al. 1986). Glutathione peroxidase (GPx) activity was measured by the method of Paglia and Valentine (1967). Reduced glutathione (GSH) was estimated by the method of Beutler et al. (1963). Malondialdehyde (MDA), a product of lipid peroxidation, was determined by the method of Buege and Aust (1978).

Statistical analysis

One-way ANOVA followed by LSD post hoc multiple comparison test was used for comparison among the four groups (SPSS/PC+ version 18, SPSS Inc. Chicago, Illinois, USA). Probability (p) <0.05 was considered statistically significant.

Results

As_2O_3-induced alterations in the serum Glucose, CK and LDH

As_2O_3 treatment significantly ($p < 0.05$) increased the serum Glucose (Fig. 1), CK (Fig. 2) and LDH (Fig. 3) when compared to the group I control. These biochemical parameters also showed statistical significance ($p < 0.05$) when compared between groups II, III and IV. The concentrations of Glucose, CK and LDH were increased with respect to the increase in concentration of As_2O_3.

As_2O_3-induced lipid peroxidation

The lipid peroxidation product MDA was significantly increased with As_2O_3 treatment when compared to the

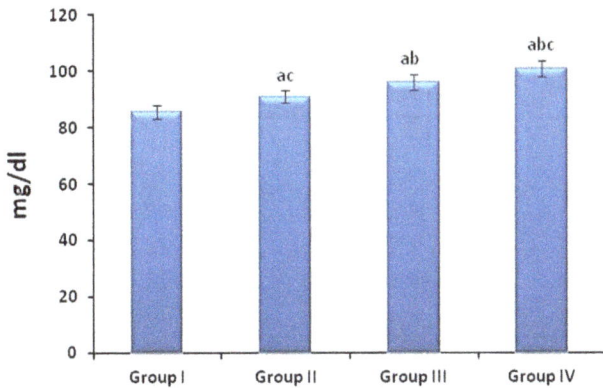

Fig. 1 Effect of As_2O_3 on Serum Glucose: Normal control (Group I), As_2O_3 2 mg/kg b.wt (Group II), As_2O_3 4 mg/kg b.wt (Group III), and As_2O_3 8 mg/kg b.wt (Group IV). Data represented as mean ± SD, $n = 6$. $p < 0.05$ was considered significant. [a]Statistical significance in comparison to normal control, [b]statistical significance in comparison to group II, and [c]statistical significance in comparison to group III

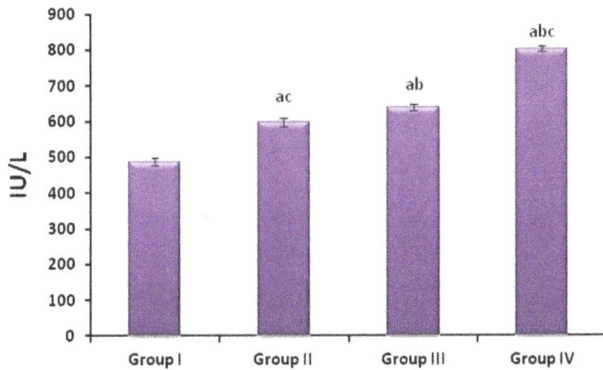

Fig. 3 Effect of As_2O_3 on Lactate Dehydrogenase: Normal control (Group I), As_2O_3 2 mg/kg b.wt (Group II), As_2O_3 4 mg/kg b.wt (Group III), and As_2O_3 8 mg/kg b.wt (Group IV). Data represented as mean ± SD, $n = 6$. $p < 0.05$ was considered significant. [a]Statistical significance in comparison to normal control, [b]statistical significance in comparison to group II, and [c]statistical significance in comparison to group III

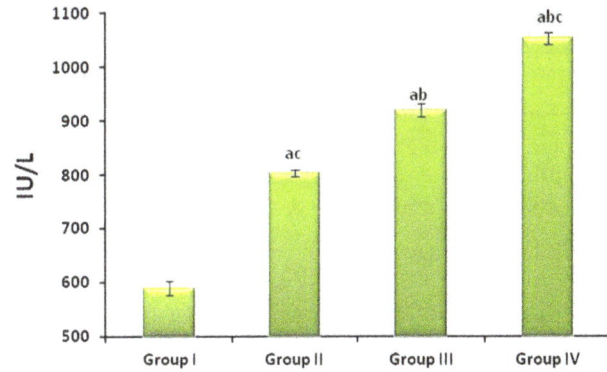

Fig. 2 Effect of As_2O_3 on Creatine Kinase: Normal control (Group I), As_2O_3 2 mg/kg b.wt (Group II), As_2O_3 4 mg/kg b.wt (Group III), and As_2O_3 8 mg/kg b.wt (Group IV). Data represented as mean ± SD, $n = 6$. $p < 0.05$ was considered significant. $p < 0.05$ was considered significant. [a]Statistical significance in comparison to normal control, [b]statistical significance in comparison to group II, and [c]statistical significance in comparison to group III

group I control. In statistical analysis, the intergroup comparison between As_2O_3-treated groups II, III and IV showed significant variation with respect to the increase in dose of As_2O_3 (Table 1).

As_2O_3-induced alterations in the antioxidant status

The tripeptide GSH was reduced significantly with respect to the control and was noticed in the arsenic-treated groups. As_2O_3 treatment also decreased the GSH-dependant antioxidant enzymes, GST and GPx, and the antiperoxidative enzymes, SOD and CAT, when compared to the group I control and between the arsenic-treated groups II, III and IV. These reductions in the antioxidant enzyme activities

were in accordance with the increase concentration of As_2O_3 (Table 1).

Discussion

As_2O_3 is an effective cancer therapeutic drug for acute promyelocytic leukaemia and has potential anticancer activity against a wide range of solid tumours (Lu et al. 2007). The adverse effects that have been noted in clinical trials of As_2O_3 are fluid retention, cardiac toxicity, hepatocellular toxicity and electrocardiographic changes (Soignet et al. 2001). The duration of the present study was selected based on the previous clinical study (Huan et al. 2000). They have reported that As_2O_3 treatment resulted in significant remission in APL patients. The As_2O_3 concentrations in our study are in the range of clinically available concentrations for anti-leukaemia treatment (Li et al. 2002).

In the present study, the treatment with As_2O_3 increased the glucose concentration in serum. Miller et al. (2002) reported that trivalent arsenic inhibits the uptake of glucose into cells, gluconeogenesis, fatty acid oxidation and further production of acetyl CoA. Pyruvate dehydrogenase, an enzyme of glucose metabolism, is susceptible to arsenic-induced reactive oxygen species (ROS) generation (Aposhian and Aposhian 2006). The thiol moiety is an important target for arsenic (Flora 2011). The increased concentration of glucose in this study, may be due to the binding of arsenic to the sulfhydryl groups of glucose metabolising enzymes, and thereby blocked the uptake of glucose. The altered blood sugar level may also due to islet cells toxicity, because arsenic administration caused severe pancreatic damage (Mukherjee et al. 2004). In our

Table 1 Effect of As_2O_3 on the Blood Antioxidant Status: Normal control (Group I), As_2O_3 2 mg/kg b.wt (Group II), As_2O_3 4 mg/kg b.wt (Group III), and As_2O_3 8 mg/kg b.wt (Group IV)

Parameters	Group I	Group II	Group III	Group IV
MDA (µM/L)	4.09 ± 0.12	4.68 ± 0.17[ac]	4.9 ± 0.09[ab]	5.05 ± 0.11[abc]
GSH (µM/gHb)	5.87 ± 0.32	5.16 ± 0.2[ac]	4.47 ± 0.32[ab]	3.84 ± 0.42[abc]
SOD (U/mgHb)	1.59 ± 0.13	1.32 ± 0.12[ac]	0.98 ± 0.14[ab]	0.75 ± 0.17[abc]
CAT (k/ml)	10.31 ± 0.45	9.12 ± 0.85[ac]	8.28 ± 0.65[ab]	7.44 ± 0.47[abc]
GPx (U/gHb)	7.94 ± 0.18	7.44 ± 0.11[ac]	7.12 ± 0.12[ab]	6.16 ± 0.1[abc]
GST (µM/min/gHb)	2.09 ± 0.11	1.64 ± 0.09[ac]	1.35 ± 0.09[ab]	0.76 ± 0.07[abc]

Data represented as mean ± SD, $n = 6$

$p < 0.05$ was considered significant

[a] Statistical significance in comparison to normal control

[b] Statistical significance in comparison to group II

[c] Statistical significance in comparison to group III

observation, the treatment with As_2O_3-induced ROS production, which may reduce insulin production by pancreatic cellular damage, leads to the increased glucose concentration in serum.

As_2O_3 administration caused myocardial damage and increased release of CK and LDH in serum (Raghu et al. 2009). In our recent report, the As_2O_3 treatment caused oxidative stress and structural aberrations in the cardiac tissue of experimental rats (Mathews et al. 2013). So the increased concentration of CK and LDH observed in this investigation may be due to the exudation of enzymes from cells to the systemic circulation because of cellular damage induced by the As_2O_3.

From our observation, it is found that the MDA production in blood is significantly increased with As_2O_3 treatment. MDA is a marker of endogenous lipid peroxidation. Liu et al. (2001) reported that the treatment with arsenic caused a significant increase in the rate of formation of ROS such as superoxide anion radical, hydroxyl radical and hydrogen peroxide. The toxic potential exerted by these compounds is through their reactivity with sulphur containing compounds and the generation of ROS (Hughes et al. 2011). Arsenic-induced MDA production could be due to the impairment of cells' natural protective system and could be directly related to the GSH depletion in blood cells (Wang et al. 2006). Higher the rate of MDA production corresponds to As_2O_3 inversely associated with GSH. Depletion of GSH results in the increased production of arsenic-induced ROS, which may enhance the lipid peroxidation as observed in the present study.

The most significant alteration in the antioxidant defence is the decrease in GSH concentration, and GSH has direct antioxidant activity (Schulz et al. 2000). In this study, As_2O_3-treated rats showed decreased concentration of GSH and GSH-dependant antioxidant enzymes GPx

and GST. This reduction is suggested to be due to the consumption of glutathione while protecting against the arsenic-induced oxidative stress, for maintaining cellular redox status (Hughes 2002). As_2O_3 administration reduced the antioxidant and antiperoxidative enzyme concentration in liver tissue of experimental rats (Mathews et al. 2012a). GPx and GST play an important role in arsenic detoxification and the arsenic-induced oxidative stress (Thompson et al. 2009). GST utilises GSH as a cofactor, and therefore, the decrease in the activity of GST after As_2O_3 treatment may suggest coming from the paucity of GSH.

The exposure to arsenic decreased the activities of antiperoxidative enzymes SOD and CAT. The decreased SOD activity in serum suggested that the accumulation of superoxide anion radical might be responsible for increased lipid peroxidation following arsenic treatment (Maiti and Chatterjee 2000). ROS can themselves reduce the activity of the antioxidant enzymes CAT and GPx (Datta et al. 2000). Reduction in the antioxidant and antiperoxidative enzymes during As_2O_3 treatment may leads to the deposition of arsenic in tissues (Mathews et al. 2012b). SOD catalyses the dismutation of superoxide anions and prevents the subsequent formation of hydroxyl radicals in blood cells (Wang et al. 2006). In the present study, the decreased SOD activity may suggest that the accumulation of superoxide anion radical; might be responsible for increased lipid peroxidation following arsenic treatment as observed by Maiti and Chatterjee (2000). Exposure to arsenic decreased the CAT activity. CAT catalyses the removal of H_2O_2 formed during the reaction catalysed by SOD (Lee and Ho 1995). In the present study, the decreased CAT activity indicates the impaired ability to detoxify H_2O_2 and may leads to the accumulation of H_2O_2 and thereby oxidative stress.

Conclusion

In the current investigation, As_2O_3 treatment at its clinically different concentrations induced toxic effects by varying the blood glucose, CK, LDH and the oxidative status. As_2O_3-induced oxidative stress and the lipid peroxidation may be due to the reduced activity of GSH and GSH-dependant antioxidant and antiperoxidative enzymes. We also suggest that further studies are necessary for identifying the cellular and molecular mechanism of toxicity of arsenic at its clinical concentrations.

Acknowledgments We thank the University Grants Commission, New Delhi for rendering financial support for the study (F. No: 39-683/2010SR) and the award of research fellowship in sciences for meritorious student to Mr. Mathews. V. Varghese (No. F.4-1/2006 (BSR)/11-29/2008(BSR)).

Conflict of interest The authors declare that they have no conflict of interest.

References

Aebi H (1984) Catalase in vitro. Methods Enzymol 105:121–126

Aposhian HV, Aposhian MM (2006) Arsenic toxicology: five questions. Chem Res Toxicol 19:1–15

Basu A, Mahata J, Gupta S, Giri AK (2001) Genetic toxicology of a paradoxical human carcinogen, arsenic: a review. Mutat Res 488:171–194

Beutler E, Duron O, Kelly BM (1963) Improved method for the determination of blood glutathione. J Lab Clin Med 61:882–888

Beutler E, Gelbart T, Pegelow C (1986) Erythrocyte glutathione synthetase deficiency leads not only to glutathione but also to glutathione-S-transferase deficiency. J Clin Invest 77:38–41

Buege JA, Aust SD (1978) The thiobarbituric acid assay. Meth Enzymol 52:306–307

Datta K, Sinha S, Chattopadhyay P (2000) Reactive oxygen species in health and disease. Natl Med J India 13:304–310

Drabkin DL, Austin JH (1932) Spectrophotometric studies. I. Spectrophotometric constants for common hemoglobin derivatives in human, dog, and rabbit blood. J Biol Chem 98:719–733

Flora SJ (2011) Arsenic-induced oxidative stress and its reversibility. Free Radic Biol Med 51:257–281

Huan SY, Yang CH, Chen YC (2000) Arsenic trioxide therapy for relapsed acute promyelocytic leukemia: an useful salvage therapy. Leuk Lymphoma 38:283–293

Hughes MF (2002) Arsenic toxicity and potential mechanisms of action. Toxicol Lett 133:1–16

Hughes MF, Beck BD, Chen Y, Lewis AS, Thomas DJ (2011) Arsenic exposure and toxicology: a historical perspective. Toxicol Sci 123:305–332

Keyse SM, Applegate LA, Tromvoukis Y, Tyrrell RM (1990) Oxidant stress leads to transcriptional activation of the human heme oxygenase gene in cultured skin fibroblasts. Mol Cell Biol 10:4967–4969

Lee TC, Ho IC (1995) Modulation of cellular antioxidant defense activities by sodium arsenite in human fibroblasts. Arch Toxicol 69:498–504

Li Y, Sun X, Wang L, Zhou Z, Kang YJ (2002) Myocardial toxicity of arsenic trioxide in a mouse model. Cardiovasc Toxicol 2:63–73

Liu SX, Athar M, Lippai I, Waldren C, Hei TK (2001) Induction of oxyradicals by arsenic: implication for mechanism of genotoxicity. Proc Natl Acad Sci USA 98:1643–1648

Lu J, Chew EH, Holmgren A (2007) Targeting thioredoxin reductase is a basis for cancer therapy by arsenic trioxide. Proc Natl Acad Sci USA 104:12288–12293

Maiti S, Chatterjee AK (2000) Differential response of cellular antioxidant mechanism of liver and kidney to arsenic exposure and its relation to dietary protein deficiency. Environ Toxicol Pharmacol 8:227–235

Mathews VV, Binu P, Sauganth Paul MV, Abhilash M, Manju A, Nair RH (2012a) Hepatoprotective efficacy of curcumin against arsenic trioxide toxicity. Asian Pac J Trop Biomed 2(2):S706–S711

Mathews VV, Paul MS, Abhilash M, Manju A, Abhilash S, Nair RH (2012b) Mitigation of hepatotoxic effects of arsenic trioxide through omega-3 fatty acid in rats. Toxicol Ind Health.

Mathews VV, Paul MV, Abhilash M, Manju A, Abhilash S, Nair RH (2013) Myocardial toxicity of acute promyelocytic leukaemia drug-arsenic trioxide. Eur Rev Med Pharmacol Sci Suppl 1:34–38

Miller WH Jr, Schipper HM, Lee JS, Singer J, Waxman S (2002) Mechanisms of action of arsenic trioxide. Cancer Res 62:3893–3903

Mukherjee S, Das D, Darbar S, Mukherjee M, Das AS, Mitra C (2004) Arsenic trioxide generates oxidative stress and islet cell toxicity in rabbits. Curr Sci 86:854–857

Paglia DE, Valentine WN (1967) Studies on the quantitative and qualitative characterization of erythrocyte glutathione peroxidase. J Lab Clin Med 70:158–169

Paoletti F, Aldinucci D, Mocali A, Caparrini A (1986) A sensitive spectrophotometric method for the determination of superoxide dismutase activity in tissue extracts. Anal Biochem 154:536–541

Raghu KG, Yadav GK, Singh R, Prathapan A, Sharma S, Bhadauria S (2009) Evaluation of adverse cardiac effects induced by arsenic trioxide, a potent anti-APL drug. J Environ Pathol Toxicol Oncol 28:241–252

Ratnaike RN (2003) Acute and chronic arsenic toxicity. Postgrad Med J 79:391–396

Rust DM, Soignet SL (2001) Risk/benefit profile of arsenic trioxide. Oncologist 6(suppl 2):29–32

Schulz JB, Lindenau J, Seyfried J, Dichgans J (2000) Glutathione, oxidative stress and neurodegeneration. Eur J Biochem 267:4904–4911

Soignet SL, Frankel SR, Douer D, Tallman MS, Kantarjian H, Calleja E, Stone RM, Kalaycio M, Scheinberg DA, Steinherz P, Sievers EL, Coutré S, Dahlberg S, Ellison R, Warrell RP Jr (2001) United States multicenter study of arsenic trioxide in relapsed acute promyelocytic leukemia. J Clin Oncol 19:3852–3860

Tchounwou PB, Wilson B, Ishaque A (1999) Important considerations in the development of public health advisories for arsenic and arsenic-containing compounds in drinking water. Rev Environ Health 14:211–229

Thompson JA, White CC, Cox DP, Chan JY, Kavanagh TJ, Fausto N, Franklin CC (2009) Distinct Nrf1/2-independent mechanisms mediate As 3+-induced glutamate-cysteine ligase subunit gene expression in murine hepatocytes. Free Radic Biol Med 46:1614–1625

Tseng CH, Chang CK, Tseng CP, Hsueh YM, Chiou HY, Tseng CC, Chen CJ (2003) Long-term arsenic exposure and ischemic heart disease in arseniasis-hyperendemic villages in Taiwan. Toxicol Lett 137:15–21

Vega L, Gonsebatt ME, Ostrosky-Wegman P (1995) Aneugenic effect of sodium arsenite on human lymphocytes in vitro: an individual susceptibility effect detected. Mutat Res 334:365–373

Wang ZY, Chen Z (2008) Acute promyelocytic leukemia: from highly fatal to highly curable. Blood 111:2505–2515

Wang L, Xu ZR, Jia XY, Jiang JF, Han XY (2006) Effects of arsenic (As-III) on lipid peroxidation, glutathione content and antioxidant enzymes in growing pigs. Asian Aust J Anim Sci 5:727–733

24

Characterization of root-associated bacteria from paddy and its growth-promotion efficacy

Yachana Jha · R. B. Subramanian

Abstract Bacteria from rhizosphere (*Bacillus pumilus*) and endorhizophere (*Pseudomonas pseudoalcaligenes*) of rice plant were isolated and evaluated for their effect on the growth-promotion efficiency on rice in greenhouse. Ability to solubilize phosphate, siderophore, indoleacetic acid (IAA), gibberellin production and utilization of ACC (1-aminocyclopropane-1-carboxylate) as sole nitrogen source were evaluated, which were produced in high concentration by *P. pseudoalcaligenes* in this present study. Inoculation of isolated microorganism resulted in the reduction of pH (from neutral to acidic) of the medium used for phosphate solubilization, and has direct relation with titratable acidity, but gluconate production showed an opposite trend. *P. pseudoalcaligenes* better helped the plant to overcome or suppress fungal pathogen infection by producing β-1, 3-glucanase and chitinase as well as also have enhanced dry weight, plant height, and root length. Based on these results, *P. pseudoalcaligenes* in this study proved a better candidature as PGPR than *B. pumilus*.

Keywords PGPR · Phosphate solubilization · ACC deaminase · Phytohormones · Siderophore · NifH gene

Y. Jha (✉)
N. V. Patel College of Pure and Applied Sciences, Sardar Patel University, V. V. Nagar, Anand, Gujarat, India
e-mail: yachanajha@gmail.com; yachanajha@ymail.com

R. B. Subramanian
BRD School of Biosciences, Sardar Patel University, Post Box No. 39, V. V. Nagar, Anand 388120, Gujarat, India
e-mail: subramanianrb@gmail.com

Introduction

Rice (*Oryza sativa*) is one of the most important stable food crops in the world. In Asia, more than two billion people get 60–70 % of their energy requirement from rice and its derived products. To sustain present food self-sufficiency and to meet future food requirements, there is a need to increase rice productivity by 3 % per annum (Thiyagarajan and Selvaraju 2001). However, the production of rice is adversely affected by a number of biotic (viruses, bacteria, fungi, nematodes, insects, etc.) and abiotic (unfavorable soil, wound, temperature, flooding, etc.) stresses (Goff 1999). The techniques such as use of resistant variety, crop rotation, chemical method, and several other control methods have been used to meet the requirement of growing population. But these techniques have several drawbacks. Chemical methods have been used since long but they damage the natural beneficial insects, environment and also contaminate the natural resources. So presence of plant growth-promoting N_2-fixing bacteria and the possibility of a significant increase in plant performance and yield under nutrient limiting conditions have been discussed for many years. In the context of increasing international concern for food and environmental quality, the use of PGPR for reducing chemical inputs in agriculture is potentially important. PGPR has been applied to crops in various forms to enhance growth, seed emergence and crop yield (Minorsky 2008). Although plants are naturally exposed to several phytopathogenic microorganisms, they exhibit tolerance to these pathogens, through various morphological, anatomical structures (cuticles, trichomes, stomata and tyloses) and biochemical mechanisms (such as phenols, phytoalexins, cyanogenic glycosides, protease inhibitors and hydrolases) (Caramori et al. 2004). The objective of present study is to characterize and elucidate the effect of isolated PGPR on plant growth promotion by producing phytohormones, siderophores, ACC (1-aminocyclopropane-1-carboxylic acid) deaminase, amplification of nifH gene and pathogenesis-related proteins (PR proteins).

Materials and methods

Characterization of plant growth-promoting mechanism

Bacillus pumilus and *Pseudomonas pseudoalcaligenes* strains were isolated from the rice field and identified (data not shown) as per our published method (Jha et al. 2011). Their growth-promotion efficiency was analyzed by their ability to solubilize phosphate, produce siderophore, indoleacetic acid (IAA), gibberellins, and utilizes ACC as sole nitrogen source and to overcome or suppress infection by producing β-1, 3-glucanase and chitinase.

Quantitative estimation of phosphate solubilization

Phosphate solubilization was estimated by Ames (1964) method by inoculating fresh culture in freshly prepared 10 % ascorbic acid mixed with cold 0.42 % ammonium molybdate in 1 N H_2SO_4 in a ratio of 1:6 and incubated on an ice bath for at least 1 h. The readings were taken at an interval of 3 days in 3 replicates.

Estimation of titratable acidity and gluconic acid production

Titratable acidity was determined by titrating 1 ml of culture filtrate against 10 mM NaOH in presence of phenolphthalein (Whitelaw et al. 1999). For estimation of gluconic acid released by cultures, 1 ml of culture supernatant was used and estimation was done by Welcher's method (1958). The result was expressed in mmol l^{-1} and carried in 3 replicates.

Estimation of Indole acetic acid and gibberellic acid production

Overnight grown cultures were inoculated in N-broth containing 0.2 % yeast extract, 1 % glucose and incubated for 24 h, and indole acetic acid was estimated by Gordon and Weber (1951) method. Gibberellic acid production was estimated by colorimetric method of Hohlbrook et al. (1961). Absorbance was measured at 254 nm and experiment was carried out in 3 replicates.

Estimation of β-1, 3-glucanase and chitinase

Bacterial cultures were inoculated in N-broth and allowed to grow for 24 h at 30 °C on shaker at 150 rpm. The bacterial culture was centrifuged at 10,000*g* for 20 min and the supernatant was used as enzyme source. β-1, 3-glucanase activity expressed as nmol min^{-1} mg^{-1} was estimated by method of Pan et al. (1991). Chitinase activity was estimated by Reissig et al. (1995) method and expressed as μmol Glc-NAc equivalents' s^{-1} g^{-1}. Experiment was carried out in 3 replicates.

Estimation of hydroxymate and catechol siderophores production

Estimation of hydroxymate type siderophores was carried out by Mayer and Abdallah's (1978) method and catechol groups was estimated by Arnow's (1937) colorimetric assay method.

ACC deaminase activity assay

ACC deaminase activity of bacterial isolates was estimated by Penrose et al. (2001) method, and the amount of F-ketobutyric acid (F-KA) generated from the cleavage of ACC was monitored using spectrophotometer. The amount of F-KA produced during this reaction was determined by comparing the absorbance at 540 nm of a sample to a standard curve of F-ketobutyrate and expressed as the amount of F-ketobutarate produced per mg of protein per hour.

Extraction of genomic DNA and PCR amplification of nifH gene

For DNA extraction, colonies from bacterial isolates were cultured in 3 ml of liquid 1/2 DYGS medium overnight at 30 °C. The cells were centrifuged and further used for DNA extraction. Genomic DNA was extracted and purified by use of the Fast DNA spin kit (Qbiogene Inc., CA, USA) according to the manufacturer's protocol. Amplification of the nifH gene from the extracted DNA was performed using the primers Pol F (5'-TGCGAYCCSAARGCBG ACTC-3') and Pol R (5'-ATSGCCATCATYTCRCCGG A-3'). Amplification was performed in 50 ml final volume containing 1 ml genomic DNA (50 ng), 20 pmol each of forward and reverse primer, PolF and PolR, a 200 mM concentration of each of dNTPs (Sigma, USA), 10XTaq polymerase buffer and 2.5 U of Taq polymerase (Sigma, USA). PCR conditions consisted of initial denaturation step at 94 °C for 4 min, 30 amplification cycles of denaturation at 94 °C for 1 min, annealing at 55 °C for 1 min and primer extension at 72 °C for 2 min; followed by a final extension at 72 °C for 5 min with MyCycler™ PCR System (BioRad, USA). Aliquots of the PCR products were analyzed in 1.5 % (w/v) agarose gels (Sigma, USA) by horizontal gel electrophoresis. PCR products were eluted from agarose gel, purified and sequenced.

Inocula preparation, seedling germination and greenhouse study

Bacteria were grown in yeast mannitol broth (YMB) and exponentially growing cells in shaken broth culture were used for inoculation. Rice seeds were surface sterilized by 70 % ethanol in a flask and were treated with 1 % sodium hypochlorite for 2 min followed by six times washing with sterile water. After that, the seeds were soaked in various PGPR broths. Seeds soaked in normal broth were treated as control. Seeds of both inoculated and controls were put in sterilized petri dishes containing filter paper (Whatman #102) and the petri dishes were kept in an incubator at 30 °C for 120 h. After soaking, the air-dried seeds were used for germination and the survival percent.

The bacterial isolates, either alone or as a mixture, were assessed for their efficiency in suppressing rice blast under greenhouse conditions. The spore suspension of *Magnaporthe grisea* with a spore load of 10^4 conidia ml^{-1} was sprayed on the plants, which caused more than 75 % infection under greenhouse conditions. Observations on the percent disease incidence of rice blast were recorded. Disease index was calculated as grades 0–5 by Sriram et al. (1999) method using the formula:

$$Disease\ index = Total\ grade \times 100/No.\ of\ sheaths\ observed \times maximum\ grade.$$

Plant obtained from germinated seeds were transferred to plastic pots containing sterilized sand-perlite (1:1) and kept in a greenhouse. The plants were irrigated with water and Hoagland nutrient solution once a week. Shoots' and roots' lengths, fresh and dry weights were determined after 4 weeks. All experiments were carried in 3 replicates.

Statistical analysis

Data were analyzed by one-way ANOVA (analysis of variance). All treatments were replicated 3 times. Differences were considered to be significant at the $P < 0.05$ level. Means were compared by Fisher's protected LSD.

Results and discussion

Rhizosphere is the most dynamic ecological niche where inter and intra species interactions of microbes, such as bacteria, fungi and protozoa, occur due to the presence of a rich and diverse microbial food source (Bais et al. 2006). The importance of rhizosphere microbial populations for maintenance of root health by nutrient uptake, and tolerance of environmental stress are now well-recognized (Bowen and Rovira 1991). Secondly, biofertilization by PGPR improve nutrient status of plant by associative nitrogen fixation, phosphorus solubilisation and siderophores production, altering the permeability and transforming nutrients in the rhizosphere thus increasing their bio-availability (Mantelin and Touraine 2004). In addition, hormonal effects occur when PGPR either produce or metabolize chemical signaling compounds that directly impact on plant growth and function (Patten and Glick 2002).

In the present study, two bacterial isolates *P. pseudoalcaligenes* and *B. pumilus* were selected from thirty-five isolates obtained from the paddy field at the botanical garden of S. P. University, Gujarat, India and were found to be efficient with reference to their phosphate solubilizing capability. Bacterial genera such as *Bacillus, Pseudomonas* and *Brevibacillus* are known to be promoting growth, and

Table 1 Phosphate released, titratable acidity and gluconic acid concentration during solubilisation of tricalcium phosphate over incubation period of 12 days by the *P. pseudoalcaligenes* ($n = 3$)

Days	pH	Phosphate released ($\mu g\ P\ ml^{-1}$)	Titratable acidity ($\times 10^{-2}$)	Gluconate ($\times 10^{-4}$ g %)
P. pseudoalcaligenes				
0	7.00 ± 0.01	70.50 ± 2.22	0	0
3	5.03 ± 0.03	225.00 ± 5.63	7.69 ± 0.10	6.42 ± 1.29
7	3.51 ± 0.01	565.50 ± 10.67	26.40 ± 0.12	4.68 ± 1.29
9	3.80 ± 0.05	726.50 ± 19.30	30.70 ± 0.10	3.62 ± 1.90
12	3.90 ± 0.05	934 ± 20.14.00	32.50 ± 0.13	2.20 ± 2.24
B. pumilus				
0	7.00 ± 0.05	72.5 ± 06.24	0	0
3	5.00 ± 0.03	211.5 ± 10.10	12.3 ± 0.15	10.20 ± 1.76
7	5.03 ± 0.01	242.5 ± 14.00	14.3 ± 0.30	10.20 ± 2.29
9	5.55 ± 0.03	385.0 ± 07.50	19.0 ± 0.10	8.92 ± 1.40
12	5.83 ± 0.02	182.5 ± 17.55	09.1 ± 0.10	6.72 ± 1.15

Values are mean of three replications. ($P \leq 0.05$; LSD test)

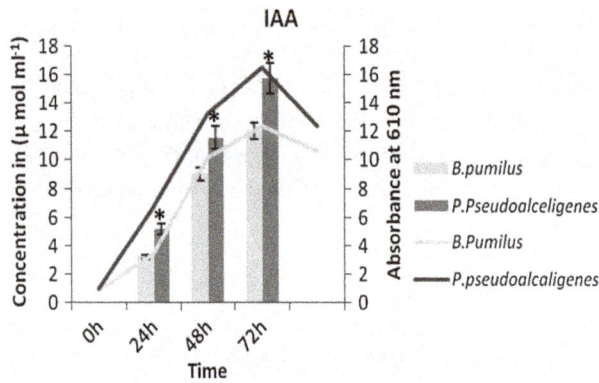

Fig. 1 IAA production by the isolates at different time interval on suitable medium ($n = 3$) in *Bar graph* and growth curve of isolates in *Line graph*

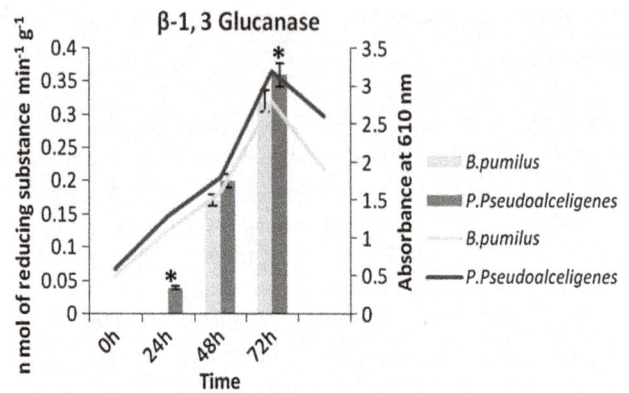

Fig. 3 β-1,3 glucanase production by *B. pumilus* and *P. pseudoalcaligenes* at different time interval on suitable medium ($n = 3$) in *Bar graph* and growth curve of isolates in *Line graph*

yield in different non-leguminous plants was also reported by Selva kumar et al. (2008). Phosphorus is one of the major nutrients, second only to nitrogen in requirement for plants. Most of phosphorus in soil is present in the form of insoluble phosphates and cannot be utilized by the plants (Pradhan and Sukla 2005). In the present study, the phosphate released by *B. pumilus* was increased by 5 times and titratable acidity by 1.5 times after 9 days of inoculation, while phosphate by *P. pseudoalcaligenes* released was increased by 13 times and titratable acidity by 4.1 times after 12 days of inoculation in the medium. Both the isolates were able to solubilize phosphorus with the production of gluconic acid as shown in Table 1. Zaidi et al. (2009) reported that the mineral phosphorus solubilization could probably be due to secretion of organic acids, such as gluconic, 2-ketogluconic. Production of organic acids for solubilisation of phosphates is a very well-known mechanism (Jones 1998); the reason for reduction in pH in present study may also be due to production of organic acids gluconate by the isolates. The reduction in pH also

has role in the phosphate solubilization and has been supported by Stumn and Morgan (1996), who reported that below pH 5, solubilization of phosphates of Ca, Al and Fe(III) increase.

Auxins and gibberelline may function as an important signal molecule in the regulation of plants growth and development. In this study, IAA and gibberellic acid production increased with time by both the isolates, production of IAA increased 4 times by *B. pumilus* and 3 times by *P. pseudoalcaligenes* (Fig. 1), while gibberellic acid increased 3 times by *B. pumilus* and only 2 times by *P. pseudoalcaligenes* in 96 h compared to initial concentration at 72 h (Fig. 2).

Siderophores are low molecular weight chemical compounds that scavenge iron, present in the environment as complexes and make this element available to the microorganisms (Neilands and Nakamura 1991). In the present study, the cultures were characterized for production of catechol and hydroxymate siderophores. *B. pumilus* produced 1.4 times higher catechol siderophore

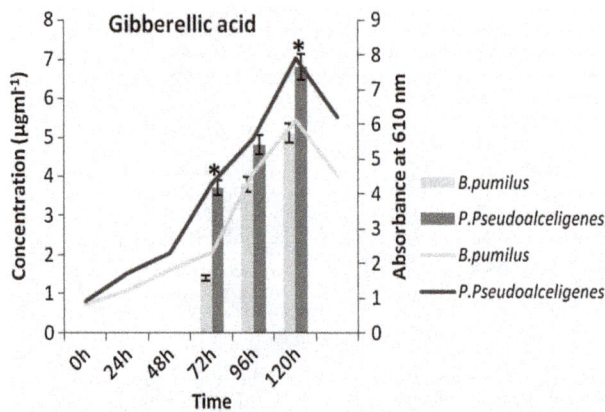

Fig. 2 GA₃ production by *B. pumilus* and *P. pseudoalcaligenes* at different time interval on suitable medium ($n = 3$) in *Bar graph* and growth curve of isolates in *Line graph*

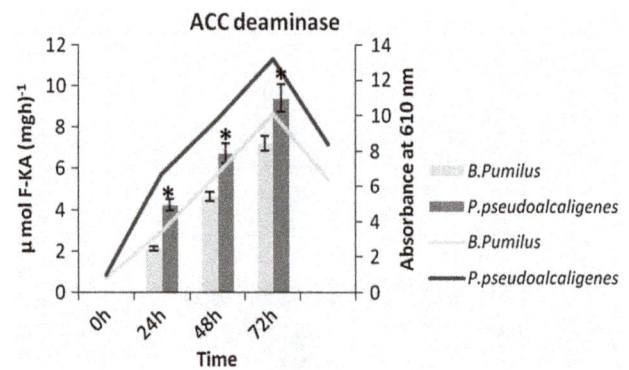

Fig. 4 ACC deaminase production by *B. pumilus* and *P. pseudoalcaligenes* at different time interval on suitable medium ($n = 3$) in *Bar graph* and growth curve of isolates in *Line graph*

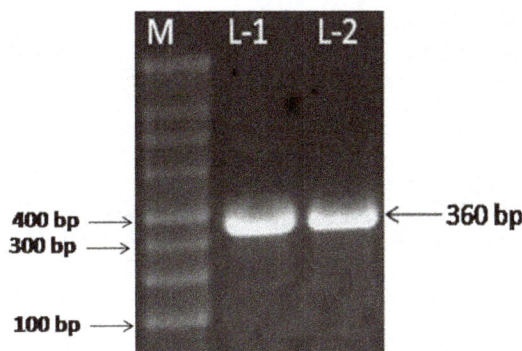

Fig. 5 Agarose gel electrophoresis of amplified NifH gene of isolates. *M* marker 100 bp DNA ladder, *L-1* and *L-2* NifH gene from *B. pumilus* and *P. pseudoalcaligenes*, respectively

Table 2 Effect of PGPR on growth parameter under glasshouse study ($n = 3$)

Mean ± SD			
	Control	Control + *B. pumilus*	Control + *P. pseudoalcaligenes*
Germination (%)	61.1 ± 0.01	65.4 ± 0.01	71.8 ± 0.04
Survival (%)	85.7 ± 0.11	89.1 ± 0.03	91.3 ± 0.02
Plant height (cm)	11.1 ± 0.04	13.6 ± 0.04	16.3 ± 0.01
Root length (cm)	2.7 ± 0.03	3.1 ± 0.01	4.0 ± 0.04
Dry weight (cm)	0.5 ± 0.02	0.7 ± 0.11	0.9 ± 0.02

Data are represented as means per pot from three pot replicates, each containing 15 transplanted plants per strain ($P \leq 0.05$; LSD test)

(12.9 µg ml^{-1}) and production of hydroxymate sidero-phore was 1.3 times higher by *P. pseudoalcaligenes* (3.75 µg ml^{-1}). These characters indicate that both isolates are better candidates for biofertilizer.

Many microorganisms produce and release lytic enzymes that can hydrolyze a wide variety of polymeric compounds, including chitin, proteins, cellulose, hemicellulose and expression of these enzymes by different microbes which can sometimes result in the suppression of plant pathogen activities directly; β-1, 3-glucanase and chitinase contribute significantly to biocontrol activities (Palumbo et al. 2006). Chitin is the most frequently occurring structural element of many invertebrates and fungi, and chitinase enzyme attacks chitin polymer. The β-1, 3-glucanase and chitinase production increased in both the isolates with time duration. β-1, 3-glucanase was 5 times high in 48 h and 9 times high at 72 h by *P. pseudoalcaligenes,* while in *B. pumilus*, it only increased by 0.5 times in same time duration (Fig. 3). Chitinase increased

17 times by *B. pumilus* at third day of inoculation and 5 times high at 24 h of incubation by *P. pseudoalcaligenes*. Its production increased by 15 times by *B. pumilus* and 3 times by *P. pseudoalcaligenes* in a 3-day time duration (data already communicated). Microorganisms, which secrete a complex of mycolytic enzymes, are considered to be potential biological control agents of plants diseases Dal Soglio et al. (1998).

In the present study, the ACC deaminase activity also increased with time duration by both the isolates. It increased 3.4 times by *B. pumilus* and 2 times by *P. pseudoalcaligenes* after 72 h (Fig. 4). Belimov et al. (2002) also reported that *B. pumilus* and *Pseudomonas putida* showed ACC deaminase from the rhizoplane of pea (*Pisum sativum*) and Indian mustard (*Brassica juncea*).

Presence of nifH gene was confirm by amplification of the structural gene for nitrogenase reductase (nifH) from the isolates, showed its potential for nitrogen fixation (Fig. 5). Xie et al. (2006) supported identification of nifH gene in the bacterial isolates from the rice field. In the greenhouse study the rice plants inoculated with isolates showed significantly higher plant height, root length and dry weight and also positive response on germination and survival percentage as shown in Table 2, and similar findings are reported by Chi et al. (2005). The present study strongly supports the development of biocontrol strategies using bacterial strains having antagonistic metabolites, to reduce the damage caused by plant pathogens. Present study showed that plants co-inoculated with PGPR and fungus *M. grisea* have disease index 38–43 % only in comparison to non-inoculated plants, where infection with fungus has 76 % disease index. The findings are supported by Ramamoorthy et al. (2002) who reported that PGPR plays a vital role in the management of various fungal diseases and Adesemoye et al. (2008) confirmed growth promotion by one representative each from both species of bacteria (*Pseudomonas* and *Bacillus*), but little variations were observed in bacterial effectiveness among parameters and crop types.

Observations were also supported by our studies on induction of defense related enzymes (Jha and Subramanian 2011) and accumulation of osmoprotectants (Jha et al. 2011) in presence of *P. pseudoalcaligenes* and *B. pumilus* alone and in combination helps the paddy under stress.

Conflict of interest We certify that there is no conflict of interest with any financial organization regarding the material discussed in the manuscript.

References

Adesemoye AO, Torbert HA, Kloepper JW (2008) Enhanced plant nutrient use efficiency with PGPR and AMF in an integrated nutrient management system. Can J Microbiol 54:876–886

Ames BN (1964) Assay of inorganic phosphate, total phosphate and phosphatases. Met Enzymol 8:115–118

Arnow LE (1937) Colorimetric determination of the components of 3,4dihydroxyphenylalanine-tyrosine mixtures. J Biol Chem 118:531–537

Bais HP, Weir TL, Perry LG, Gilroy S, Vivanco JM (2006) The role of root exudates in rhizosphere interactions with plants and other organisms. Annu Rev Plant Biol 57:233–266

Belimov AA, Safronova VI, Mimura T (2002) Response of spring rape (Brassica napus var. oveifera L.) to inoculation with plant growth-promoting rhizobacteria containing 1-aminocyclopropane-1-carboxylate deaminase depends on nutrient status of the plant. Can J Microbiol 48:189–199

Bowen GD, Rovira AD (1991) The rhizosphere and its management to improve plant growth. Adv Agron 66:1–102

Caramori SS, Lima CS, Fernandes KF (2004) Biochemical characterization of selected plant species from Brasilian savanas. Braz Arch Biol Tech 47:253–259

Chi F, Shen S-H, Cheng H-P, Jing Y-X, Yanni YG, Dazzo FB (2005) Ascending migration of endophytic rhizobia, from roots to leaves, inside rice plants and assessment of benefits to rice growth physiology. Appl Environ Microbiol 71:7271–7278

Dal Soglio FK, Bertagnolli BL, Sinclair JB, Yu GZ, Eastbum DM (1998) Production of chitinolytic enzymes and endoglucanase in the soybean rhizosphere in the presence of Trichoderma harzianum and Rhizoctonia solani. Biol Control 12:111–117

Goff SA (1999) Rice as a model for cereal genomics. Curr Opin Plant Biol 2:86–89

Gordon SA, Weber RP (1951) Colorimetric estimation of indoleacetic acid. Plant Physiol 2:192–195

Holbrook A, Edge W, Bailey F (1961) Spectrophotometric method for determination of gibberellic acid. Adv Chem Ser 28:159–167

Jha Y, Subramanian RB (2011) Endophytic Pseudomonas pseudoalcaligenes shows better response against the Magnaporthe grisea than a rhizospheric Bacillus pumilus in Oryza sativa (Rice). Arch Phytopathol PFL 44:592–604

Jha Y, Subramanian RB, Patel S (2011) Combination of endophytic and rhizospheric plant growth promoting rhizobacteria in Oryza sativa shows higher accumulation of osmoprotectant against saline stress. Acta Physiol Plant 33:797–802

Jones DL (1998) Organic acids in the rhizosphere—a critical review. Plant Soil 205:25–44

Mantelin S, Touraine B (2004) Plant growth-promoting bacteria and nitrate availability: impacts on root development and nitrate uptake. J Exp Bot 55:27–34

Mayer JM, Abdallah MA (1978) The florescent pigment of Pseudomonas fluorescens Biosynthesis, purification and physical-chemical properties. J Gen Microbiol 107:319–332

Minorsky PV (2008) On the inside. Plant Physiol 146:323–324

Neilands JB, Nakamura K (1991) In: Winkelmann G (ed) CRC handbook of microbial iron chelates. CRC Press, Florida, pp 1–14

Palumbo JD, Baker JL, Mahoney NE (2006) Isolation of bacterial antagonists of Aspergillus flavus from almonds. Microbial Ecol 52:45–52

Pan SQ, Ye XS, Kuc J (1991) Association of β-1,3 glucanase activity and isoform pattern with systemic resistance to blue mold in tobacco induced by stem injection with Peronospora tabacina or leaf inoculation with tobacco mosaic virus. J Physiol Mol Plant Pathol 39:25–30

Patten CL, Glick BR (2002) Role of Pseudomonas putida indoleacetic acid in development of the host plant root system. App Environ Microbiol 68:3795–3801

Penrose DM, Barbara M, Glick BR (2001) Determination of ACC to assess the effect of ACC-deaminase-containing bacteria on roots of canola seedlings. Can J Microbiol 47:77–80

Pradhan N, Sukla LB (2005) Solubilization of inorganic phosphate by fungi isolated from agriculture soil. Afr J Biotechnol 5:850–854

Ramamoorthy V, Raguchander T, Samiyappan R (2002) Induction of defense-related proteins in tomato roots treated with Pseudomonas fluorescens Pf1 and Fusarium oxysporum f. sp. lycopersici. Plant Soil 239:55–68

Reissig Jl, Strominger LF, Leloir J (1995) A modified colorimetric method for the estimation of N-acetylamino sugars. J Biol Chem 217:959–966

Selva kumar G, Kundu S, Gupta AD, Shouche YS, Gupta HS (2008) Isolation and characterization of nonrhizobial plant growth promoting bacteria from nodules of Kudzu (Pueraria thunbergiana) and their effect on wheat seedling growth. Curr Microbiol 56:134–139

Sriram PP, Shin YC, Park CS, Chung YR (1999) Biological control of fusarium wilts of cucumber by chitinolytic bacteria. Phytopathol 89:92–99

Stumn W, Morgan JJ (1996) Aquatic chemistry. Wiley, New York, pp 404–409

Thiyagarajan TM, Selvaraju R (2001) Water saving in rice cultivation in India. In: Proceedings of an international workshop on water saving rice production systems. Nanjing University, China, pp 15–45

Welcher FJ (1958) The analytical uses of ethylene diamine tetraacetic acid (EDTA). D. Van Nostrand company, Inc., Princeto

Whitelaw MA, Harden TJ, Helyar KR (1999) Phosphate solubilisation in solution culture by the soil fungus Penicillium radicum. Soil Biol Biochem 31:655–665

Xie GH, Cui Z, Yu CJ, Yan J, Hai W, Steinberger Y (2006) Identification of nif genes in N2-fixing bacterial strains isolated from rice fields along the Yangtze River Plain. J Basic Microbiol 46:56–63

Zaidi A, Khan MS, Ahemad M, Oves M (2009) Plant growth promotion by phosphate solubilizing bacteria. Acta Microbiol Immunol Hung 56:263–284

25

Influence of insecticides flubendiamide and spinosad on biological activities in tropical black and red clay soils

G. Jaffer Mohiddin · M. Srinivasulu ·
K. Subramanyam · M. Madakka · D. Meghana ·
V. Rangaswamy

Abstract A laboratory experiment has been conducted to investigate the ecological toxicity of flubendiamide and spinosad at their recommended field rates and higher rates (1.0, 2.5, 5.0, 7.5, 10.0 kg ha^{-1}) on cellulase, invertase and amylase in black and red clay soils after 10, 20, 30 and 40-day exposure under controlled conditions in groundnut (*Arachis hypogaea* L.) soils of Anantapur District, Andhra Pradesh, India. Flubendiamide and spinosad were stimulatory to the activities of cellulase, invertase and amylase at lower concentrations at 10-day interval. The striking stimulation in soil enzyme activities noticed at 2.5 kg ha^{-1}, persists for 20 days in both soils. Overall, the higher concentrations (5.0–10.0 kg ha^{-1}) of flubendiamide, and spinosad were toxic or innocuous to cellulase, invertase and amylase activities, respectively. The results of the present study thus, clearly, indicate that application of the insecticides in cultivation of groundnut, at field application rates improved the activities of cellulase, invertase and amylase in soils.

Keywords Enzyme activities · Flubendiamide · Groundnut (*Arachis hypogaea* L.) soils · Spinosad

Introduction

In modern agriculture, it has become a common trend to apply different groups of pesticides, either simultaneously or in succession, for effective control of a variety of pests (Quazi et al. 2011). Pesticides are deliberately introduced into agricultural systems with various formulations to protect crops against weeds, insects, fungi and other pests (Yang et al. 2007; Moorman 1989; Singh et al. 1999; Bhuyan et al. 1992; Chu et al. 2008). However, much of the applied pesticides will finally reach the soil often leading to a combined contamination of pesticide residues in the soil environment (Chu et al. 2008), which may affect the growth and activity of soil microbial communities (Singh and Singh 2005), and in turn affect the enzyme activities.

G. J. Mohiddin (✉) · M. Srinivasulu · D. Meghana ·
V. Rangaswamy
Department of Microbiology, Sri Krishnadevaraya University,
Anantapur 515 055, Andhra Pradesh, India
e-mail: jaffermicro@gmail.com

M. Srinivasulu
e-mail: mandalasrinivasulu@yahoo.in

D. Meghana
e-mail: meghanadasetty@gmail.com

V. Rangaswamy
e-mail: rangamanjula@yahoo.com

M. Madakka
Department of Biotechnology and Bioinformatics, Yogi Vemana
University, Kadapa 516 003, Andhra Pradesh, India
e-mail: mekapogu@gmail.com

K. Subramanyam
Plant Molecular Biology Unit, Department of Biotechnology and
Genetic Engineering, Bharathidasan University, Tiruchirappalli
620024, Tamil Nadu, India
e-mail: prasamshika2@gmail.com

M. Srinivasulu
Radioactive Waste Management (Bioremediation) Lab, Division
of Advanced Nuclear Engineering, Pohang University of Science
and Technology, Pohang-si, Republic of Korea

Present Address:
G. J. Mohiddin
Department of Life Sciences and Agriculture, Universidad de las
Fuezas Armada, Sangolqui, Quito, Ecuador, South America

Increasing use of pesticides in agriculture led to the development of soil microbial testing programme for examination of the side effects (Swaminathan et al. 2009). The testing programmes include measurement of activities of soil enzymes, and physicochemical properties.

The economy of India is largely dependent on agricultural production. Better harvest requires rigorous cultivation, irrigation, fertilizers and pesticides to protect plants from pests and plant diseases. In India, 15–20 % of all produce is destroyed by pests (Bhalerao and Puranik 2009). Groundnut (*Arachis hypogaea* L.) is one of the most important cash crops grown in Indian agricultural soils with the highest yield among the oil seeds' crops (Singh and Singh 2005; Menon et al. 2004; Bera et al. 2002) and is the primary source of edible oil in India (Ramesh babu et al. 2002). India is a world leader in groundnut farming, with 6.0 million hectares of the cultivated area during the year 2010–11 (USDA 2011). Within India, Andhra Pradesh State ranks first in area and production (Hegde and Kiresur 1999). Among different regions of Andhra Pradesh, Anantapur District, a semi-arid region relies on groundnut cultivation, predominantly (Anonymous 2011). In spite of its high range of cultivation, groundnut productivity is low, fluctuating around 9 q/ha on average, and an annual yield loss of Rs. 150 crores due to pests has been reported (Loganathan et al. 2002). Among the pesticides, insecticides of flubendiamide and spinosad are in the current list of modern pesticides in Indian agriculture used to control incidence of pest attack over groundnut crop.

Soil is a natural system containing microbes which are the driving force behind many soil processes, including transformation of organic matter, nutrient release and degradation of xenobiotics (Zabaloy et al. 2008). Many studies have shown that biological parameters have been used to assess soil quality and health as affected by agricultural practices (Gianfreda et al. 2005; Truu et al. 2008; Garcia-Ruiz, et al. 2009). In this respect, soil enzymes can be used as potential indicators of soil quality for sustainable management because they are sensitive to ecological stress and land management practices (Tejada 2009). Flubendiamide represents a novel class of insecticides with extremely high activity against a broad spectrum of lepidopterous insects (Tohnishi et al. 2005). Spinosad is a biologically derived insecticide that consists of two active compounds, spinosyns A and D, produced by fermentation culture of an actinomycete isolated from soil (*Saccharopolyspora spinosa* Mertz and Yao). Structurally, these compounds are macrolides and contain a unique tetracycling system to which two different sugars are attached (Kirst et al. 1992). Negative impact of pesticides on soil enzymes activities has been widely reported throughout the literature (Ismail et al. 1998; Menon et al. 2005)

unfortunately no reports were available on these two new insecticides on enzyme activities.

The quorum-sensing systems allow bacteria to monitor their environment for the presence of other bacteria and to respond to fluctuations in the number and/or species present by altering particular behaviors. Most quorum-sensing systems are species- or group-specific, which presumably prevents confusion in mixed-species environments. However, some quorum-sensing circuits control behaviors that involve interactions among bacterial species. These quorum-sensing circuits can involve both intra- and interspecies communication mechanisms. Finally, anti-quorum-sensing strategies are present in both bacteria and eukaryotes and these are apparently designed to combat bacteria that rely on cell-cell communication and for the successful adaptation to particular niches. Many enzymes of both microbial or plant origins have been recognized to be able to transform pollutants at a detectable rate and potentially suitable to restore polluted environment. The main enzymatic classes involved in such a process are hydrolases, dehalogenases, and oxidoreductases. Amide, ester and peptidic bonds undergo hydrolytic cleavage by amidases, esterases and proteases in several xenobiotic compounds and may lead to products with little or no toxicity. Hydrolases responsible for the cleavage of pesticides are among the best studied groups of enzymes. Most of these hydrolases are extracellular enzymes, except for the cell wall-bound enzymes of penicillium and arthrobacter sp., which hydrolyze barban and propham.

Cellulase can catalyze hydrolysis of 1, 4, beta- D-glycosidic bonds of cellulose and is also an important indicator for carbon circulation. Invertase is known to be a very stable and persistent enzyme, and its association with soil components is well documented (Kiss et al. 1978). Amylase plays an important role in biochemical reactions and nutrient cycling. Apparently, it has become necessary to determine the effects of agronomically needed pesticides (flubendiamide and spinosad), applied at recommended levels and at higher doses, in order to establish the significance, in terms of biogeochemical reactions and nutrient cycling. Hence the present study was carried out to determine the influence of insecticides on the activity of cellulase, invertase and amylase in two groundnut soils of Anantapur district, Andhra Pradesh, India from December 2, 2010 to July 15, 2011.

Materials and methods

Soils

Black and red clay soils were used in the present study. Soil samples taken from groundnut-cultivated fields of Anantapur district, Andhra Pradesh, India, were chosen with a known history of pesticides use, from a depth of 12 cm, air-dried and sieved through 2 mm sieve before usage. Mineral matter of soil samples such as sand, silt, and clay contents were analyzed with use of different sizes of sieves by following the method of Alexander (1961). Cent percent water-holding capacity of soil samples was measured by finding amount of distilled water added to both the soil samples to get saturation point and then 60 % water-holding capacity of soil was calculated by the Johnson and Ulrich method (1960). Soil pH was measured at 1:1.25 soil-to-water ratio in a Systronics digital pH meter with calomel glass electrode assembly. Organic carbon content in soil samples was estimated by Walkley–Black method, and the organic matter was calculated by multiplying the values with 1.72 (Jackson 1971). Electrical conductivity of soil samples after addition of 100 ml distilled water to 1 g soil samples was measured by a conductivity bridge. Total nitrogen content in soil samples was determined by the method of micro-Kjeldahl method (Jackson 1971). Content of inorganic ammonium–nitrogen in soil samples after extraction of 1 M KCl by Nesslerization method (Jackson 1971), contents of nitrite–nitrogen (Barnes and Folkard 1951) and contents of nitrate–nitrogen by Brucine method (Ranney and Bartlett 1972) after extraction with water were determined, respectively. Physicochemical characteristics of the two soils are listed in Table 1.

Insecticides

In order to determine the influence of selected insecticides on the microbial activities, commercial grades of flubendiamide and spinosad were obtained from Bayer's Science India.

Soil treatment

The soil ecosystem stimulating non-flooded portions of the soil samples were added in test tubes (25 × 150 mm) and moistened with water in order to maintain at 60 % water-holding capacity. Same model was used previously to elucidate the effect of insecticides on microbial activities by Mohiddin et al. (2011).

Cellulase (EC 3.2.1.4), invertase activity (EC 3.2.1.26) and amylase activity (EC 3.2.1.1)

Five-gram portion of the soil samples was weighed and dispersed into sterile test tubes (25 × 150 mm). Stock solutions from selected insecticides were added to the rate of 10, 25, 50, 75 and 100 $\mu g \ g^{-1}$ soil equivalent to field application rates of 1.0, 2.5, 5.0, 7.5 and 10.0 kg ha^{-1} respectively. Soil samples without insecticide treatment served as controls. Soil samples were mixed thoroughly for uniform distribution of insecticide that was added. Triplicates were maintained for each treatment at room temperature (28 ± 4 °C) with 60 % water-holding capacity throughout the incubation period. After desired intervals of incubation, soil samples were extracted in distilled water for estimation of enzyme activities. Similar model was used earlier by (Singaram and Kamalakumari 2010; Mohiddin et al. 2010).

In order to determine cellulase enzyme activity in soils, the method employed for the assay of cellulase was developed by Cole (1977) and followed by Tu (1981a, b). The soil samples were transferred to 100 ml Erlenmeyer flasks and were treated with 1 ml of toluene to arrest the enzyme activity. After 15 min, 10 ml of carboxy methyl cellulose (CMC) 1 % was used as a substrate followed by 10 ml of acetate buffer (pH 5.9) and incubated for 24 h to determine the reducing sugar content in the filtrate (Deng and Tabatabai 1994). In another experiment, cellulase activity was determined at 10, 20, 30 and 40 days of soil incubation. Testing samples were passed through Whatman No. 1 filter paper and the filtrate was assayed for the amount of glucose by the Nelson method (1944) in a Spectronic 20 D spectrophotometer.

The method employed for assay of invertase was developed by Cole (1977) and followed by (Tu 1981a, b). The soil samples were transferred to 100 ml Erlenmeyer flasks and treated with 1 ml toluene to arrest the enzyme activity. After 15 min, 6 ml of 18 mM sucrose was added to the soil samples and incubated for 24 and 48 h; the testing samples were passed through Whatman No. 1 filter paper and the filtrate was assayed for the amount of glucose

Table 1 Physicochemical properties of the soils

Properties	Black clay soil	Red clay soil
Sand (%)	68.45	53.25
Silt (%)	21.45	27.12
Clay (%)	10.0	19.8
pH[a]	7.8	6.7
Water holding capacity (ml g^{-1} soil)	0.7	0.4
Electrical conductivity (mmhos)	258	232
Organic matter (%)[b]	1.34	0.74
Total nitrogen (%)[c]	0.086	0.038
NH$_4^+$–N ($\mu g \ g^{-1}$ soil)[d]	6.96	6.01
NO$_2^-$–N ($\mu g \ g^{-1}$ soil)[e]	0.58	0.42
NO$_3^-$–N ($\mu g \ g^{-1}$ soil)[f]	0.94	0.73

[a] 1:1.25 = soil:water slurry

[b] Walkley–Black method (Johnson and Ulrich 1960)

[c] Micro-Kjeldahl method (Johnson and Ulrich 1960)

[d] Nesslerization method (Johnson and Ulrich 1960)

[e] Diazotization method (Ranney and Bartlett 1972)

[f] Brucine method (Barnes and Folkard 1951)

by the Nelson method (1944) in a Spectronic 20 D spectrophotometer.

The method employed for the assay of amylase was developed by Cole (1977) and followed by Tu (1981a, b). The soil samples were transferred to 100 ml Erlenmeyer flasks and treated with 1 ml toluene to arrest the enzyme activity. After 15 min, 6 ml of 0.2 M of acetate phosphate buffer (5.5 pH) containing 2 % starch was added to each of the testing samples and closed with cotton plugs. After 24 and 72 h of incubation, the testing samples were made up to a volume of 50 ml with sterile distilled water and passed through Whatman No. 1 filter paper and the filtrate was assayed for the amount of glucose by Nelson's method (1944) in a Spectronic 20 D spectrophotometer.

Statistical analysis

The concentration of the cellulase, invertase and amylase was calculated based on soil weight (oven dried). Data were analyzed using one-way ANOVA, and the differences contrasted using Duncan's multiple range test (DMRT) (Megharaj et al. 1999; Mohiddin et al. 2011). All statistical analyses were performed at $P \leq 0.05$ using SPSS statistical software package.

Results and discussion

The black and red clay soils are predominantly used for the cultivation of groundnut (*Arachis hypogaea* L.) in the Anantapur district of Andhra Pradesh, India. The major constraints in the groundnut crop are insects and fungi pests. For this reason, pesticides are frequently used for crop protection. Continuous and indiscriminate use of these pesticides causes a major risk of soil health. Hence, these soils were selected to study the effect of insecticides on

enzyme activities. In general, the organic matter content is high in black soil. Therefore, the biological activity was also pronounced more in black soil than in red soil under the influence of insecticides. There have been many reports of the effects of pesticides on soil enzyme activities (Anonymous 2011; Loganathan et al. 2002) and it has been

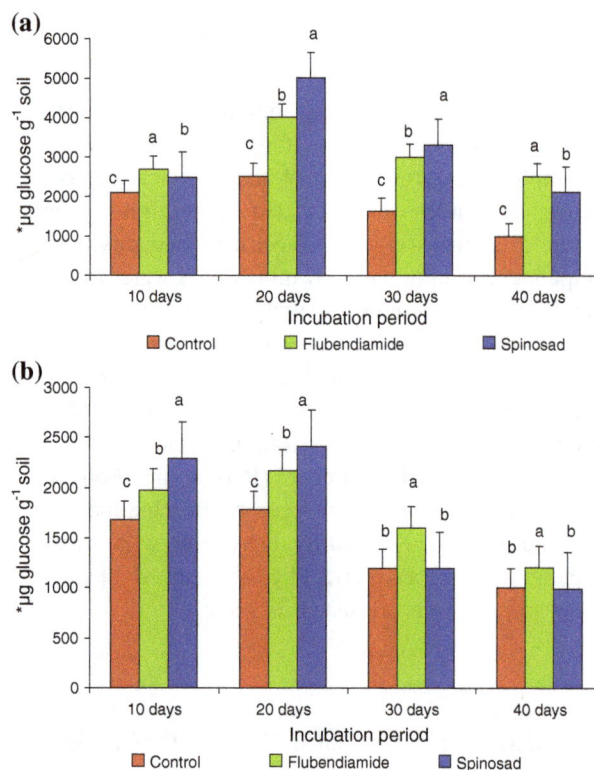

Fig. 1 Influence of flubendiamide and spinosad on cellulase*activity in **a** black clay and **b** red clay soil at 2.5 kg ha^{-1}. *µg glucose per gram soil formed after 24 h incubation with Carboxy methyl cellulose (CMC). The values are the mean ± SE for each incubation period, are not significantly different ($P \leq 0.05$) from each other according to Duncan's multiple range (DMR) test

Table 2 Activity of cellulase under the impact of different concentrations of flubendiamide and spinosad in black and red clay soils for 24 h after 10 days

Concentration of insecticides (kg ha^{-1})	Black clay soil		Red clay soil	
	Flubendiamide 24 h	Spinosad 24 h	Flubendiamide 24 h	Spinosad 24 h
0.0	2,090 ± 5.773 c (100)	2,090 ± 5.773 c (100)	1,680 ± 11.547 c (100)	1,680 ± 11.547 d (100)
1.0	2,400 ± 57.735 b (115)	2,330 ± 17.320 b (134)	1,720 ± 5.773 b (102)	1,880 ± 11.547 c (112)
2.5	2,700 ± 2.886 a (129)	2,490 ± 2.886 a (172)	1,980 ± 0.577 a (118)	2,290 ± 4.041 a (136)
5.0	1,700 ± 17.320 d (81)	1,350 ± 5.773 d (134)	1,400 ± 17.320 d (83.3)	2,200 ± 57.735 b (130)
7.5	1,390 ± 4.041 e (66)	1,190 ± 0.577 f (128)	1,090 ± 2.309 e (65)	1,200 ± 25.980 e (71)
10.0	1,100 ± 5.773 f (52)	1,250 ± 3.464 e (60)	990 ± 1.732 f (97)	1,120 ± 14.433 f (67)

µg glucose per gram soil formed after 24 h of incubation with 1 % carboxy methyl cellulose (CMC)

Each column is mean ± SE for six concentrations in each group; columns not sharing a common letter (a, b, c, d, e and f) differ significantly with each other ($P \leq 0.05$; DMRT)

observed that the responses of soil enzymes to different pesticides are not the same. As a new pesticide, unfortunately, there is no information available regarding the influence of flubendiamide and spinosad on soil enzyme activities. Of course, when the flubendiamide and spinosad concentration was increased, the potential hazard to soil would increase. Soil enzyme activities are more sensitive to the environment. They reflect the soil quality more quickly and directly (Srinivasulu et al. 2012).

Since enzyme activity has been considered as a very sensitive indicator, any disturbance due to biotic or environmental stresses in the soil ecosystem may affect soil biological properties. Our analysis revealed that cellulase activity was significantly increased from 0.1 to 2.5 kg ha^{-1} whereas the activity was decreased at higher concentrations (5.0–10.0 kg ha^{-1}) of pesticides in both soils (Table 2). The cellulase activity was significantly enhanced at 2.5 kg ha^{-1} level in both soils for flubendiamide and spinosad and showed individual increments of cellulase activity ranging from a low increase 15–29, 11–19 and 2–18, 12–36 % in comparison to control (Table 2). The

stimulatory concentration (2.5 kg ha^{-1}) induces the highest cellulase activity after 20, 30 and 40 days of incubation in black clay soils (Fig. 1a) with flubendiamide and spinosad when compared to control. Whereas in red clay soil a similar trend was followed by flubendiamide, induces the highest cellulase activity after 20, 30 and 40 days of incubation but spinosad showed a variable pattern was observed at 30 and 40 days, the cellulase enzyme activity remained same with control (Fig. 1b). The relatively low activity of cellulase might result from the toxic effect of flubendiamide and spinosad on soil microorganisms, which in turn produces cellulase. The inhibition of cellulase activity by flubendiamide and spinosad could be attributed to the properties of flubendiamide and spinosad. Similar type of reports were identified by (Ramudu et al. 2011; Mohiddin et al. 2010) chlorothalonil and propiconazole, imidacloprid, and acephate. Similar observations were made by Katayama and Kuwatsuka (1991) and Jaya Madhuri and Rangaswamy (2002) on the cellulase activity. Analogous report was obtained by Ismail et al. (1996a, b) on application of metolachor to Malaysian soil. Gigliotti

Table 3 Activity of invertase under the impact of different concentrations of flubendiamide and spinosad in black clay soil for 24 and 48 h after 10 days

Concentration of insecticides (kg ha^{-1})	Flubendiamide		Spinosad	
	24 h	48 h	24 h	48 h
0.0	900 ± 0.577 c (100)	950 ± 2.309 e (100)	900 ± 0.577 c (100)	950 ± 2.309 e (100)
1.0	915 ± 2.886 b (102)	960 ± 5.773 d (101)	900 ± 0.577 b (100)	1,780 ± 11.547 c (187)
2.5	1,520 ± 5.773 a (169)	1,650 ± 4.618 a (174)	1,620 ± 4.618 a (180)	1,900 ± 17.320 a (200)
5.0	800 ± 5.773 d (89)	1,310 ± 1.732 b (138)	600 ± 0.577 d (67)	1,850 ± 5.773 b (195)
7.5	750 ± 3.464 e (83)	980 ± 11.547 c (103)	440 ± 23.094 e (49)	1,060 ± 3.464 d (111)
10.0	700 ± 1.732 f (78)	800 ± 5.773 f (84)	400 ± 2.886 f (44)	900 ± 0.577 f (95)

µg glucose per gram soil formed after 24 h of incubation with 18 Mm sucrose

Each column is mean ± SE for six concentrations in each group; columns not sharing a common letter (a, b, c, d, e and f) differ significantly with each other ($P \leq 0.05$; DMRT)

Table 4 Activity of invertase under the impact of different concentrations of flubendiamide and spinosad in red clay soil for 24 and 48 h after 10 days

Concentration of insecticides (kg ha^{-1})	Flubendiamide		Spinosad	
	24 h	48 h	24 h	48 h
0.0	420 ± 4.618 f (100)	800 ± 8.660 d (100)	420 ± 4.618 d (100)	800 ± 5.773 f (100)
1.0	600 ± 0.577 d (143)	1,020 ± 4.618 b (127)	580 ± 11.547 b (138)	1,320 ± 4.041 d (165)
2.5	760 ± 4.041 a (181)	1,120 ± 11.547 a (140)	600 ± 0.577 a (143)	1,800 ± 0.577 a (225)
5.0	700 ± 1.732 b (166)	920 ± 4.618 c (115)	589 ± 11.547 b (140)	1,720 ± 11.547 c (215)
7.5	650 ± 28.867 c (155)	780 ± 11.547 e (97)	490 ± 2.886 c (117)	1,620 ± 5.773 b (202)
10.0	620 ± 11.547 e (148)	670 ± 17.320 f (84)	350 ± 10.392 e (83)	1,300 ± 2.886 e (162)

µg glucose per gram soil formed after 24 h of incubation with 18 Mm sucrose

Each column is mean ± SE for six concentrations in each group; columns not sharing a common letter (a, b, c, d, e and f) differ significantly with each other ($P \leq 0.05$; DMRT)

et al. (1998) also reported that bensulfurn methyl at 16 and 160 μg/g inhibited cellulase activity in soil samples. In a diverse study made by Gherbawy and Abdelzaher (1999),

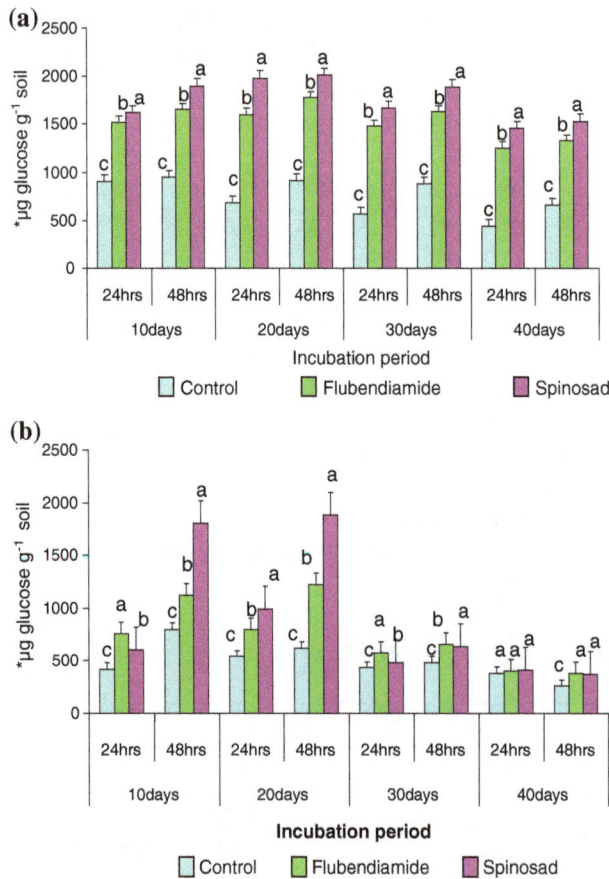

Fig. 2 Influence of flubendiamide and spinosad on invertase*activity in **a** black clay and **b** red clay soil at 2.5 kg ha^{-1}. *μg glucose per gram soil formed after 24 and 48 h incubation with 18 Mm sucrose. The values are the mean ± SE for each incubation period, are not significantly different ($P \leq 0.05$) from each other according to Duncan's multiple range (DMR) test

alteration in the activity of cellulase by metalaxyl was marked in pure fungal cultures. Similar results were obtained by Arinz and Yubedee (2000) that kelthane and fenvalerate caused inhibition to enzyme activity.

Invertase activity was depressed in flubendiamide and spinosad treated soils throughout the experiment when compared to the controls in both soils incubated for 10 days (Tables 3, 4) The maximum activity was observed at 2.5 kg ha^{-1} (stimulatory) for flubendiamide and spinosad showed individual increments of invertase activity ranged from a low increase 2–69, 1–74 % and 0–80, 87–100 % for black clay soil and for red clay soil, 43–81, 27–40 % and 38–43, 65–125 % received 2.5 kg ha^{-1} respectively in comparison to control at 24 and for 48 h (Tables 3, 4). The results reveal that invertase enzyme is rather sensitive to flubendiamide and spinosad. Figure 2 showed the variation of invertase activity after flubendiamide and spinosad application. Although enzyme activities of samples were lower than that of control, significant differences ($P \leq 0.05$) were found among the enzyme activities between treated soil samples and the control (Tables 3, 4). With the increase in incubation periods, the stimulated enzyme activities were also increased up to 20-days further increase in the incubation decrease in the enzyme activity was noticed (Fig. 2a, b). Our results appeared to be consistent with previous reports, in which it is demonstrated that pesticides stimulated invertase activity of soils (Ramudu et al. 2011; Sannino and Gianfreda 2001; Srinivasulu and Rangaswamy 2006). Rate of invertase activity followed the same trend of initial stimulation followed by inhibition as reported by Rangaswamy and Venkateswarlu (1992). On the contrary, Tu (1995) affirmed initial inhibition followed by recovery with five insecticides in sandy loam soil.

Amylase activity (Tables 5, 6) showed a variable pattern in response to different insecticide concentration after

Table 5 Activity of amylase under the impact of different concentrations of flubendiamide and spinosad in red clay soil for 24 and 72 h after 10 days

Concentration of insecticides (kg ha^{-1})	Red clay soil			
	Flubendiamide		Spinosad	
	24 h	72 h	24 h	72 h
0.0	180 ± 2.886 d (100)	260 ± 2.886 d (100)	180 ± 2.886 d (100)	260 ± 2.886 c (100)
1.0	290 ± 2.309 b (161)	340 ± 5.773 b (131)	250 ± 5.773 b (139)	300 ± 0.577 b (115)
2.5	420 ± 1.732 a (233)	520 ± 0.577 a (200)	270 ± 1.732 a (150)	480 ± 2.309 a (185)
5.0	200 ± 2.309 c (111)	280 ± 17.320 c (108)	200 ± 0.577 c (111)	300 ± 2.309 b (115)
7.5	160 ± 2.309 e (89)	200 ± 0.577 e (77)	205 ± 2.886 c (114)	265 ± 2.886 c (102)
10.0	120 ± 5.773 f (67)	180 ± 11.547 f (69)	180 ± 2.886 d (100)	250 ± 5.773 d (96)

μg glucose per gram soil formed after 24 and 72 h of incubation with 2 % starch

Each column is mean ± SE for six concentrations in each group; columns not sharing a common letter (a, b, c, d and e) differ significantly with each other ($P \leq 0.05$; DMRT)

Table 6 Activity of amylase under the impact of different concentrations of flubendiamide and spinosad in black clay soil for 24 and 72 h after 10 days

Concentration of insecticides (kg ha^{-1})	Black clay soil			
	Flubendiamide		Spinosad	
	24 h	72 h	24 h	72 h
0.0	310 ± 5.773 d (100)	380 ± 2.886 e (100)	310 ± 5.773 b (100)	380 ± 2.886 c (100)
1.0	420 ± 0.577 b (135)	490 ± 5.773 c (129)	310 ± 11.547 b (100)	450 ± 17.320 d (118)
2.5	500 ± 8.660 a (161)	600 ± 11.547 a (158)	450 ± 5.773 a (145)	570 ± 1.154 a (150)
5.0	450 ± 2.886 b (145)	520 ± 1.154 b (137)	300 ± 0.577 b (97)	400 ± 5.773 b (105)
7.5	350 ± 17.320 c (113)	460 ± 2.309 d (121)	250 ± 2.886 c (81)	305 ± 1.154 e (80)
10.0	210 ± 5.773 e (67)	320 ± 5.773 f (84)	240 ± 5.773 c (77)	280 ± 1.154 f (74)

μg glucose per gram soil formed after 24 and 72 h of incubation with 2 % starch

Each column is mean ± SE for six concentrations in each group; columns not sharing a common letter (a, b, c, d and e) differ significantly with each other ($P \leq 0.05$; DMRT)

10 days of incubation. Amylase activity increased under lower doses and decreased under higher doses compared to the controls in black and red clay soils. The maximum activity was observed at 2.5 kg ha^{-1} (stimulatory) for flubendiamide, spinosad. Amylase activity showed an individual increment of 35–61, 29–58, 0–45, 8–50 %, in black clay soil and 61–133, 31–100, 39–50, 15–85 %, in comparison to control at 24 and for 72 h received 2.5 and 5.0 kg ha^{-1} respectively in red clay soil. With the increase in incubation periods, the stimulated enzyme activities were also increased up to 20-days further increase in the incubation decrease in the enzyme activity was noticed (Fig. 3). Our results were in contrast with the several researcher works (Srinivasulu and Rangaswamy 2006; Mohiddin et al. 2010; Tu 1981a, b, 1982, 1988), triazophos, a phosphorothioate triazole is stimulated for amylase at 5 and 10 mg/kg incubated for 3 days in an organic soil. As per the observation made by the Prasad and Mathur (1983) the amylase activity increased during germination in both control, and Cuman treated seeds at 0.25, 0.5, 0.75 and 1 % respectively. Interaction effects on soil enzyme activities, including amylase activity received least attention. There were only isolated reports on interaction effects between two chemical compounds in axenic culture studies with algae, cyanobacteria and fungi (Megharaj et al. 1989; Stratton and Corke 1982a, b). Kennedy and Arathan (2002) reported that application of carbofuran at 1 and 1.5 kg ha^{-1} significantly reduced the activity of soil enzymes, viz., alpha -amylase, beta -glucosidase, cellulase, urease and phosphatase up to 30 days after carbofuran application. However, application of carbofuran at the recommended level (0.5 kg a.i. ha^{-1}) had no significant effect upon the activity of soil enzymes, which are biologically significant as they play an important role not only in the soil chemical and biological properties but also affect the nutrient availability to plants. Rate of amylase

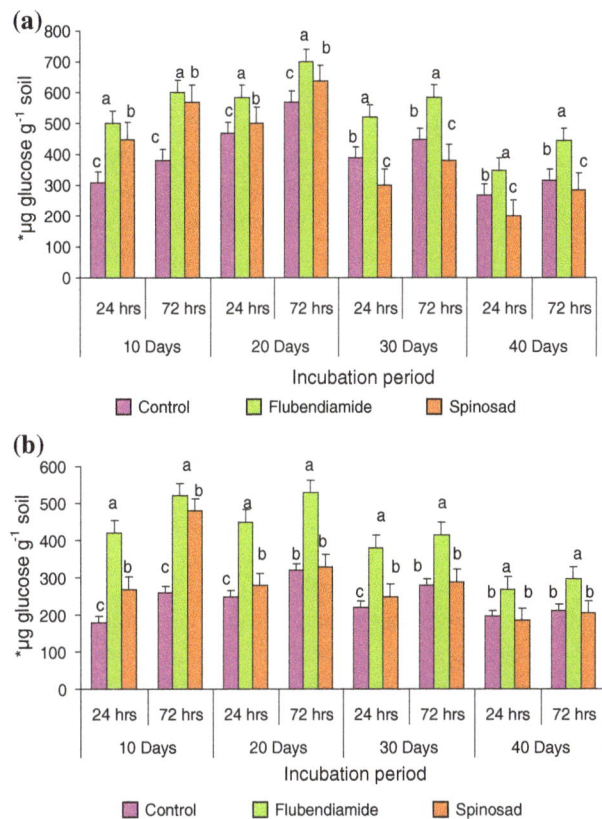

Fig. 3 Influence of flubendiamide and spinosad on amylase*activity in **a** black clay and **b** red clay soil at 2.5 kg ha^{-1}. *μg glucose per gram soil formed after 24 and 72 h incubation with 2 % starch. The values are the mean ± SE for each incubation period, are not significantly different ($P \leq 0.05$) from each other according to Duncan's multiple range (DMR) test

activity followed the same trend of initial stimulation followed by inhibition as reported by Rangaswamy and Venkateswarlu (1992) and Vijay Gundi et al. (2007). Thus, far, no information has been available regarding the influence of flubendiamide and spinosad on these soil enzyme activities. At the same time, much more should be done to understand the influence of flubendiamide and spinosad on soil enzymes clearly. Hence further investigation is needed to evaluate the influence of insecticides on the enzyme activities in agricultural soils which are important and affect nutrient cycling and fertility of soils.

Conclusions

Results from this study indicated that the cellulase enzyme activity was profoundly increased up to 2.5 kg ha^{-1} where as at higher concentrations (5.0–10.0 kg ha^{-1}) of pesticide concentration the enzyme activity were dramatically decreased in both the soils except spinosad at 5.0 kg ha^{-1} in red clay soil. Invertase enzyme activity was decreased from 5.0–10.0 kg ha^{-1} level when compared to control in black clay soil where as the enzyme activity at 1.0–2.5 kg ha^{-1} level. In red clay soil the invertase enzyme activity was stimulated up to 10.0 kg ha^{-1} when compared to control except fubendiamide for 48 h and for spinosad for 24 h at 7.5–10.0 kg ha^{-1}. Amylase enzyme activity showed a stimulatory activity up to 5.0 kg ha^{-1} further increase in the pesticide concentration repression in the enzyme activity was noticed in both soils. Overall soil enzymes were affected by the application of flubendiamide and spinosad at higher concentrations (5.0–10.0 kg ha^{-1}). However, as an important agent for the control of plant pathogens, flubendiamide and spinosad is often used at rates much greater than the recommended dosage.

Overall, flubendiamide and spinosad at a normal field dose (1.0–2.5 kg ha^{-1}) would not pose a threat to soil enzymes among them spinosad is more effective than flubendiamide in inducing the cellulase and invertase with exception of amylase enzyme activities at normal field rates (1.0–2.5 kg ha^{-1}). When flubendiamide and spinosad concentration was increased (5.0–10.0 kg ha^{-1}), however, the threat to soil, cellulase, invertase and amylase increased. A very few reports are available on the influence of insecticides on these enzyme activities cellulase, invertase, and amylase.

Acknowledgments The authors are grateful to the University Grants Commission UGC-SAP, New Delhi, India, for financial assistance (UGC S.LR. No. F.3-25/2009) and at the same time we are very much thankful to the Department of Microbiology, Sri Krishnadevaraya University for providing all the necessary facilities in fulfilling our research.

Conflict of interest The author(s) declare(s) that there is no conflict of interests regarding the publication of this article.

References

Alexander M (1961) Introduction to soil microbiology. Wiley Estern Ltd, New Delhi

Anonymous (2011) Agriculture production plan for Anantapur District, Andhra Pradesh, India. Department of Agriculture, Anantapur, Andhra Pradesh, India

Arinz AE, Yubedee AG (2000) Effect of fungicides on *Fusarium* grain rot and enzyme production in maize (*Zea mays* L.). Glob J Appl Sci 6(4):629–634

Barnes H, Folkard BR (1951) The determination of nitrite. Analyst 76:599–603

Bera SK, Dash P, Singh SP, Dash MM (2002) Bacterial Pod rot: a new threat to groundnut under rice-based cropping system in Orissa, India. Int Arac News Lett 22:40–41

Bhalerao TS, Puranik PR (2009) Biodegradation of organochlorine pesticide, endosulfan, by a fungal soil isolate, *Aspergillus niger*. Int Biodeterior Biodegrad 59:315–319

Bhuyan S, Sau S, Adhya K, Sethunathan TK (1992) Accelerated aerobic degradation of γ-hexaclorocyclohexane in suspensions of nonflooded soils pretreated with hexaclorocyclohexane. Biol Fertil Soils 12:279–284

Chu XF, Hua P, Xuedong W, Xiao S, Bo Min F, Yunlong Y (2008) Degradation of chlorpyrifos alone and in combination with chlorothalonil and their effects on soil microbial populations. J Environ Sci 20:464–469

Cole MA (1977) Lead inhibition of enzyme synthesis in soil. Appl Environ Microbiol 33:262–268

Deng S, Tabatabai M (1994) Colorimetric determination of reducing sugars in soils. Soil Biol Biochem 26:473–477

Garcia-Ruiz R, Ochoa V, Vinegla B, Hinojosa MB, Pena-Santiago R, Liebanas G, Linares JC, Carreira JA (2009) Soil enzymes, nematode community and selected physico-chemical properties as soil quality indicators in organic and conventional olive oil farming: influence of seasonality and site features. Appl Soil Ecol 41:305–314

Gherbawy YA, Abdel Zaher HMA (1999) Isolation of fungi from tomato rhizosphere and evaluation of the effect of some fungicides and biological agents on the production of cellulose enzymes. Czech Mycol 51:157–170

Gianfreda L, Rao MA, Piotrowska A, Palumbo G, Colombo C (2005) Soil enzyme activities as affected by anthropogenic alterations: intensive agricultural practices and organic pollution. Sci Total Environ 341:265–279

Gigliotti C, Allievi L, Salandi C, Ferrari F, Farini A (1998) Microbiol ecotoxicity and persistence in soil of the herbicide bensulfuron methyl. J Environ Sci Health 33(4):381–398

Hegde DM, Kiresur V (1999) Oilseeds. Changing paradigms. In: Venkataramani G (ed) Survey of Indian agriculture. The Hindu, Chennai, pp 67–72

Ismail BS, Fugon D, Omar O (1996a) Effect of metolachlor on soil enzymes in Malaysian soil. J Environ Sci Health 31(6):1267–1278

Ismail BS, Omar O, Ingon O (1996b) Effects of metolachlor on the activities of four soil enzymes. Microbios 87(353):239–248

Ismail BS, Yapp KF, Omar U (1998) Effects of metsulfuron methyl on amylase, urease and protease activities in two soils. Aus J Soil Res 36:449–456

Jackson ML (1971) Soil chemical analysis. Prentice Hall India, New Delhi

Jaya Madhuri R, Rangaswamy V (2002) Influence of selected insecticides on phosphatase activity in groundnut (*Arachis hypogaea* L.) soils. J Env Biol 23(4):393–397

Johnson CM, Ulrich A (1960) Determination of moisture in plant tissues. Calif Agric Bull 766:112–115

Katayama A, Kuwatsuka S (1991) Effects of pesticides on cellulose degradation in soil under upland and flooded conditions. Soil Sci Plant Nutr 37:1–6

Kennedy ZJ, Arathan SS (2002) Influence of carbofuran on the activity of soil enzymes in submerged rice soil ecosystem. In: Rajac RC (ed) Biotechnology of microbes and sustainable utilization. India, pp 322–326

Kirst HA, Michel KH, Mynderse JS, Choco EH, Yao RC, Nakatsukasa WM (1992) Discovery, isolation and structure elucidation of a family of structurally unique fermentation derived tetracyclic macrolides, in synthesis and chemistry of Agrochemicals III. In: Basker DR, Fenyes JG, Steffens JJ (eds) American Chemical Society, Washington, pp 214–215

Kiss S, Dragan-Bularda M, Radulescu D (1978) Soil polysaccharidases: activity and agricultural importance. In: Burns RG (ed) Soil enzymes. Academic Press, London, pp 117–147

Loganathan M, Sundara babu PC, Balasubramanyam G (2002) Efficacy of biopesiticides against Spdeoptera litura (Fab.) on groundnut (Arachis hypogaea L.). Mad Agric J 89(7–9):521–524

Megharaj M, Venkateswarlu K, Rao AS (1989) Interaction effects of insecticides combinations towards the growth of Scenedesmus bijugatus and Synechococcus elongates. Plant Soil 114:159–163

Megharaj M, Singleton I, Kookana R, Naidu R (1999) Persistence and effects of fenamiphos on native algal populations and enzymatic activities in soil. Soil Biol Biochem 31:1549–1553

Menon P, Gopal M, Prasad P (2004) Influence of two insecticides chloripyriphos and quinolphos on arginine ammonification and mineralizable nitrogen in two tropical soil types. J Agric Food Chem 24(52):7370–7376

Menon P, Gopal M, Prasad P (2005) Effects of chlorpyrifos and quinalphos on mineralization in soils of diverse genesis under differing management systems. Biol Fertil Soils 27:430–438

Mohiddin GJ, Srinivasulu M, Madakka M, Rangaswamy V (2010) Influence of insecticides on the activity of amylase and cellulase in groundnut (Arachis hypogaea L.) soil. Ecol Environ Conserv 3(16):383–388

Mohiddin GJ, Srinivasulu M, Madakka M, Rangaswamy V (2011) Influence of selected insecticides on enzyme activities in groundnut (Arachis hypogaea L.) soils. Dyn Soil Dyn Plant 1(5):65–69

Moorman TB (1989) A review of pesticide effects on microorganisms and microbial processes related to soil fertility. J Prod Agric 2(1):4–23

Nelson N (1944) A photometric adaptation of Somogyi method for determination of glucose. J Biol Chem 153:375–380

Prasad BN, Mathur SN (1983) Effect of metasystox and cumin-l on seed germination, reducing sugar content and amylase activity in Vigna mungo (L.) Hepper. Ind J Plant Physiol 2(24):209–213

Quazi S, Datta R, Sarkar D (2011) Effects of soil types and forms of arsenical pesticide on rice growth and development. Int J Environ Sci Technol 8:445–460

Ramesh Babu K, Santharam G, Chandrasekharan S (2002) Bioefficacy of imidacloprid against leaf miner Aproaerema modicella deventor on groundnut (Arachis hypogaea L.). Pestology 6(26):13–16

Ramudu AC, Mohiddin GJ, Srinivasulu M, Madakka M, Rangaswamy V (2011) Impact of fungicides chlorothalonil and propiconazole on microbial activities in groundnut (Arachis hypogaea L.) soils. ISRN Microbiol, p 7

Rangaswamy V, Venkateswarlu K (1992) Activities of amylase and invertase as influenced by the application of monocrotophos, quinalphos, cypermethrin and fenvalerate to groundnut soils. Chemosphere 25(4):525–530

Ranney TA, Bartlett RJ (1972) Rapid field determination of nitrate in natural waters. Commun Soil Sci Plant Anal 3:183–186

Sannino F, Gianfreda L (2001) Pesticide influence on soil enzymatic activities. Chemosphere 45:417–425

Singaram P, Kamalakumari K (2010) Effect of continuous application of different levels of fertilizers and farm yard manure on enzyme dynamics of soil. Mad Agric J 87:364–365

Singh J, Singh DK (2005) Bacterial, azotobacter, actinomycetes, and fungal population in soil after diazinon, imidacloprid, and lindane treatments in groundnut (Arachis hypogaea L.) fields. J Environ Sci Heal B 40(5):785–800

Singh BK, Kuhad RC, Singh A, Lal R, Triapthi KK (1999) Biochemical and molecular basis of pesticide degradation by microorganisms. Crit Rev Biotechnol 19:197–225

Skujins J (1978) History of abiotic soil enzyme research. In: Burns RG (ed) Soil Enzyme. Academic Press, London, pp 1–49

Srinivasulu M, Rangaswamy V (2006) Activities of invertase and cellulase as influenced by the application of tridemorph and captan to groundnut (Arachis hypogaea) soil. Afr J Biotechnol 2(5):175–180

Srinivasulu M, Mohiddin GJ, Rangaswamy V (2012) Effect of insecticides alone and in combination with fungicides on nitrification and phosphatase activity in two groundnut (Arachis hypogaea L.) soils. Environ Geochem Health 3(34):365–374

Stratton GW, Corke CT (1982a) Toxicity of the insecticide permethrin and some degradation products towards algae and cyanobacteria. Environ Poll Ser A 29:71–80

Stratton GW, Corke CT (1982b) Comparative fungitoxicity of the insecticide permethrin and ten degradation products. Pestic Sci 13:679–685

Swaminathan P, Prabharan D, Uma L (2009) Fate of few pesticides metabolizing enzymes in the marine Cyanobacterium Phormidium Valderianum BDU 20041 in perspective with chlorpyrifos exposure. Pestic Biochem Physiol 94:68–72

Tejada M (2009) Evolution of soil biological properties after addition of glyphosate, diflufencian and glyphosate + diflufencian herbicides. Chemosphere 76:365–373

Tohnishi M, Nakao H, Furuya T, Seo A, Kodama H, Tsubata K, Fujioka S, Kodama H, Hirooka T, Nishimatsu T (2005) Flubendiamide, a novel insecticide highly active against lepidopterous insect pests. J Pestic Sci 30:354–360

Truu M, Truu J, Ivask M (2008) Soil microbiological and biochemical properties for assessing the effect of agricultural management practices in Estonian cultivated soils. Eur J Soil Biol 44:231–237

Tu CM (1981a) Effect of pesticides on activity of enzymes and microorganisms in a clay loam soil. J Env Sci Health 16:179–181

Tu CM (1981b) Effect of some pesticides on enzyme activities in an organic soil. Bull Environ Contam Toxicol 27:109–114

Tu CM (1982) Influence of pesticides on activities of amylase, invertase and level of adenosine triphosphate in organic soil. Chemosphere 2:909–914

Tu CM (1988) Effect of selected pesticides on activities of amylase, invertase and microbial respiration in sandy soil. Chemosphere 17:159–163

Tu CM (1995) Effect of five insecticides on microbial and enzymatic activities in sandy soil. J Environ Sci Health 30:289–306

United States Department of Agriculture (USDA) (2011) Foreign Agricultural Service (FAS)

Vijay Gundi AKB, Viswanath B, Subhosh Chandra M, Narahari Kumar V, Rajasekhar Reddy B (2007) Activities of cellulose and amylase in soils as influenced by insecticide interactions. Ecotoxicol Environ Safety 68:278–285

Yang CL, Sun TH, He WX, Zhou QX, Chen S (2007) Single and joint effects of pesticides and mercury on soil urease. J Environ Sci 19(2):210–216

Zabaloy MC, Garland JL, Gomez MA (2008) An integrated approach to evaluate the impacts of the herbicides glyphosate, 2,4-D and metsulfuron-methyl on soil microbial communities in the Pampas region, Argentina. Appl Soil Ecol 40:1–12

26

Over-expression, purification and isotopic labeling of a tag-less human glucose-dependent insulinotropic polypeptide (hGIP)

Rakesh C. Chandarana · Vikrant · Ashok K. Varma ·
Anil Saran · Evans C. Coutinho · Jacinta S. D'Souza

Abstract Glucose-dependent insulinotropic polypeptide (GIP), a gut peptide released in response to food intake brings about secretion of insulin in a glucose-dependent manner upon binding to its receptor, GIPR. GIP–GIPR has emerged as a new vista for anti-diabetic drug discovery and their interaction is being probed at the atomic level to aid rational drug design. In order to probe this interaction on cells, the current study attempts towards expressing ^{15}N-labeled GIP using classical molecular biology tools. We have developed a methodology to obtain GIP devoid of extra amino acid(s); a prerequisite to the intended inter-action study. The synthetic GIP cDNA with a Factor Xa protease site at the N-terminus of GIP was inserted in the vector pET32a(+); the fusion protein thus expressed was eventually cleaved to obtain GIP. After successful Factor Xa cleavage, the cleaved GIP was confirmed by western blot. Subsequently, the (^{15}N)GIP was obtained using the aforementioned procedure and confirmed by MALDI-TOF.

Keywords Glucose-dependent insulinotropic polypeptide · Affinity chromatography · Isotopic labeling · Recombinant fusion protein · Factor Xa protease cleavage site · Diabetes mellitus

R. C. Chandarana · A. Saran · E. C. Coutinho
Department of Pharmaceutical Chemistry, Bombay College of
Pharmacy, Kalina, Santacruz (E), Mumbai 400098, India

Vikrant · A. K. Varma
Structural and Molecular Biology Laboratory, Tata Memorial
Centre, Advanced Centre for Treatment, Research and Education
in Cancer, Kharghar, Navi Mumbai 410 210, India

J. S. D'Souza (✉)
UM-DAE Centre for Excellence in Basic Sciences, Kalina
Campus, Santacruz (E), Mumbai 400098, India
e-mail: jacinta@cbs.ac.in

Introduction

Glucose-dependent insulinotropic polypeptide (GIP) is a peptide hormone released into the blood stream in response to glucose and nutrients absorption from the intestine along with glucagon like polypeptide-1 (GLP-1) (Drucker 2006; McIntosh et al. 2009). Both peptides are known to exert varied physiological actions on different body tissues with the pancreas being the major effector. These peptides interact with their respective G-protein-coupled receptors present on the cell surface and activate the adenylate cyclase signaling pathway. It brings about insulin secretion from the pancreas in a glucose-dependent manner and thus the name GIP. This pathway of insulin secretion is known as entero-insular axis and the peptides as incretins. Due to the glucose-dependent action, both GIP and GLP-1 have attracted immense attention for the design of novel anti-diabetics. In order to identify the key determinants of their interaction, attempts have been made to study their struc-tures (Parthier et al. 2007; Runge et al. 2008; Underwood et al. 2010). In particular, studies aimed at probing the interaction of the peptides with the N-terminus domain of their receptors is of interest. Currently, nuclear magnetic resonance (NMR) is the only technique that permits determination of the structure of biomolecules at atomic-level resolution in near-physiological conditions (solution state) (Billeter et al. 2008; Banci et al. 2010). With isotopic labeling, determining complex structures and studying the interaction dynamics of biomolecules in the presence of other biomolecules and within the cell is now feasible (Stockman 2002; Takeuchi and Wagner 2006; Banci et al. 2010). However, much of the success of structural biology depends on the availability of biomolecules in its utmost pure and native state. Moreover, techniques such as X-ray crystallography and NMR require proteins in mM concentration. With the advent of recombinant DNA technology (rDNA) and techniques of gene synthesis, affinity tag-based purification and subsequent removal of the tag by protease cleavage, it is now possible to obtain proteins in their native states with higher yields (Arnau et al. 2006).

Expression of isotopic hGIP had not been reported till date. The current study reports a method of isotopically (^{15}N)-labeled and unlabeled recombinant over-expression of hGIP in pET32a(+) vector, with complete tag removal by Factor Xa cleavage and purification of the native hGIP. The peptide obtained here has been characterized by western blotting using hGIP-specific antibody and the labeling has been confirmed by MALDI.

Materials and methods

All the materials were obtained from Sigma-Aldrich India Ltd., Merck-Millipore India, Himedia (India) and were of molecular biology grade.

Designing the DNA construct for *hgip* gene

To overcome the problem of codon bias and facilitate complete removal of the tags, the cDNA encoding hGIP was commercially gene synthesized (First Base, Singapore) with a nucleotide sequence coding for an Factor Xa protease cleavage at 5′ and a stop codon at the 3′ ends. The cDNA was cloned into pET32a(+) and subsequently transformed into *Escherichia coli* BL21(DE3) and *E. coli* DH5α strains along with appropriate positive and negative controls. The colonies grown in Luria–Bertani (LB) agar plates containing ampicillin was checked for the presence of the insert by colony PCR and the amplicons were electrophoresed on a 2 % agarose gel containing ethidium bromide. Glycerol stocks of the positive clones were prepared and cryopreserved at −80 °C. Plasmids were isolated from the cells containing the construct and sequenced using pET32a(+)-based primers.

Over-expression and purification of pET-GIP fusion protein

The *E. coli* BL21(DE3) cells harboring the construct were grown to an OD of 0.6 in 10 ml LB medium containing ampicillin and induced using 1 mM IPTG. In order to determine an optimum time of expression, induction was carried out for 1, 3 and 18 h. The cells collected at different time-points were harvested by centrifugation at 5,000*g*/ 4 °C and the expression of pET-GIP fusion protein was checked by electrophoresing on a denaturing gel. The solubility of the fusion protein was ascertained by

disrupting the cell lysate under non-denaturing conditions and electrophoresing the supernatant (~100 μg total protein/lane) on a denaturing gel. The pET-GIP fusion protein was confirmed using an antibody to the hexa-histidine portion of the tag. Upon observing over-expression of the 23 kDa pET-GIP polypeptide on SDS-PAGE, the conditions were scaled up to 1 l (culture volume) for 3 h and the cells harvested by centrifugation at 5,000*g* for 10 min at 4 °C. The cell pellets were stored at −80 °C until the next step of purification. At the time of purification, the pellet was thawed and the cells were re-suspended in lysis buffer (20 mM Tris–HCl buffer pH 7.5, 300 mM NaCl, 1 mM PMSF, 10 % glycerol and 10 mM imidazole) and ruptured by sonication. The supernatant was collected after centrifuging the cell lysate at 10,000*g* for 30 min at 4 °C and mixed with Ni–NTA beads that were pre-washed with the lysis buffer, in a 50-ml centrifuge tube. The tube was kept on a cell mixer and the binding of the fusion protein to the beads was carried out for 2 h at 4 °C. After binding, the slurry was loaded on a BioRad elution column and the flow-through was collected. The column was washed with lysis buffer containing 20 mM imidazole to remove the non-specific binding of proteins to the column. Elution of the pET-GIP fusion protein was then carried out using imidazole gradient from 20 mM to 2 M in the same buffer and the various fractions were electrophoresed on a denaturing gel. The purity of the fusion protein was assessed by silver staining of the gel. Subsequent purification of pET-GIP fusion protein was carried out by the batch elution method where the protein was eluted using 5 ml of lysis buffer containing 100 mM, 500 mM and 1 M imidazole. The fraction containing the protein was dialyzed against the Factor Xa digestion buffer −20 mM Tris–HCl pH 6.5, 50 mM NaCl, 1 mM CaCl$_2$ at 4 °C.

Factor Xa cleavage of pET-GIP fusion protein

The cleavage of the purified pET-GIP fusion protein obtained in the Factor Xa digestion buffer after dialysis was optimized by varying parameters such as temperature and enzyme concentration. The cleaved protein was electrophoresed on a denaturing gel along with the uncleaved control and molecular weight marker and different fragments were observed. The fragments obtained upon cleavage were also characterized by western blot using hGIP-specific antibody to confirm the presence of the peptide.

Purification of GIP from the mixture by gel filtration

The 5-kDa hGIP was purified at room temperature by separating it from the mixture containing the residual tag (18 kDa), other contaminants (>11 kDa) and Factor Xa

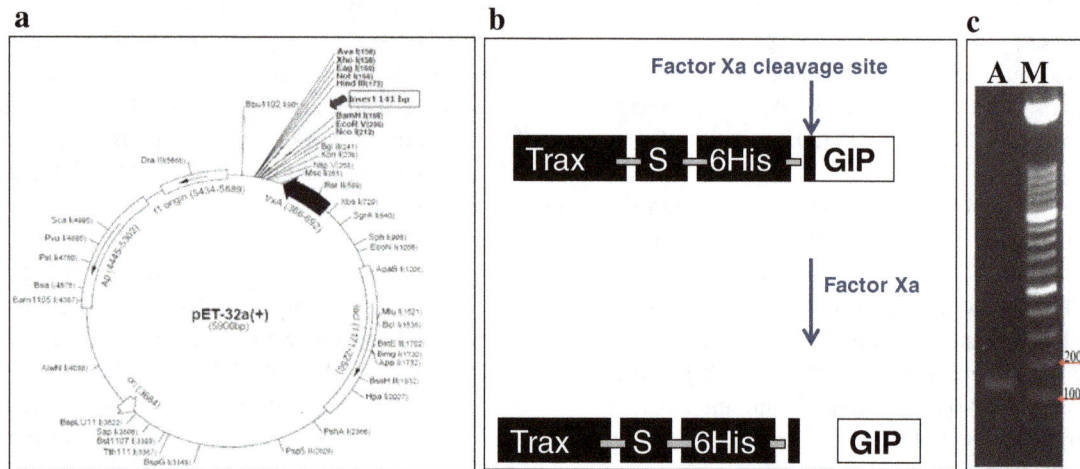

Fig. 1 Design of the construct resulting in the cloning of the GIP ORF. **a** pET32a(+) plasmid (obtained from First Base, Singapore) showing the position (between *Bam*HI and *Hin*dIII) at which the h*gip* insert was cloned. **b** A cartoon depiction of the Factor Xa cleavage site. Note that upon cleavage the released hGIP is untagged. **c** PCR of the h*gip* amplicon of ∼125 bp; lane labeled *M* is the standard molecular weight marker, lane labeled *A* is the h*gip* amplicon

enzyme (55 kDa) by gel filtration using a superdex 75 column on a GE AKTA FPLC system in 20 mM Tris–HCl buffer (pH 7) 150 mM NaCl. The elution was monitored by UV detector (280 nm) and elutes were collected as 2 ml fractions. The fractions showing the protein were electrophoresed on 15 % SDS-PAGE and the separation was checked.

Isotopic (^{15}N) labeling and characterization of hGIP by MALDI

Upon optimizing the basic strategies for over-expression, tag removal and purification of hGIP, isotopic labeling of the peptide was carried out by growing the *E. coli* BL21(DE3) cells up to OD_{600} of 0.6 in the minimal medium (M9) containing isotopic (^{15}N) ammonium chloride as the nitrogen source and induced using 1 mM IPTG. Subsequent steps of purification were carried out in the manner as reported for unlabeled hGIP.

The isotopic peptide (0.5 µl corresponding to 0.25 µg concentration) was mixed with the non-isotopic hGIP (0.5 µl corresponding to 0.4 µg concentration) in 1 µl of Sinapinic acid as matrix and spotted on a ground steel MALDI-TOF plate (Bruker Daltonics, USA). This mixture was allowed to dry at room temperature and the plate was then loaded into the instrument. Subsequent to adjusting the appropriate parameters, the spot was fired using the linear mode at 22 ± 2 °C, 27 kV and a microchannel plate detector. The molecular weight was determined using an external standard (insulin, ubiquitin i, cytochrome *c*, myoglobin that covered a mass range of ∼5,000–17,500 Da) at an error of ±100 ppm, the software for deconvolution being Flex analysis, Bruker Daltonics.

Results and discussion

The need to study the structure of biomolecules at near-physiological conditions is mandatory for rational drug design; hGIP being one of them. For this purpose, cellular and molecular biologists as well as pharmacologists have felt the need to obtain decent concentrations and yields of hGIP in their studies. For long, such experimentalists have been using either the synthetic peptide or purifying it in bulk from natural sources (Pederson and Brown 1976; Alaña et al. 2004). Besides, acquiring (^{15}N)hGIP or doubly labeled (^{15}N, ^{13}C)hGIP synthetically for structural and/or receptor-binding studies has proven to be cost intensive. The aim of the present study was to obtain an economic and constant source of unlabeled and labeled hGIP in reasonable yields using recombinant DNA technology. The pET series of vectors have T7 promoters that over-express proteins in *E. coli* BL21(DE3) cells (Studier and Moffatt 1986). Apart from the hexa-histidine affinity tag, pET32a(+) also has a localization tag (S-tag) and a solubilization tag (ThioredoxinA, TrxA) (Tsunoda et al. 2005). To prevent the possibility of the protein routing into inclusion bodies, the idea of tagging it with a TrxA tag was sought after. The S-tag would aid in future receptor-binding studies under physiological conditions. The current study has directed efforts towards obtaining a constant source of unlabeled as well as labeled hGIP. For the ultimate need of obtaining a tag-less GIP, a construct was designed suitably using pET32a(+) as the backbone vector. The ThioredoxinA (TrxA) tag was followed by the Streptavidin (S) tag, the hexa-histidine (6XHis) tag and a Factor Xa cleavage site (IEGR). The GIP sequence was

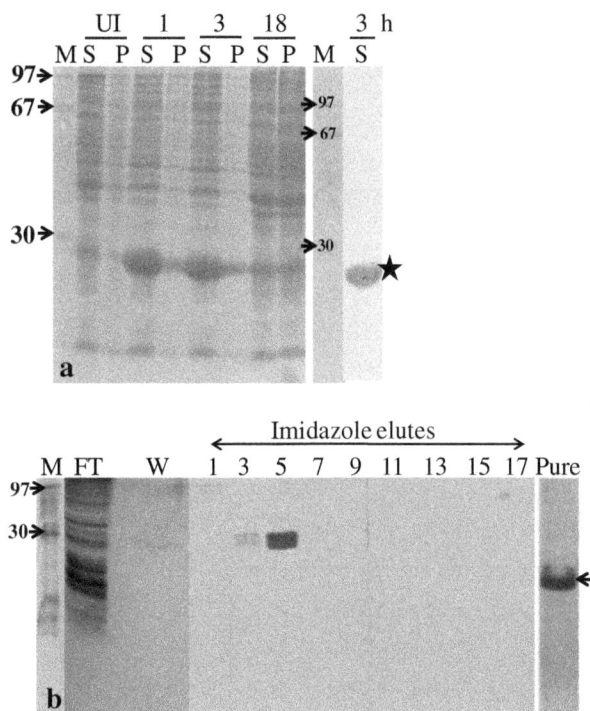

Fig. 2 Induction of the fusion protein (Trx-S-6XHis-fXa-GIP). **a** Kinetics of the over-expression of the fusion protein, induction with IPTG and detection of the 6XHis tag using an anti-hexa-histidine antibody (denoted by a *star* in the last lane of this figure). **b** Affinity (IMaC) based purification of the fusion protein using gradient elution (20–2,000 mM imidazole). Note the purification of the fusion protein in elutes 3 and 5. A silver stained gel of the purified protein (see *arrowhead*). Lanes labeled *M* protein molecular weight standards, *P* pellets, *S* supernatants obtained after extraction, *UI* uninduced control, *numbers* time-points (in hours) of induction, *FT* flow-through, *W* wash obtained after the proteins have been bound to the column; numbers against *arrows*, molecular weights in kDa; pure, purified protein (~5 μg)

introduced after this cleavage site so that the tag-less GIP alone is released upon Factor Xa cleavage. The synthetic sequence (Fig. 1a, b) so designed was then synthesized commercially by First Base, Singapore. The resultant recombinant vector (pET32a-*fXa-gip*) was used to transform *E. coli* BL21(DE3) competent cells so as to yield colonies on LB agar plates containing ampicillin as the antibiotic of selection. Subsequent characterization by colony PCR using insert-specific and vector-specific primers showed desired amplicons on 2 % agarose gel indicating the presence of the appropriate insert in all the colonies (Fig. 1c). The sequencing of the plasmids isolated from three such positive clones confirmed the presence of the correct sequence of the insert.

Over-expression and purification of pET-GIP fusion protein

Escherichia coli BL21(DE3) strains containing the plasmid (pET32a-fXa-gip) were induced with 1 mM IPTG. An intense band of fusion protein (Trx-S-6XHis-fXa-GIP) was observed at ~23 kDa; the intensity of this band increased with the duration of induction (Fig. 2a). The optimum time for induction was found to be 3 h, beyond which the protein degraded. As expected, due entirely to the TrxA tag, the fusion protein was found to be soluble as it was released into the aqueous medium by non-denaturing cell disruption (Fig. 2a; S for supernatant fraction that contained the fusion protein and not in the pellet P). The presence of the 6XHis tag, as confirmed using an anti-6XHis antibody on a western blot of the fusion protein (Fig. 2a; last lane) permitted affinity-based purification by IMaC using Ni–NTA matrix.

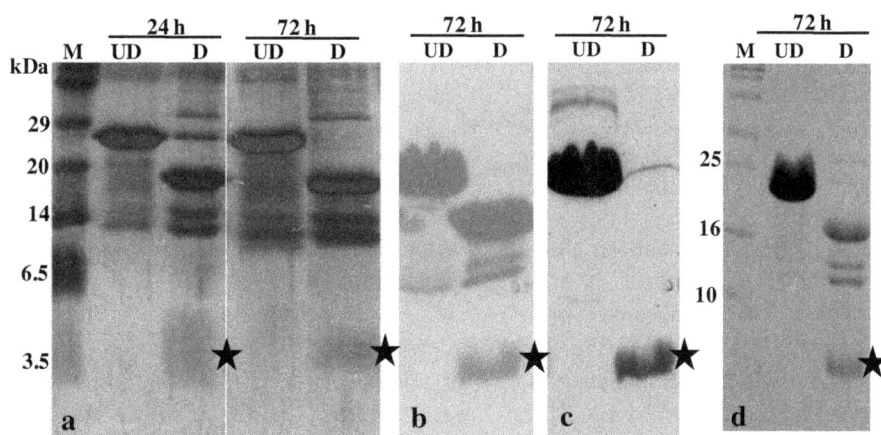

Fig. 3 Factor Xa digestion and characterization by western blotting. **a** The fusion protein (75 μg) was digested (lanes labeled *D*) with Factor Xa and the digested products were electrophoresed on a denaturing gel and silver stained; **b** Ponceau S staining of the nitrocellulose membrane after transferring the digested products (starting amount used was 200 μg) by the western blotting procedure; **c** immunoblot for the detection of unlabeled hGIP (starting amount used was 150 μg), and **d** labeled hGIP, using anti-hGIP antibodies. *UD* and *D* labels in all the figures refer to undigested and digested, respectively. The *stars* in all the figures show the position of the digested product, viz. hGIP

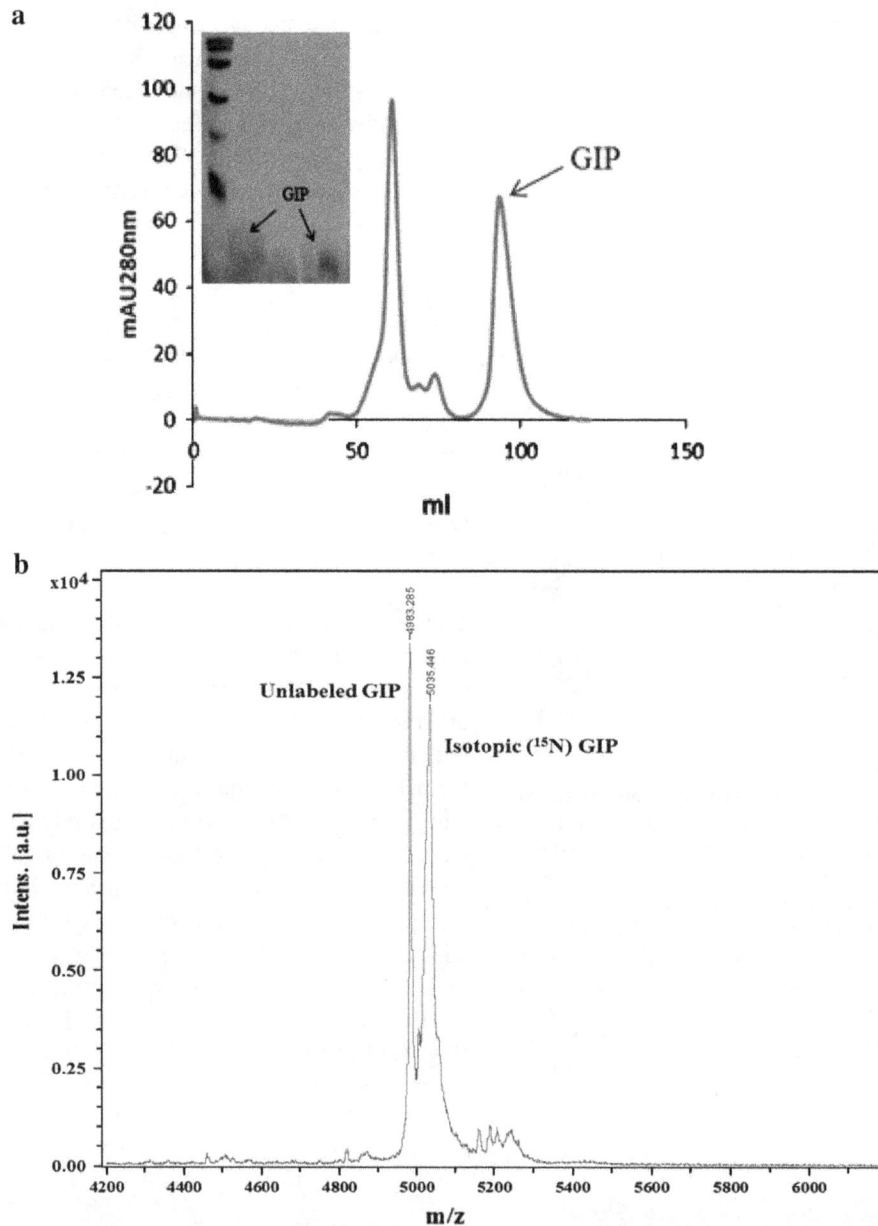

Fig. 4 a Isotopically labeled fusion protein was digested with Factor Xa and the GIP was purified using gel filtration chromatography on Superdex 75 column on an FPLC system (GE, AKTA). The *inset* is the purified protein that was obtained after digestion. **b** Molecular weight determination of unlabeled and isotopic (^{15}N)GIP by MALDI of unlabeled and isotopic (^{15}N)GIP

The purification of pET-hGIP fusion protein carried out using Ni–NTA matrix and an imidazole gradient (20 mM–2 M) showed the presence of pure 23 kDa fusion protein in the elutes when electrophoresed on 15 % SDS-PAGE (Fig. 2b). The protein eluted in the fraction corresponding to ∼100–150 mM imidazole and subsequent batch purification of the ^{15}N-labeled fusion protein at 100, 250 and 1,000 mM imidazole resulted in elution of maximum protein at 100 mM (Fig. 2b).The silver stained gel showed ∼99 % purity and absence of any contaminants. The yield of the fusion protein as estimated by Bradford was found to be ∼30 mg/l culture volume.

Factor Xa cleavage of the fusion protein, purification by gel filtration and MALDI-TOF analysis

After obtaining the purified fusion protein, the conditions for release of the GIP peptide after Factor Xa cleavage were explored. The digestion with Factor Xa was found to be optimum at pH 7.5, 4 °C, enzyme concentration of 1 unit of protease and digestion time being 72 h (Fig. 3a). Upon electrophoreses of the cleaved fragments on a 15 % SDS-PAGE, distinct bands of M_r ∼18 kDa (corresponding to the residual tags) and ∼5 kDa (corresponding to hGIP; Fig. 3a) were observed. Another doublet of M_r ∼14 kDa was also seen and possibly is a result of degradation of the fusion protein. In order to ascertain the digested product as hGIP ($\sim$5 kDa), a western blot of the digested products was probed with the anti-hGIP antibody (Fig. 3b, c; hGIP is marked with a star). This confirmed that the band was of hGIP and also permitted for further purification of the peptide.

The over-expression of the pET-hGIP fusion protein was reproducible when the cells harboring the construct were cultured in minimal media (M9) containing isotopic (^{15}N) ammonium chloride and induced using 1 mM IPTG. Subsequently, the purification of pET-GIP fusion protein was achieved successfully by affinity chromatography on a Ni–NTA column. Although the labeled protein eluted at a higher imidazole concentration (250 mM imidazole), the conditions of digestion with Factor Xa were reproducible. Further purification of hGIP was evident from the chromatogram where different fragments of the digested pET-GIP protein separated differentially by gel filtration on a Superdex 75 column, the tag and the doublet eluted in the 22nd–38th fraction (44–74 ml) and the hGIP at 41st–54th fraction (82–108 ml) showing a peak in the 47th fraction (94 ml). When the fraction corresponding to the hGIP peak was loaded on 15 % SDS-PAGE it showed the hGIP fragment in purified form (inset of Fig. 4a). The peptide was concentrated by ultrafiltration using amicon column mwco 3 kDa. The characterization of the labeled hGIP by MALDI showed the desired peak precisely at 4.98 kDa corresponding to the unlabeled peptide and also a peak at 5.44 kDa corresponding to the molecular weight of the ^{15}N labeled peptide. This confirmed the isotopic labeling of the hGIP. The yield of the hGIP peptide was $\sim$1 mg/l of the culture.

It may be noted that GIP (full-length) or truncated versions have thus far been produced synthetically (Fehmann and Göke 1995). Such synthetic peptides have been used for various binding as well as physiological studies. However, these have been useful in proton-NMR studies and thus far no such method has been developed to obtain isotopic (^{15}N)hGIP. Hence, this is the first report of ^{15}N-labeled hGIP and we present a clone that harbors a fusion gene containing the h*gip* gene that can be reproducibly used to produce the fusion protein at $\sim$30 mg/l and upon digestion and further purification, a peptide of almost 99 % purity with a yield of $\sim$1 mg/l.

Acknowledgments R.C.C. and E.C. wishes to thank the Department of Science and Technology, New Delhi, India, for the grant SR/SO/BB-10/2005 dtd. 8-9-2006. A.S. wishes to thank the Indian National Science Academy, India, for financial support. J.S.D. wishes to thank the Department of Atomic Energy, India for financial support. E.C, J.D and R.C.C wish to thank Prof. B. J. Rao, Department of Biological Sciences, Tata Institute of Fundamental Research, India for permitting the use of his laboratory during the initial part of this work.

Conflict of interest None.

References

Alaña I, Hewage CM, Malthouse JPG et al (2004) NMR structure of the glucose-dependent insulinotropic polypeptide fragment, GIP(1–30)amide. Biochem Biophys Res Commun 325:281–286.

Arnau J, Lauritzen C, Petersen GE, Pedersen J (2006) Current strategies for the use of affinity tags and tag removal for the purification of recombinant proteins. Protein Expr Purif 48:1–13

Banci L, Bertini I, Luchinat C, Mori M (2010) NMR in structural proteomics and beyond. Prog Nucl Magn Reson Spectrosc 56:247–266.

Billeter M, Wagner G, Wüthrich K (2008) Solution NMR structure determination of proteins revisited. J Biomol NMR 42:155–158.

Drucker DJ (2006) The biology of incretin hormones. Cell Metab 3:153–165.

Fehmann HC, Göke B (1995) Characterization of GIP(1–30) and GIP(1–42) as stimulators of proinsulin gene transcription. Peptides 16:1149–1152

McIntosh CHS, Widenmaier S, Kim SJ (2009) Glucose-dependent insulinotropic polypeptide (gastric inhibitory polypeptide; GIP). Vitam Horm 80:409–471.

Parthier C, Kleinschmidt M, Neumann P et al (2007) Crystal structure of the incretin-bound extracellular domain of a G protein-coupled receptor. Proc Natl Acad Sci USA 104:13942–13947.

Pederson RA, Brown JC (1976) The insulinotropic action of gastric inhibitory polypeptide in the perfused isolated rat pancreas. Endocrinology 99:780–785

Runge S, Thøgersen H, Madsen K et al (2008) Crystal structure of the ligand-bound glucagon-like peptide-1 receptor extracellular domain. J Biol Chem 283:11340–11347.

Stockman B (2002) NMR screening techniques in drug discovery and drug design. Prog Nucl Magn Reson 41:187–231

Studier FW, Moffatt BA (1986) Use of bacteriophage T7 RNA polymerase to direct selective high-level expression of cloned genes. J Mol Biol 189:113–130

Takeuchi K, Wagner G (2006) NMR studies of protein interactions. Curr Opin Struct Biol 16:109–117.

Tsunoda Y, Sakai N, Kikuchi K et al (2005) Improving expression and solubility of rice proteins produced as fusion proteins in *Escherichia coli*. Protein Expr Purif 42:268–277.

Underwood CR, Garibay P, Knudsen LB et al (2010) Crystal structure of glucagon-like peptide-1 in complex with the extracellular domain of the glucagon-like peptide-1 receptor. J Biol Chem 285:723–730.

Permissions

List of Contributors

Vidhya Lakshmi Das
School of Biosciences, Mahatma Gandhi University, PD Hills (PO), Kottayam 686 560, Kerala, India

Roshmi Thomas
School of Biosciences, Mahatma Gandhi University, PD Hills (PO), Kottayam 686 560, Kerala, India

Rintu T. Varghese
Plant Molecular Biology, Rajiv Gandhi Centre for Biotechnology, Thycaud (PO), Poojappura, Thiruvananthapuram 695 014, Kerala, India

E. V. Soniya
Plant Molecular Biology, Rajiv Gandhi Centre for Biotechnology, Thycaud (PO), Poojappura, Thiruvananthapuram 695 014, Kerala, India

Jyothis Mathew
School of Biosciences, Mahatma Gandhi University, PD Hills (PO), Kottayam 686 560, Kerala, India

E. K. Radhakrishnan
School of Biosciences, Mahatma Gandhi University, PD Hills (PO), Kottayam 686 560, Kerala, India

Yanling Cai
Division for Nanotechnology and Functional Materials, Department of Engineering Sciences, The A ° ngstro¨m Laboratory, Uppsala University, Box 534, 75121 Uppsala, Sweden

Maria Strømme
Division for Nanotechnology and Functional Materials, Department of Engineering Sciences, The A ° ngstro¨m Laboratory, Uppsala University, Box 534, 75121 Uppsala, Sweden

Ken Welch
Division for Nanotechnology and Functional Materials, Department of Engineering Sciences, The A ° ngstro¨m Laboratory, Uppsala University, Box 534, 75121 Uppsala, Sweden

Sourav Bhattacharya
Department of Microbiology, Karpagam University, Coimbatore 641021, Tamil Nadu, India

Arijit Das
Department of Microbiology, Genohelix Biolabs, Centre for Advanced Studies in Biosciences, Jain University, Bangalore 560019, Karnataka, India

Kuruvalli Prashanthi
Department of Biotechnology, Genohelix Biolabs, Centre for Advanced Studies in Biosciences, Jain University, Bangalore 560019, Karnataka, India

Muthusamy Palaniswamy
Department of Microbiology, Karpagam University, Coimbatore 641021, Tamil Nadu, India

Jayaraman Angayarkanni
Department of Microbial Biotechnology, Bharathiar University, Coimbatore 641046, Tamil Nadu, India

Samiah H. S. Al-Mijalli
Biology Department, College of Sciences, Nora Bent AbdulRahman University, Riyadh, Saudi Arabia

Priyanka Siwach
Department of Biotechnology, Chaudhary Devi Lal University, Sirsa, Haryana, India

Anita Rani Gill
Department of Biotechnology, Chaudhary Devi Lal University, Sirsa, Haryana, India

Pallavi Singh
Laboratory of Plant Tissue Culture and Stress Physiology, Department of Plant Physiology, Institute of Agricultural Sciences, Banaras Hindu University, Varanasi, India

Padmanabh Dwivedi
Laboratory of Plant Tissue Culture and Stress Physiology, Department of Plant Physiology, Institute of Agricultural Sciences, Banaras Hindu University, Varanasi, India

R. J. Rukarwa
School of Agricultural Sciences, Makerere University, P.O. Box 7062, Kampala, Uganda

S. B. Mukasa
School of Agricultural Sciences, Makerere University, P.O. Box 7062, Kampala, Uganda

B. Odongo
African Institute for Capacity Development, P.O. Box 46179, Nairobi GPO 00100, Kenya

G. Ssemakula
National Crop Resources Research Institute (NaCRRI), P.O. Box 7084, Namulonge, Kampala, Uganda

M. Ghislain
International Potato Center, P.O. Box 25171, Nairobi 00603, Kenya

A. K. Mukherjee
Central Institute for Cotton Research, PB 2, Shankar Nagar PO, Nagpur 440010, Maharashtra, India
Central Rice Research Institute, Cuttack, Odisha, India

A. Sampath Kumar
Central Institute for Cotton Research, PB 2, Shankar Nagar PO, Nagpur 440010, Maharashtra, India

S. Kranthi
Central Institute for Cotton Research, PB 2, Shankar Nagar PO, Nagpur 440010, Maharashtra, India

P. K. Mukherjee
Central Institute for Cotton Research, PB 2, Shankar Nagar PO, Nagpur 440010, Maharashtra, India

Ajit Kumar
Environment Toxicology Unit, Centre for Advanced Studies in Zoology, University of Rajasthan, Jaipur 302004, India
Centre for Bioinformatics, M.D. University, Rohtak 124001, India

Narain Bhoot
Environment Toxicology Unit, Centre for Advanced Studies in Zoology, University of Rajasthan, Jaipur 302004, India

I. Soni
Environment Toxicology Unit, Centre for Advanced Studies in Zoology, University of Rajasthan, Jaipur 302004, India

P. J. John
Environment Toxicology Unit, Centre for Advanced Studies in Zoology, University of Rajasthan, Jaipur 302004, India

Anjali Bose
BRD School of Biosciences, Sardar Patel Maidan, Sardar Patel University, Satellite Campus, Vadtal Road, P.O. Box 39, Vallabh Vidyanagar 388 120, Gujarat, India

Haresh Keharia
BRD School of Biosciences, Sardar Patel Maidan, Sardar Patel University, Satellite Campus, Vadtal Road, P.O. Box 39, Vallabh Vidyanagar 388 120, Gujarat, India

K. M. Kumar
School of Biosciences and Technology, VIT University, Vellore 632014, Tamil Nadu, India

P. Anitha
School of Biosciences and Technology, VIT University, Vellore 632014, Tamil Nadu, India

V. Sivasakthi
School of Biosciences and Technology, VIT University, Vellore 632014, Tamil Nadu, India

Susmita Bag
School of Biosciences and Technology, VIT University, Vellore 632014, Tamil Nadu, India

P. Lavanya
School of Biosciences and Technology, VIT University, Vellore 632014, Tamil Nadu, India

Anand Anbarasu
School of Biosciences and Technology, VIT University, Vellore 632014, Tamil Nadu, India

Sudha Ramaiah
School of Biosciences and Technology, VIT University, Vellore 632014, Tamil Nadu, India

Francois N. Niyonzima
Department of Biochemistry, Center for Post Graduate Studies, Jain University, Bangalore 560011, India

Sunil S. More
Department of Biochemistry, Center for Post Graduate Studies, Jain University, Bangalore 560011, India

Arthala Praveen Kumar
Applied Microbiology Laboratory, Department of Virology, Sri Venkateswara University, Tirupati 517 502, India

Avilala Janardhan
Applied Microbiology Laboratory, Department of Virology, Sri Venkateswara University, Tirupati 517 502, India

Seela Radha
Applied Microbiology Laboratory, Department of Virology, Sri Venkateswara University, Tirupati 517 502, India

Buddolla Viswanath
Applied Microbiology Laboratory, Department of Virology, Sri Venkateswara University, Tirupati 517 502, India

Golla Narasimha
Applied Microbiology Laboratory, Department of Virology, Sri Venkateswara University, Tirupati 517 502, India

Subhasish Dutta
Department of Biotechnology, National Institute of Technology Durgapur, Mahatma Gandhi Avenue, Durgapur 713209, India

Bikram Basak
Department of Biotechnology, National Institute of Technology Durgapur, Mahatma Gandhi Avenue, Durgapur 713209, India

Biswanath Bhunia
Department of Bio Engineering, National Institute of Technology Agartala, Barjala, Tripura 799055, India

Samayita Chakraborty
Department of Biotechnology, National Institute of Technology Durgapur, Mahatma Gandhi Avenue, Durgapur 713209, India

Apurba Dey
Department of Biotechnology, National Institute of Technology Durgapur, Mahatma Gandhi Avenue, Durgapur 713209, India

Merina Paul Das
Department of Industrial Biotechnology, Bharath University, Chennai 600073, Tamil Nadu, India

Santosh Kumar
Department of Industrial Biotechnology, Bharath University, Chennai 600073, Tamil Nadu, India

N. Ashwini
Department of Microbiology, Centre for PG Studies, Jain University, 18/3, 9th Main, Jayanagar 3rd Block, Bangalore 560011, India

S. Srividya
Department of Microbiology, Centre for PG Studies, Jain University, 18/3, 9th Main, Jayanagar 3rd Block, Bangalore 560011, India

Ram Prasad Metuku
Department of Microbiology, Kakatiya University, Hanamkonda, Warangal 506009, India

Shivakrishna Pabba
Department of Microbiology, Kakatiya University, Hanamkonda, Warangal 506009, India

Samatha Burra
Department of Microbiology, Kakatiya University, Hanamkonda, Warangal 506009, India

N S. V. S. S. S. L. Hima Bindu
Department of Microbiology, Kakatiya University, Hanamkonda, Warangal 506009, India

Krishna Gudikandula
Department of Microbiology, Kakatiya University, Hanamkonda, Warangal 506009, India

M. A. Singara Charya
Department of Microbiology, Kakatiya University, Hanamkonda, Warangal 506009, India

B. Jasim
School of Biosciences, Mahatma Gandhi University, Priyadharshini Hills PO, Kottayam Dist, Kerala 686560, India

Aswathy Agnes Joseph
School of Biosciences, Mahatma Gandhi University, Priyadharshini Hills PO, Kottayam Dist, Kerala 686560, India

C. Jimtha John
School of Biosciences, Mahatma Gandhi University, Priyadharshini Hills PO, Kottayam Dist, Kerala 686560, India

Jyothis Mathew
School of Biosciences, Mahatma Gandhi University, Priyadharshini Hills PO, Kottayam Dist, Kerala 686560, India

E. K. Radhakrishnan
School of Biosciences, Mahatma Gandhi University, Priyadharshini Hills PO, Kottayam Dist, Kerala 686560, India

Sourabh Jain
Uttarakhand Technical University, Dehradun, India
National Research Centre on Plant Biotechnology, IARI Campus, Pusa, New Delhi, India

Arun Bhatt
Department of Crop Improvement, Uttarakhand University of Horticulture and Forestry, Uttarakhand, India

V. Mohanasrinivasan
School of Biosciences and Technology, VIT University, Vellore 14, Tamil Nadu, India

Mudit Mishra
School of Biosciences and Technology, VIT University, Vellore 14, Tamil Nadu, India

Jeny Singh Paliwal
School of Biosciences and Technology, VIT University, Vellore 14, Tamil Nadu, India

Suneet Kr. Singh
School of Biosciences and Technology, VIT University, Vellore 14, Tamil Nadu, India

E. Selvarajan
School of Biosciences and Technology, VIT University, Vellore 14, Tamil Nadu, India

V. Suganthi
School of Biosciences and Technology, VIT University, Vellore 14, Tamil Nadu, India

C. Subathra Devi
School of Biosciences and Technology, VIT University, Vellore 14, Tamil Nadu, India

Dina Morshedi
Department of Industrial and Environmental Biotechnology, National Institute of Genetic Engineering and Biotechnology, Shahrak-e Pajoohesh, km 15, Tehran-Karaj Highway,P. O. Box: 14965/161, Tehran, Iran

Farhang Aliakbari
Department of Industrial and Environmental Biotechnology, National Institute of Genetic Engineering and Biotechnology, Shahrak-e Pajoohesh, km 15, Tehran-Karaj Highway,P. O. Box: 14965/161, Tehran, Iran
Department of Biotechnology, Semnan University of Medical Sciences, Semnan, Iran

Hamid Reza Nouri
Department of Industrial and Environmental Biotechnology, National Institute of Genetic Engineering and Biotechnology, Shahrak-e Pajoohesh, km 15, Tehran-Karaj Highway,P. O. Box: 14965/161, Tehran, Iran

Majid Lotfinia
Department of Stem Cells and Developmental Biology, CellScience Research Center, Royan Institute for Stem Cell Biology and Technology, ACECR, Tehran, Iran
Department of Biochemistry, Pasteur Institute of Iran, Tehran, Iran

Jafar Fallahi
Department of Industrial and Environmental Biotechnology, National Institute of Genetic Engineering and Biotechnology, Shahrak-e Pajoohesh, km 15, Tehran-Karaj Highway,P. O. Box: 14965/161, Tehran, Iran

Jatin Srivastava
Department of Applied Sciences, Faculty of Environmental Science, Himalayan Institute of Technology and Management, BKT, NH 24, Lucknow 227005, UP, India

Harish Chandra
Department of Medicinal and Aromatic Plants, School of Agriculture and Allied Sciences, High Altitude Plant Physiology Research Center, H.N.B. Garhwal University, Srinagar, Uttrakhand, India

Anant R. Nautiyal
Department of Medicinal and Aromatic Plants, School of Agriculture and Allied Sciences, High Altitude Plant Physiology Research Center, H.N.B. Garhwal University, Srinagar, Uttrakhand, India

Swinder J. S. Kalra
Department of Chemistry, Dayanand Anglo Vedic College, Civil Lines, Kanpur, UP, India

Mathews V. Varghese
School of Biosciences, Mahatma Gandhi University, P.D Hills P.O, Kottayam 686560, Kerala, IndiaAlex Manju

M. Abhilash
School of Biosciences, Mahatma Gandhi University, P.D Hills P.O, Kottayam 686560, Kerala, IndiaAlex Manju

M. V. Sauganth Paul
School of Biosciences, Mahatma Gandhi University, P.D Hills P.O, Kottayam 686560, Kerala, IndiaAlex Manju

S. Abhilash
School of Biosciences, Mahatma Gandhi University, P.D Hills P.O, Kottayam 686560, Kerala, IndiaAlex Manju

R. Harikumaran Nair
School of Biosciences, Mahatma Gandhi University, P.D Hills P.O, Kottayam 686560, Kerala, IndiaAlex Manju

Yachana Jha
N. V. Patel College of Pure and Applied Sciences, Sardar Patel University, V. V. Nagar, Anand, Gujarat, India

R. B. Subramanian
BRD School of Biosciences, Sardar Patel University, Post Box No. 39, V. V. Nagar, Anand 388120, Gujarat, India

G. Jaffer Mohiddin
Department of Microbiology, Sri Krishnadevaraya University, Anantapur 515 055, Andhra Pradesh, India
Department of Life Sciences and Agriculture, Universidad de las Fuezas Armada, Sangolqui, Quito, Ecuador, South America

M. Srinivasulu
Department of Microbiology, Sri Krishnadevaraya University, Anantapur 515 055, Andhra Pradesh, India
Radioactive Waste Management (Bioremediation) Lab, Division of Advanced Nuclear Engineering, Pohang University of Science and Technology, Pohang-si, Republic of Korea

K. Subramanyam
Plant Molecular Biology Unit, Department of Biotechnology and Genetic Engineering, Bharathidasan University, Tiruchirappalli 620024, Tamil Nadu, India

M. Madakka
Department of Biotechnology and Bioinformatics, Yogi Vemana University, Kadapa 516 003, Andhra Pradesh, India

D. Meghana
Department of Microbiology, Sri Krishnadevaraya University, Anantapur 515 055, Andhra Pradesh, India

V. Rangaswamy
Department of Microbiology, Sri Krishnadevaraya University, Anantapur 515 055, Andhra Pradesh, India

Rakesh C. Chandarana
Department of Pharmaceutical Chemistry, Bombay College of Pharmacy, Kalina, Santacruz (E), Mumbai 400098, India

Vikrant
Structural and Molecular Biology Laboratory, Tata Memorial Centre, Advanced Centre for Treatment, Research and Education in Cancer, Kharghar, Navi Mumbai 410 210, India

Ashok K. Varma
Structural and Molecular Biology Laboratory, Tata Memorial Centre, Advanced Centre for Treatment, Research and Education in Cancer, Kharghar, Navi Mumbai 410 210, India

Anil Saran
Department of Pharmaceutical Chemistry, Bombay College of Pharmacy, Kalina, Santacruz (E), Mumbai 400098, India

Evans C. Coutinho
Department of Pharmaceutical Chemistry, Bombay College of Pharmacy, Kalina, Santacruz (E), Mumbai 400098, India

Jacinta S. D'Souza
UM-DAE Centre for Excellence in Basic Sciences, Kalina Campus, Santacruz (E), Mumbai 400098, India

www.ingramcontent.com/pod-product-compliance
Lightning Source LLC
Chambersburg PA
CBHW080623200326
41458CB00013B/4486